Noise and Man

Noise and Man

WILLIAM BURNS, C.B.E., D.Sc., F.R.C.P.

Professor of Physiology in the University of London at
Charing Cross Hospital Medical School
Honorary Consultant Otologist to
the Charing Cross Group of Hospitals

J B LIPPINCOTT COMPANY, PHILADELPHIA

To Margaret

RA 772 N7 B8 1973 cp. 2

First published in the United States of America 1969
Second edition 1973

© William Burns 1968, 1973

Printed in Great Britain
Library of Congress Catalog Card No. 72-4874
ISBN 0-397-58098-3

Published in the United States by
J. B. Lippincott Company
Philadelphia

Contents

Abbreviations

In the text and references the following abbreviations are used

AAOO	American Academy of Ophthalmology and Otolaryngology
ANSI	American National Standards Institute
ASHRAE	American Society of Heating, Refrigeration and Air Conditioning Engineers
BRS	Building Research Station
BSI	British Standards Institution
CHABA	Committee on Hearing, Bioacoustics and Biomechanics
FAA	Federal Aviation Administration
ICAO	International Civil Aviation Organisation
IEC	International Electrotechnical Commission
ISO	International Organization for Standardization
MRC	Medical Research Council
NASA	National Aeronautics and Space Administration
NPL	National Physical Laboratory

Preface to the first edition

There is at present growing concern about the occurrence of un-wanted sounds, commonly called noise, and their possible effects upon man. Despite frequent conferences and symposia, and the exist-ence of an extensive literature on noise, the necessary information is to some extent elusive. This introduction to the subject of noise and its effects on man attempts to provide the basic information and point the way to fuller treatments of the several aspects of the subject.

Many problems arise from noise: annoyance; interference with conversation, leisure or sleep; effects on the efficiency of work; and potentially harmful effects, particularly on hearing. These problems may arise in a wide variety of situations, where they may have to be considered, at least initially, by those who could not reasonably be expected to be expert in the subject. Audiologists, engineers, medical officers in industry, medical practitioners, otologists, physicists and health physicists, safety officers in industry, psychologists, public health officers in the municipal fields, or administrators—for instance in local government—may all find themselves with prob-lems containing elements outside their previous training and experi-ence; I trust that this book may be of some help in easing their entry into a rather wide and often confusing field. I have also had in mind the enquiring non-specialist reader who wishes to have a short factual account of this topical subject.

I have naturally been influenced by the scientific outlook and writings of a number of workers in this field. Thus I would like to record my indebtedness to Dr Hallowell Davis, and his co-author in the classical text *Hearing*, Dr S. S. Stevens; and to Dr G. von Békésy for his writings on many aspects of auditory function. At a personal level, I acknowledge the help, encouragement and kindness of Dr Davis, extending over many years.

Numerous collaborators, colleagues and friends have in the course of scientific associations and discussions benefited this book, directly or indirectly. Dr D. E. Broadbent, Dr R. R. A. Coles, Mr C. Copeland, Dr M. E. Delany, Dr D. H. Eldredge, Dr R. Hinchcliffe, Dr J. J. Knight, Dr A. E. Knowler, Dr K. D. Kryter, Dr T. S. Littler,

Mr N. W. Ramsey, Dr D. W. Robinson, Dr W. Rudmose, Dr R. W. B. Stephens, Dr W. Taylor and Dr J. C. Webster have all put me in their debt for information or advice of different kinds. I have found the practical aspect of acoustics to be greatly illuminated by the writings of Dr L. L. Beranek, and have derived much help from them, as also from Dr I. J. Hirsh's publications, particularly on the measurement of hearing, and from Dr H. E. von Gierke's work.

I am greatly indebted to a large number of authors and publishers for their kind permission to use, either directly or by adaptation, illustrations from their work. These are gratefully acknowledged and are identified here or by the reference attached to the figure. I am especially grateful to Dr A. E. Knowler and Mr J. G. Roberts of the UK Civil Aviation Authority for their advice on the text and illustrations of Chapter 13 and for official permission to reproduce Figs. 13.1, 13.4 and 13.5, and to Messrs Rolls-Royce for the illustrations of turbojet and turbofan engines in the same chapter. I am also particularly grateful for the help of HM Stationery Office in enabling me to adapt figures from the report of the Committee on the Problem of Noise (Chapters 8 and 13). Mr J. Kuehn and Mr W. T. Gracey of B. & K. Laboratories have provided examples of frequency analyses of noise, and Mr W. Coates of the Royal Institution of Great Britain kindly arranged the loan of apparatus shown in Fig. 2.4.

I am particularly grateful to the British Standards Institution, and specifically to Messrs A. D. Falk and J. S. Vickers, for the invaluable assistance they have given in providing material. Mr H. P. Roberts provided technical information for Fig. 2.1 from the British Railways Board. Dr H. Johansen and Dr D. Bentz have drawn my attention to Danish legislation on the sale of fireworks as a precaution against damage to hearing.

The illustrations have been drawn by Mr John Stokes; the photographic work, and much good advice, I owe to Miss P. M. Turnbull, head of the department of Medical Illustration of Charing Cross Hospital and Medical School; Mr K. Sparke, chief technician of my department has provided preliminary drawings for a number of the figures.

The arduous work of producing many drafts in typescript, together with the provision of much assistance on the manuscript in general, has been the work of Miss J. M. Edwards. To all these I offer my thanks.

Finally, I express my indebtedness to John Murray, for the very great assistance I have had from the patient and persistent exercise of their professional skill and competence as publishers.

W.B.

1
Introduction

The conditions of life in countries where technology is in an advanced condition are obviously very different from those in societies based largely on agricultural pursuits. In general, agricultural society persisted in Britain and in other Western countries until the early part of the nineteenth century. At this time the great expansion in the utilisation of coal and iron ore, and the rapid and widespread introduction of mechanical methods for production of goods and equipment and for their transportation, created new conditions of living. The social and material environment of this period has been amply explored by historians, and its imprints, to various degrees, can be seen to this day in every country with a history of heavy industrialisation. Incongruously established in the countryside, the new coalmines, ironworks, factories and their attendant railway developments attracted round them the crowded housing characteristic of the period; and to the atmospheric pollution, slag-heaps and visible industrial waste was added a new factor, noise. Contemporary accounts of the new scenes, sounds and smells of the industrial revolution are plentiful and display various reactions of admiration, awe, doubt or dismay in the minds of the beholders. The physical impact of the proximity of heavy moving machinery, after centuries of principally agricultural pursuits, must have been formidable. There are numbers of contemporary illustrations of the new technology (Figs. 1.1, 1.2), emphasising ponderous machines, factory buildings and smoke-filled skies, and the implication of noise in many of these scenes is obvious.

The industrial revolution, in the main, rested on the availability of coal. For land-based power, for ship propulsion, for railway and even road haulage, the ubiquitous steam engine was fed from coal-fired boilers of riveted construction. This phase lasted for nearly a century; but its end came with the arrival of new technical advances, including the evolution of high-pressure water-tube boilers with elaborate combustion systems and welded construction, supplying steam to turbines; with the adoption of the internal-combustion engine, especially in the diesel form; and finally with the availability

1.1 The black country around Wolverhampton, 1866. (Radio Times Hulton Picture Library.)

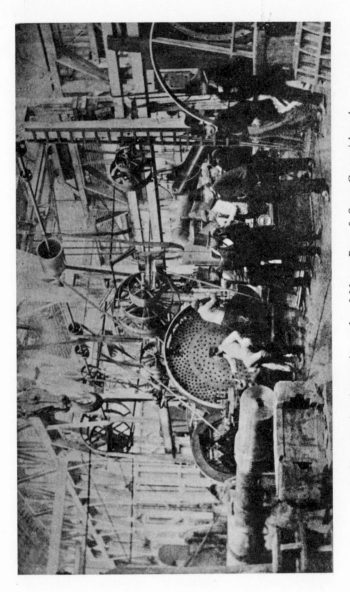

1.2 Heavy engineering at the works of Messrs Penn & Sons, Greenwich and Deptford. [From Barry, *Dockyard Economy and Naval Power* (Sampson Low, London, 1863). (Radio Times Hulton Picture Library.)]

of electric power at an economic rate. The steam locomotive itself, one of the most significant and historic of all the machines used by man, is now almost extinct and superseded by other types of haulage possessing superior economics and greater convenience. Thus the reign of the traditional prime movers of the nineteenth century came to an end, but the phase has left a permanent mark in the terminology of clinical medicine in the term 'boiler-maker's ear'. This term signifies a particular form of hearing impairment in which there is deafness to high tones, as a consequence of the noise of hammering, particularly associated with riveting, which was then the usual way of joining steel plates, and an essential feature of the construction of the boilers of those times.

This is not to say that man-made noise was unknown before the days of the industrial revolution. Wherever metal was shaped by hammering to make ornaments, utensils, armour or weapons; where armies joined in battle; or where gunpowder was used, loud sounds must have been produced. Some interesting contemporary references to these exist. C. H. Parry (1825) describes cases in which hearing was stated to have been impaired, temporarily or permanently, due to noise. In one incident Admiral Lord Rodney is described as being almost entirely deaf for fourteen days following the firing of eighty broadsides from his ship, H.M.S. *Formidable*, in the year 1782. In another instance an officer is stated to have been rendered permanently deaf by the nearby discharge of a cannon at the battle of Copenhagen. There is little doubt that these contemporary opinions on the damaging effect of gunfire noise on hearing were well founded.

We can distinguish between natural sounds and man-made sounds. In general, the more primitive the society the less, in loudness and frequency of occurrence, is man-made sound likely to be. Thus the natural sounds of wind and water, of thunder or the roar of a forest fire, tend to give way in the civilised environment to the sounds made by man, especially in urban communities. Among the greatest sources of noise noticeable out-of-doors now is road traffic. There may be disturbance to communities from industrial noise, for in their immediate proximity many processes and types of machinery produce very loud sounds. We shall apply the term noise to describe sounds which are unwanted and possibly also loud and objectionable. The criteria are thus subjective. The very nature of these definitions presupposes a very wide range of reactions by different people to the same sound, but if the sound is sufficiently loud or long-lasting, or both, or if it has some peculiarity in quality or time pattern, it will be found disagreeable by some people. By and large the louder the noise the greater the number of people who will find it objectionable; with certain noises, a larger proportion of those exposed will be likely to object strongly.

For many years the daily lives of people, particularly in urban communities, have been more and more invaded by noise. Road traffic noise, which had been becoming steadily more noticeable, was accepted without very much complaint until recently. However, the introduction of the turbojet engine into commercial airline operation evoked great antagonism from populations in the vicinity of airports used by these aeroplanes, and it is possible that this particular situation has drawn attention to the existence of other sources of noise, previously tolerated, and may have stimulated a more critical public attitude towards noise in general. The difficulty about aircraft noise, as we shall see later, is the large and often fairly densely populated ground areas in the vicinity of airports which may be subjected to it. At present, this is a consequence of the siting of airports, most of which date back to the era of piston and not jet engines; and of the particular noise characteristics and pattern of landing and take-off flight paths of current large jet-engined transport aircraft. Take-off and landing procedures have in fact already been determined by noise reduction requirements, so that noise has become a limiting factor in the operation of commercial aircraft near the ground. In addition penalties of increased weight and reduction in engine thrust have up till now had to be accepted to secure some reduction in noise output.

In the field of surface transport, road traffic noise now produces serious community disturbance. As well as the actual numbers of vehicles involved, the upward trend in speed or weight or both together has increased noise, to which the almost universal adoption of the diesel engine in commercial road vehicles has contributed. By comparison, rail noise is not in general a source of disturbance: it is intermittent and the areas and locations of the tracks are circumscribed.

Industrial processes are bound to include noisy situations, especially in heavy industry, but these in a sense are perhaps a more contained problem, amenable to design and arrangement of plant in the future. Industry is capable of causing annoyance to communities, or hearing loss in those immediately concerned with noisy operations. Community disturbance should be controlled by area planning, so that industrially noisy areas are segregated from residential areas. Occupational hearing loss can be controlled by appropriate means. The increasing use of automation will doubtless reduce the number of persons actively engaged in noisy situations, but a hard core of intrinsically noisy conditions is likely to remain and must be considered specifically. In contemplating the incidence of noise it must be remembered that conditions can vary very much in the same locality, in the same industry and even in the same factory; we cannot stigmatise a whole industry because of a single noisy situation involving perhaps one isolated process. Similarly, the peace and quiet of hill pastures enjoyed by the shepherd is not shared by his fellow workers on the land whose daily tasks are assisted by the familiar tractor or by other more exotic forms of agricultural machinery. Indeed, significant exposure to noise may be sustained in the course of recreations: the noise of firearms is harmful to hearing; and certainly the loudness of certain types of music argues considerable fortitude on the part of its devotees. This further emphasises the subjective nature of reactions to loud sound, which may depend largely upon the viewpoint of the listener.

In these general considerations we have accepted that certain noises can be objectionable to certain people; one might even add, in certain circumstances. This supposes that the noise is a nuisance in some way; does it annoy, irritate, distract from some task, interfere with sleep, or the understanding of spoken or telephone conversation, or is it simply so loud as to be distressing, or even painful to the ears? Which of these effects are important, and in what conditions? How does hearing actually suffer from exposure to noise, and what are the relations between the noise and the possible effects? These questions, and others, are the ones which naturally and repeatedly arise whenever noise, as a factor in people's lives, is considered.

Fortunately, in the very recent past, a sudden public realisation of the existence of an almost catastrophic deterioration of the habitability of urban and even country areas, and of sea coasts, has occurred. The cause is the presence of physical and chemical agents which are by-products of unrestrained technological development, and of these, noise is one. The whole range of factors, physical, social and economic, involved in protecting civilisation from its undesirable consequences must be taken into account in the acoustic field as in others, before balanced judgements can be reached on the necessary restraints to protect living conditions.

We shall examine the known effects of noise on man, to assess their importance in various circumstances and to come to some conclusions about the desirable limitation of noise to avoid ill effects. But we must first establish certain basic facts about the nature of sound, and the way in which it activates the mechanism of hearing.

2

Physical properties of sound

In this and the ensuing two chapters, a brief account is given of sound as a physical entity. A reasonable appreciation of the quantitative aspects of sound is essential for any systematic correlation between sound and its possible biological effects. There will doubtless be many directions in which readers may wish to pursue the subject further and several texts are available. In particular, *Noise and Vibration Control*, edited by L. L. Beranek (1971), describes sound and its measurement directed to those interested in its practical applications. The same author's textbook, *Acoustics* (Beranek, 1954), is also orientated in a practical manner. A shorter textbook of sound is that of Hall and Matthews (1965). Other sources are Kinsler and Frey, *Fundamentals of Acoustics* (1962), and an advanced comprehensive text is Stephens and Bate, *Acoustics and Vibrational Physics* (1966).

The most obvious feature of sound as a physical phenomenon is its ability to travel in a material medium. The noticeable lapse of time between the lightning flash and the noise of thunder is a reminder that the sound is travelling at a slower speed than that of light. The sound is in this example passing through the air; from common experience we know that sound can be carried also in liquids and in solids. Sound is defined (BSI, 1969) thus: first, as a 'mechanical disturbance, propagated in an elastic medium, of such a character as to be capable of exciting the sensation of hearing'. An alternative subjective definition is also given: 'sound is the sensation of hearing excited by mechanical disturbance.' The common subjective attributes of sound—its pitch, its loudness and its quality—must have their counterparts in the physical nature of the sounds.

Reverting for the moment to common experience of a subjective nature, we can obtain more information by asking ourselves what sound feels like. Consider the impression we receive of loud noises: the departure from an airport of a jet aircraft, or the sound of the more boisterous forms of 'pop' music at close range, particularly where drums are a conspicuous feature. In each case we have a sensation of obviously loud sound, but also a feeling of movement particularly in the chest area, as though the sound were exerting pressure on the body generally, and setting up vibrations which we can actually feel. This pressure effect may be noticed on inanimate objects also; a thunderclap may rattle windows, and so can the noise of a jet engine. The 'elastic medium' in this case is the atmospheric air, and the particular aspect of the 'mechanical disturbance', which is being detected by our ears or other parts of the body, consists of small variations in the normal atmospheric pressure. These variations are the local evidence of the arrival of sound waves from their particular sources. We should now study in more detail the nature of sound waves, and how they may be described quantitatively. Since our main interest is the transmission of sound in air, we may begin by considering the atmosphere which carries these waves.

The atmosphere

Our atmospheric air is a mixture of gases, of which by far the most plentiful are nitrogen and oxygen. The two are mixed but not chemically combined, and there is about four times as much nitrogen, by volume, as there is oxygen. For the purposes of our discussion of air as an 'elastic medium' through which sound may travel, the chemical constitution may be neglected, and the air merely regarded as a gas. Air as a gaseous medium possesses the property of being able to exert a pressure, measurable, for example, by the height of the mercury column of a barometer (Fig. 2.1A). Ascent to higher altitudes gives a decrease in total atmospheric pressure but no significant change in chemical composition. The atmospheric pressure of course varies with time and place even at the same altitude, and its study forms a fundamental part of the science of meteorology. Because of this variability in atmospheric pressure, it is necessary to assume a standard value for calculation purposes.

ATMOSPHERIC PRESSURE   We can profitably consider in more detail the nature of the atmospheric pressure. If an airtight vessel has most of its air removed, the atmospheric pressure will be exerted on the outside only, and may crush the vessel if it is not sufficiently strong. The atmospheric pressure is in fact utilised to operate various mechanisms: for example, if a piston in a cylinder is subjected to unequal pressures on its two faces, it will be displaced towards the side with the lower pressure. This effect is easily achieved, for instance, if the cylinder is evacuated on one side of the piston, while the other side is acted upon by the atmospheric

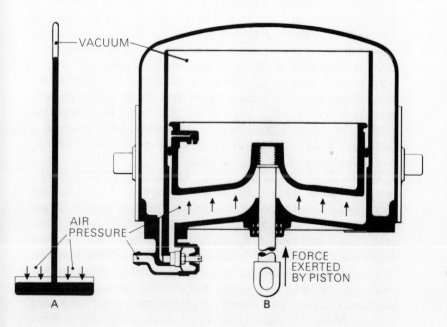

2.1 Demonstrations of atmospheric pressure. A: Mercury barometer. The atmospheric pressure supports the column, and its height indicates the actual pressure; the standard atmosphere is equal to a barometric height of 760 mm of mercury. The meteorological 'bar' equals about 750 mm of mercury or 1 million dyn/cm² or 100000 N/m². The N/m² is the acoustic unit of pressure in the SI system used in this book. B: Railway vacuum brake cylinder. Greater air pressure below the piston causes it to rise, so applying the brakes. (Details by courtesy of British Railways Board.)

pressure (Fig. 2.1B). The atmospheric pressure is appreciable and is measured like other pressures as force per unit area. In general use, force and mass are not clearly distinguished, so that kilogrammes per square centimetre (kg/cm²) or pounds per square inch (lb/in²) are common measures of pressure. The height of the mercury barometer column is used in meteorology; an average atmospheric pressure at sea level is about 1 kg/cm², or 14·7 lb/in². In scientific usage, including that of acoustics, pressure is measured in specific units of force per unit area, such as newtons[1] per square metre (N/m²), or dynes per square centimetre (dyn/cm²). In this book the Système International d'Unités (SI) is used, with the addition of certain references to other units. The relevant SI units, with equivalents in other systems, are given in Appendix B, and a glossary of acoustical terms by courtesy of the British Standards Institution is given in Appendix A. When a pressure is specified for general purposes, it is usual to give the excess pressure over the prevailing atmospheric pressure. Thus a motor tyre gauge may register a pressure of 1·75 kg/cm², or 25 lb/in², but the implication is that outside the tyre is a pressure of 1 atmosphere, while inside is a pressure of 1 atmosphere plus the indicated pressure, or as near to it as the accuracy of the gauge permits.

The basis of this all-pervading, non-directional atmospheric pressure rests on the fact that the air is composed of gas molecules which possess mass and are in continual random motion. These molecules are the smallest units of a gas normally found freely existing, and their random motion is facilitated by the fact that there is free space around them. In the course of their motion, however, they collide with one another and with any solid boundaries of the volume occupied by the gas. We may consider the implications of mass and velocity a little further. In physics, mass is defined as that quantity which, when acted upon by a force, is accelerated in direct proportion to that force. Weight is a property of mass acted upon by the force of gravity. Air can be weighed by finding the weight of an airtight vessel before and after removing its contained air. A suitcase of average size would be about 85 grammes (g) or 3 ounces lighter if it could be evacuated. The air in a living-room might weigh about 70 kg in the usual conditions of temperature and pressure. On this evidence we can be assured that air molecules possess mass.

We noted above that air molecules are in constant random motion; they will thus possess momentum, since in physical terms mass multiplied by velocity is equal to momentum. When the molecules collide with the wall of a containing vessel or any boundary surface there is a change in direction of the motion of the molecules and therefore change of momentum. Since rate of change

[1] After Sir Isaac Newton (1642–1727).

of momentum with time is equal to mass times acceleration, which has the same dimensions as force, in consequence the impact of the molecules will exert a force on the wall. Where pressure exists in excess of the prevailing atmospheric pressure, e.g. inside a motor tyre, more molecules have been forced into the interior of the tyre by the air pump than would have occupied this space by atmospheric pressure alone. Thus, on the same area, there are more impacts on the inner surface of the tyre wall in unit time than there are on the outer surface. In any closed chamber containing gas the same effect, that is increased pressure, can be achieved without adding more molecules by raising the velocity of the molecules. This can be done by increasing the temperature of the enclosed gas.

Sound pressure waves

In dealing with sound waves, we noted that they produce pressure fluctuations above and below the ambient, or prevailing, air pressure. It is usual to describe the behaviour of the air by imagining the volume of air to be divided up into innumerable little compartments or packets. Each is imagined to be large, however, compared with the average distance between the molecules and thus to consist of a considerable number of molecules. These little compartments of air we call 'particles' and we assume that although the contained molecules are in random motion, this does not result in any mass movement of the air, and so the particles are not moving relative to one another when no sound is being carried at that point. This condition of lack of particle disturbance when there is lack of sound does not preclude the whole mass of particles from being moved bodily as in a wind or convection current of air.

Now we must return to the physical requirements for transmission, specifically to the criteria for an elastic medium. For the transmission of an elastic wave, the medium must possess two qualities, elasticity and mass. The principles governing the spread of elastic waves are quite simple: the particles of the medium should have the ability to restore themselves to their original condition if displaced; and they should have mass and therefore momentum so that in colliding with adjacent particles they change the motion of the latter.

That air, and thus its constituent particles, has mass we have just established. With regard to elasticity, air is compressible, and the degree of compression is proportional to the applied force. This spring-like quality of air is utilised in a variety of engineering applications. The pneumatic tyre is an example, as is the use of air as a spring medium for the suspension of road or railway (Koffman & Jarvis, 1964) vehicles. Nobody who has used a hand- or foot-operated air pump can be unaware of this quality. The requirements of mass and elasticity are thus present in air and so the transmission of elastic waves is possible.

SOUND WAVES: MECHANICAL ANALOGY   The general principles of sound waves can be visualised by means of mechanical analogies, perhaps the best of which is that provided by a long coil spring. The spring is usually wound so that the turns at rest are completely closed up, and in the condition of moderate extension suitable for the demonstration, the turns are separated by a small interval, and free to move to some extent without touching one another (Fig. 2.2). These conditions are achieved for demonstration purposes by suspending the spring by threads, while applying slight extension. The turns can be regarded as the particles of air, one turn being able to displace an adjacent turn by movement towards it, until eventually

2.2   Coil spring, seen stationary in this illustration, for demonstrating wave motion.

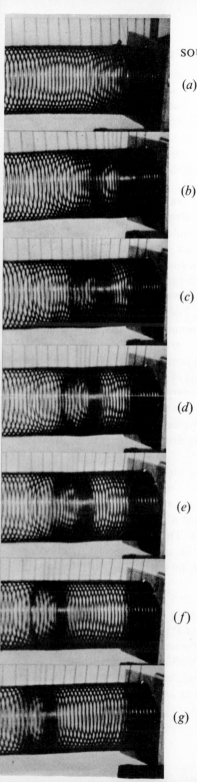

(a)

(b)

(c)

(d)

(e)

(f)

(g)

the turns may collide. This can be illustrated by compressing one end of the spring. Let us suppose that the end turn is thus rapidly compressed and released, so returning it to the position it originally occupied. Viewed in a series of film frames (Fig. 2.3), compression is seen to involve the end turn first, then adjacent turns, so that the region of compression continues to travel along the spring and eventually reaches the other end. If the movements of the spring are caused to die out at the end of their travel by dissipating the energy of the system through some form of damping, for instance by the introduction of friction, that will be the end of the visible disturbance. If not, the compression wave will be reflected and will travel back towards its place of origin.

This wave motion can be sustained if we cause the driving member to oscillate continuously. We can, in fact, by a suitable mechanical arrangement, set up a particular type of oscillating movement known as simple harmonic motion in the end of the spring, and study its effects. This type of motion is fundamental in acoustics and requires further consideration.

2.3 Wave travelling in a coil spring. Selected frames, a to g, from a film taken at 64 frames per second, show the wave as it passes from right to left along the spring.

SIMPLE HARMONIC MOTION   This is the name given to the regular
to-and-fro oscillations which in certain circumstances can be closely
approached by the movement of the blades of a tuning fork, or by
a pendulum swinging through a small arc. Simple harmonic motion
is seen when a small unit of a solid, fluid or gaseous medium, which
we have called a particle, is caused to vibrate along a line of motion
by being displaced from its normal position and then released, provided the force required to displace the particle is proportional to
the displacement. In the case of the tuning fork, when the blades
are struck, they are displaced and subsequently vibrate back and
forth; a slow motion cine picture would show that a blade has a
regular motion between limits on either side of the resting position.
When in motion the blade is for an instant stationary at these limit
positions, and between these, it accelerates to a maximum velocity
at the mid position and then slows down again to the stationary
condition at the other limit position. The acceleration and decelera-
tion are in a regular and orderly manner. In the case of the tuning
fork the limits of the displacement gradually become less until
the blade is stationary in its normal position again, due to dissipation of mechanical energy in the metal of the fork as heat, and in the
air as sound. If, on the other hand, energy is fed into the system by
driving the blades so that they are given a little push at appropriate
intervals, the same amplitude of displacement can be maintained
indefinitely. Tuning forks electrically driven in this way are used in
many applications.

If a pen were attached to a tuning fork blade and made to write
on a sheet of paper which is moving at a uniform speed in a direction
at right angles to the direction of motion of the blade, a waveform
would result which is approximately a graph against time of the
simple harmonic motion (Fig. 2.4). The graph is actually a sine
wave, so called from the curve of the trigonometrical function of
that name. The pressure changes in the simplest sounds, known as
pure tones, are of this type.

Simple harmonic motion is described precisely by the projection
of the motion of a point moving round a circle at a constant rate.
In detail (Fig. 2.5) we consider point P travelling at uniform speed
round the circumference of a circle whose radius is r, centre O and
diameter AOB. If a line vertical to AOB is drawn through the centre
O (XOY) as the point P travels round the circle, its vertical distance
above or below the diameter AOB at any instant can be indicated

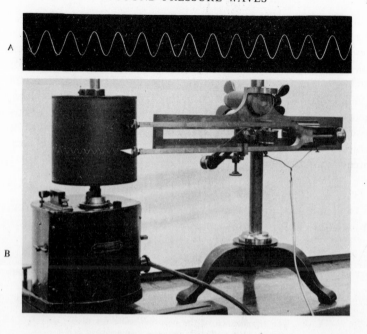

2.4 Trace of simple harmonic motion (A) produced by a large electrically driven tuning fork writing on paper mounted on a rotating drum (B).

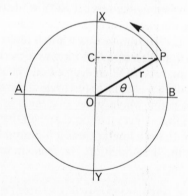

2.5 Derivation of simple harmonic motion.

by its projection on the line XOY, as at the instant indicated by OC. The position of the point P is denoted by the angle θ which the radius r, or rotating vector, makes with the diameter AOB; $\theta = 0°$ when P is at B. A rotation of 360° is known as one cycle. When θ is 0° or 180° the point C coincides with the centre O, and the displacement is zero. When θ is 90° or 270° point P is at X or Y and the displacement is maximal, and is known as the amplitude, symbol a. For any position of P indicated in degrees of angle θ the displacement of its projection from O, as a proportion of the amplitude will be indicated by the relation

$$\text{displacement} = a \sin \theta \qquad (2.1)$$

The relation of the angle θ to displacement of its projection from O is indicated, for various values of θ in Table 2.1.

TABLE 2.1

Angle θ	0°	45°	90°	225°
Displacement	0	$0.707a$	a	$-0.707a$

The significance of the negative value for displacement at $\theta = 225°$ is that the displacement is on the other side of the zero position O.

The graph of the displacement of the projection of point P on the vertical axis of Fig. 2.5 can be plotted as a function of the angle θ, or fractions of a cycle. The curve which results is as if a pen were moving up and down along the vertical XOY while the paper passes across in the direction BA. These are in fact the conditions which produced the tracing of the tuning fork which is illustrated in Fig. 2.4.

In considering simple harmonic motion we described the angle θ made by the rotating vector OP, in degrees, and in consequence one complete rotation of 360° is equal to one cycle of the simple harmonic motion, after which the whole series of events is repeated. Each cycle will occupy a finite interval of time, and all the cycles will be of the same duration. In most applications it is desirable to include the factor of time. In the time notation, the angle moved through will be given by

$$\theta = \text{angular velocity } (\omega) \times \text{time } (t) \qquad (2.2)$$

Angular velocity is expressed in radians per second, where one radian is the angle subtended at the centre of a circle by an arc whose length is equal to the radius, hence 2π radians $= 360°$. The equation (Equation 2.1) for deriving displacement, modified by substituting radians per second for degrees, from Equation 2.3, is then

$$\text{displacement} = a \sin \omega t \tag{2.3}$$

where a = amplitude
  ω = angular velocity in rad/s
  t = time elapsed since P was at point B in Fig. 2.5.

EFFECT OF SINUSOIDAL DRIVE ON COIL SPRING ANALOGY If a suitable mechanical arrangement is used to impart a sinusoidal movement to the end of the coil spring, waves of compression and expansion can be seen to travel down the spring. As in the case of the single compression wave of Fig. 2.3, in addition to the compression, expansion regions with greater spacing than in the resting condition can be seen travelling between regions of compression. These originate at the source, because when the driving member moves back again, it separates the turns, only to close them again on the next compressive stroke. Thus any particular turn irrespective of its position is vibrating back and forth with respect to its normal position. The wave travels continually forward, but not the individual turns. The speed of travel of the wave in the spring corresponds to the speed of sound. The wavelength is the distance between any two corresponding points on two adjacent waves, for example, between two successive points of maximum compression. The degree of compression and opening out of the spring is equivalent to the increase or decrease of the pressure in the sound waves, relative to the ambient or surrounding atmospheric pressure.

The behaviour of the air particles in sound waves may now be appreciated more clearly by leaving the mechanical analogy and considering the compressional or *longitudinal* waves of airborne sound.

Qualitative picture of sound in air

To transfer the performance of the coil spring to the actual transmission of sound in air, we must remember the physical constitution of gases. We recall that the molecules of the atmospheric air are in continual motion, with the inevitable result that they are colliding with one another very frequently, and altering their direction of motion in consequence. The velocity of these molecules in their short erratic movements is, we recall, dependent on temperature; in normal atmospheric conditions it is about 1600 kilometres per hour (km/h) or 1000 miles per hour (mile/h). The distance travelled between collisions is very short and any given molecule will sustain about one thousand million collisions in one second. The concept of a stationary molecule, acted upon by moving surfaces to produce motions which set up sound waves, is therefore not a true representation of the facts. As we have noted, it is customary in acoustic physics to speak of the atmospheric air as being divisible arbitrarily into groups of molecules, known as particles. The merit of the term is that it is more satisfactory to describe a little volume of air, occupied by molecules in random motion as having velocity or displacement, since it represents the mean or collective behaviour of all the molecules in such an imaginary compartment.

Bearing these facts in mind, we can transfer the coil spring analogy to sound waves in air with very little change. The coil spring now becomes a tube containing air. The driving member can conveniently be a loudspeaker diaphragm, which can be made to perform to-and-fro movements, with simple harmonic motion (Fig. 2.6A). Sound waves will then travel down the tube, and in order to avoid the formation of reflected waves at the other end, some termination of a sound-absorbing character should be provided at the end of the tube remote from the diaphragm. We now have a situation in which we can consider the result of movement of the loudspeaker diaphragm in simple harmonic motion. Let us suppose it first moves from its position of rest towards the air in the tube. This will push all the particles in immediate contact with it further into the tube, and they in turn, by collision with their immediate neighbours, will displace them also. The result is that initially the particles in the immediate vicinity of the diaphragm are accelerated. The resulting wave of increased pressure moves towards the right of the diagram. The diaphragm will now reverse its direction. The particles in contact with it at once follow the movement, again in a manner similar to that shown by the coil spring. This produces a reduction in pressure (i.e. a rarefaction), which in turn passes down the tube in the wake of the compression, only to be succeeded by further compressions and rarefactions. Although the waves move down the tube, the particles only move back and forth through a small amplitude, like the turns of the spring. If a pressure-measuring device is inserted at any point in the tube it will register an oscillation of pressure which will, in the example, have the characteristics of a sinusoidal wave if the pressure is plotted against time. The velocity of the waves is the velocity of sound. The frequency of the sound is the number of times which each complete cycle repeats itself in

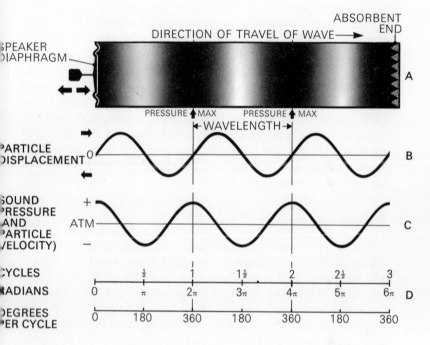

2.6 Sinusoidal wave motion. (A) shows a vibrating diaphragm setting up sound waves in a tube, which travel towards the right of the diagram. The events in A, B, C and D are all synchronised. Darker areas in the tube indicate regions of increased pressure. Note that the same curve (C) describes sound pressure and particle velocity. ATM = atmospheric pressure.

unit time (normally taken as one second) as we saw in connection with the derivation of simple harmonic motion. The shape of the sinusoidal pressure wave is virtually unchanged during transmission. The magnitude of the air particle movements in the illustration is dependent on the amplitude of the oscillations of the diaphragm.

The different parts A, B, C and D of Fig. 2.6 show the way in which particle displacement, particle velocity and pressure in the sound wave are related in time and space in this plane sinusoidal wave motion. The variation with time, throughout each cycle, of all the quantities can be described by sinusoidal waves. All four parts A, B, C and D of the diagram cover the same period of time. In A the loudspeaker diaphragm, while in motion, is shown at the instant when it passes through its midposition, travelling from left to right. At this instant the particles in contact with it are in the position of zero displacement as indicated in B, and the cycle just starting, indicated by zero on the scale of cycles, degrees or radians, shown in D. At this same instant, however, particle velocity and pressure, which run together (i.e. are in phase), are at a maximum, as shown in C. Particle velocity and pressure are, in fact, one-quarter of a cycle (or 90° or $\pi/2$ radians) in advance of displacement, and this phase relation will persist throughout the series of waves generated by the diaphragm and shown passing along the tube to the right (scale D). Study of A, B, C and D will show the relationship of events throughout three complete cycles.

Quantitative aspects of sound

We must now consider how sound can be specified quantitatively. There are two main dimensions of sound which must be covered in any statement of the physical quantities associated with any particular sound. These are, assuming the sound to be a pure tone, the magnitude of the disturbance in the air, and its frequency. For more complex sounds the same basic principles apply, as we shall discuss later. First, let us consider the quantitative description of a pure tone.

The dimension of frequency, as we have seen, is not difficult to specify, and this is almost invariably given in the number of complete cycles occurring in one second, abbreviated to 'c/s' or 'cps', or hertz (Hz).[1]

The other dimension, the magnitude of the vibrational disturbance, is amenable to evaluation in a number of different ways. To appreciate these we must return to the qualitative statements previously made about the nature of sound waves in air, and illustrated in Fig. 2.6. The measurable aspects of sound in common use are: *particle displacement*; *particle velocity*; *particle acceleration*; *sound*

[1] After H. R. Hertz (1857–1894).

pressure; and the magnitude of the disturbance can be described as *sound intensity* or rate at which energy is transmitted through a specified area. The variety of these measures may be somewhat disconcerting at first sight, but they are, however, all aspects of the physical events which we have followed from the coil spring analogy onwards, and they are all systematically related to one another. Sound pressure, sound intensity and frequency concern us principally here.

In practice, the measurement of the frequency of sound presents no particular problems, while the magnitude of the sound is most easily measured as sound pressure. This is, in addition, appropriate since the human ear is a pressure-sensitive mechanism, sensing the regions of greatest sound pressure fluctuation as those giving the maximum loudness, for any particular sound. Finally, provided certain conditions are satisfied, sound intensity can be derived directly from sound pressure, since it is proportional to the square of the latter.

Sound pressure: units

Sound pressure, as we have noted, fluctuates in a pure tone in such a way that for half the time of a complete cycle it is above and half below atmospheric pressure. When specifying the sound pressure we could use the amplitude, which is the peak value of each half cycle, but in fact it is customary to use the average pressure throughout the cycle, ignoring the fact that for half the cycle the pressure is positive and the other half negative, with respect to the prevailing or ambient pressure. This average pressure is obtained by squaring the values of the sound pressure at a large number of instants throughout one cycle (or a whole number of cycles), which eliminates the negative values, averaging these values, and taking the square root of the average. The resulting value is the so-called *root mean square* (RMS) or *effective* value. For pure tones this value is equal to 0·707 times the amplitude of the wave. The RMS value is applicable to other waveforms than pure tones, and is important inasmuch as it is the pressure related to the power being transmitted by the sound wave, as we shall consider later.

For the measurement of sound pressure we require a unit to denote a force applied over a specified area. The pressures with which we are concerned in ordinary sounds are infinitesimally small, and in the course of transmission over appreciable distances a pure tone will remain unchanged in waveform. Large disturbances due to such events as explosions or the movement through the air of projectiles or aircraft travelling faster than the local speed of sound, are in a different category. These waves alter their shape while travelling, and are known as shock waves. The normal sound waves with which we are immediately concerned only set up exceedingly small sound pressures. The faintest audible sounds are measurable in fractions of one-millionth of an atmospheric pressure, and the range of magnitude between these sound pressures and the values for sounds painfully loud is of the order of one million-fold. For this and other reasons it is usual to express the magnitudes of sounds in terms of ratios relative to some reference value, for example, of pressure or of intensity. These ratios are not specified in direct numbers, but in logarithmic ratios called bels.[1] It has been found convenient to subdivide the bel into ten steps, and this is the basis of the decibel scale, which we shall discuss further later. Actual measurement of sound pressures is based on a pressure scale in which the basic unit is a pressure of one-millionth of an atmospheric pressure of nearly 750 millimetres (mm) of mercury (Hg) measured with the mercury in the barometer at a temperature of $0°C$.[2] This pressure is known by various names according to the system of units used. A good explanatory term, much used in the USA is the microbar; the prefix 'micro', as is usual, indicating one millionth. In the centimetre–gramme–second (cgs) system of units 1 microbar is equal to 1 dyne per square centimetre (dyn/cm^2), and in the metre–kilogramme–second (mks) system the same pressure would be 0·1 newton per square metre (N/m^2). As noted above the metre–kilogramme–second units on which the Système International (SI) is based will be used here with, in some instances, the cgs equivalent or other equivalents (see Appendix B, p. 379).

Sound intensity : units

For some purposes the specification of sound on an energy basis is necessary, and thus the magnitude of a given sound may be stated in terms of intensity. The distinction between pressure and intensity

[1] After Alexander Graham Bell (1847–1922).
[2] °C, formerly Centigrade, now Celsius, after Anders Celsius (1701–1744). Here the mercury is taken to be acted upon by standard gravity.

must be clearly emphasised. The sound intensity is equal to the square of the sound pressure divided by a factor, the characteristic impedance. This is the ratio of the effective sound pressure to the effective particle velocity at a point in a progressive plane sound wave. The value of the characteristic impedance in air is the product of the density of the air and the speed of sound in the air, and so depends on the prevailing air temperature and atmospheric pressure. The watt[1] (symbol W), as in other power applications, is used in the quantitative expression of intensity:

$$I = \frac{p^2}{\rho c} \qquad (2.4)$$

where I = sound intensity in W/m^2

p = RMS sound pressure in N/m^2

ρc = characteristic impedance of air (density of air × speed of sound in air) in mks rayls[2]

(At a temperature of 18°C and a barometric pressure of 0·750 m Hg, ρc is nearly 409 mks rayls. In the cgs system ρc in similar conditions would be nearly 40·9 rayls, p expressed in dyn/cm^2, and I in W/cm^2).

The form of Equation 2·4 will be familiar to many, since it is analogous to the formula relating electrical power, voltage and resistance:

$$P = \frac{E^2}{R} \qquad (2.5)$$

where P = watts

E = volts

R = ohms

The sound intensity at a point in a travelling sound wave is the average rate of flow of sound energy per unit area, this area being perpendicular to the direction of propagation of the sound. With these essential facts in mind we can proceed to consider sound in realistic situations.

---

[1] After James Watt (1736–1819).
[2] After Lord Rayleigh (1842–1919).

3
Types of sound

So far, sound has been presented in terms of a quantitative description of the behaviour of sound waves consisting of pure tones, travelling in unobstructed space. This situation is not often found. More realistic circumstances, somewhat more complex, must now be discussed.

Sound consisting of numbers of frequencies

The sounds normally heard are made up of numbers of different frequencies. The possible range of frequencies of sound is large. In the usual notation of numbers of cycles per second, the whole range or spectrum of frequencies of sound is roughly classified into three bands of frequencies by the criterion of audibility to the human sense of hearing. Thus we speak of the frequencies we are able to hear as consisting of audible or sonic frequencies. Frequencies too low to excite the sensation of hearing are known as infrasonic frequencies, and those too high as ultrasonic frequencies. The actual location of these bands in the sound spectrum is not very rigidly defined, but, as we shall see later, the sonic range for young people is usually taken to be from about 20 to 20000 Hz. Other mammals, such as cats, dogs and bats, can hear sounds of much higher frequency than is possible in man; and the lower limit of frequency perceptible to human hearing varies with the intensity of sound, since a pulsating sensation persists at frequencies too low to be regarded as having a musical pitch, and the greater the pressure fluctuations the lower the frequency we can appreciate.

Confining ourselves to the consideration of the sounds having frequencies within the sonic range, we find that a pure tone is a very rare phenomenon. Individual 'notes' struck on the piano, for instance, or blown on a wind instrument, which we identify as middle C, or A or some other recognised musical notation for a particular musical note, are far from being pure tones, but, on the contrary, are complicated waveforms which are composed of numbers of frequencies. The same applies to notes sung by the human voice or the musical hum of some machines. Such sounds are usually called complex tones. An oscillogram, that is a graph of sound pressure against time, of a vowel sound sung by a human voice, is shown in Fig. 3.1A. Great variations in waveform occur between voices and even in the same voice under different conditions. The difference between this waveform and that of Fig. 2.4A, which is effectively a sine wave, is obvious.

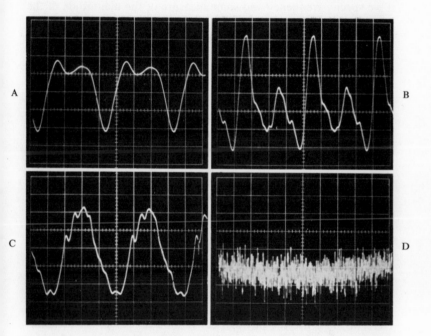

3.1   Oscillograms of pressure against time for different sounds. A: A vowel sound sung by a human voice. B: Alto saxophone. C: Clarinet. The frequency in each case is approximately the same, being about 250 Hz. D: Oscillogram of random noise, electronically generated. Time on horizontal scale: 1 division $= \frac{1}{1000}$ s. (Dr A. F. Lewis and Dr G. B. May.)

The pattern of the oscillogram of the human voice can be seen to repeat itself at regular intervals, although it is obviously not a sine wave. The frequency of repetition of the pattern of Fig. 3.1A is near to 250 Hz, the pitch of the note being a little lower than middle C, the frequency of which is normally 261·6 Hz. Fig. 3.1B and C, show about the same frequency on the alto saxophone and clarinet respectively. They can also be seen to be repetitive complex waves of approximately the same frequency of occurrence, but again different from the oscillogram of the human voice. The physical nature of these tones is described in the next section.

Complex tones : periodic waveforms

The physical description of a pure tone can be completely achieved if we specify frequency and some measure of the magnitude of the sound, such as the size of the pressure waves. The complex tone, however, needs additional description. The waveform of the pressure wave may be complex, but in the types of sound from instruments or voice, which we are considering, it is a stable and regularly repeated pattern. The frequency of repetition of the pattern is the fundamental frequency. In the case of the voice and instrumental sounds of Fig. 3.1 the fundamental is about 250 Hz, but other frequencies are present, which give the characteristic shape of each. Such waveforms may be mathematically analysed by the method of Fourier analysis. This method demonstrates that any regularly repeating waveform can be resolved into a number of pure tones, the frequency of each being a whole or integral multiple of the fundamental frequency. These integral multiple frequencies are known as harmonics, and the frequencies are said to be harmonically related. In the terminology of acoustic physics and of music there is a difference in the meaning of the word harmonic. In physics, for example, the harmonic which is twice the frequency of the fundamental is known as the second harmonic; in music as the first harmonic, and similarly for higher multiples of the fundamental. Here we shall use the physical convention in which, for a fundamental of n Hz, the second harmonic is of frequency $2n$ Hz. Commonplace sounds may contain considerable numbers of harmonics, and these may be conveniently displayed, as we shall see, in a simple plot of the sound pressure of each frequency. Such a description is known generally as a frequency spectrum. To sum up, therefore, the concept of Fourier analysis means that, from a waveform or sound pressure on the vertical axis and time on the horizontal axis, it is possible to say what pure tones, in terms of frequency, amplitude and phase, are making up the complex waveform; and moreover, having determined in this way the frequency spectrum, the waveform could be reproduced by combining in the appropriate phase relations the various pure tones of different sound pressure and frequency. An illustration of the synthesis of a waveform very unlike that of a sine wave is given in Fig. 3.2. Here a square wave is analysed, showing that a recognisable approach to it can be achieved by three sine waves added together, while 15 sine

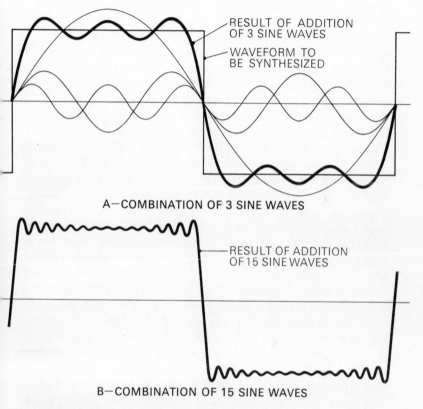

A—COMBINATION OF 3 SINE WAVES

B—COMBINATION OF 15 SINE WAVES

3.2 Synthesis of a complex wave from a number of sine waves. At any given time, the amplitude of the complex wave is derived by adding the amplitudes of the individual sine waves. Formation of a square wave: A, from 3 sine waves; B, from 15 sine waves. (After Wood, A. B., *A Textbook of Sound* (3rd Ed. revised). London, G. Bell.)

waves of suitable frequency, intensity and phase relations show a waveform which is beginning to approximate to a square wave. It should, however, be stated here that, within certain circumstances, the ear disregards the phase relations of different components, for instance in a steady complex tone.

Non-periodic waveforms: noise

Some sound waves, unlike those just considered, do not repeat themselves in a regular way, but are random in their fluctuations. Such waves may be caused by a very large number of pure tones, not harmonically related, occurring together. The total pressure due to the various pure tones at a given moment could theoretically be the sum of their maximum pressures if it so happened that each coincided at a particular moment of time. The greater the number of frequencies, the less likely is this to occur, so that an undulating irregular waveform considerably less in maximum pressure than the sum of the individual frequencies is the result.

When the components of different frequency are very numerous and not harmonically related to one another we find the type of sound which the physicist knows as noise. This is not to be confused with the subjective definition of the same word. To the physicist, noise may mean, in addition to the usual definition of unwanted sound, random electrical or acoustic disturbances. Acoustic noise, therefore, in this sense is sound with a continuous spectrum, that is, with a very great number of frequencies so close together that they are not easily distinguishable as separate entities. Such noise is produced when gas escapes at high pressure to the atmosphere. An excellent example is the sound of steam escaping from the safety valve of a steam locomotive, the noise of a waterfall or surf breaking on a beach. Some types of jet engine exhaust, or even some machinery noise such as that of many weaving looms operating together, produce continuous spectrum noise. An oscillogram of such a spectrum, generated electronically, is shown in Fig. 3.1D.

A particular type of continuous spectrum noise is white noise. This is so called in analogy with white light, which contains all the wavelengths of the visible spectrum. White noise therefore is sound having a continuous spectrum in which the sound pressure in bands of frequency one cycle per second wide is substantially constant over an appreciable range of frequencies.

Not infrequently continuous spectrum noise contains also quite powerful pure tones, which may be heard against the background of roaring and hissing characteristic of the wide band element. This occurs conspicuously in some types of jet engine. In machinery noises which are nominally pure tones and which sound musically pitched, harmonics also occur, so that once again we find that a pure tone in isolation is not a common form of sound.

TIME CHARACTERISTICS OF NOISE   So far, we have not considered the effect of variations in sound pressure with time. This is an important feature of many noises in practical situations, and it can occur in various ways. A noise may be discontinuous, and the intermissions may be of a regular or irregular nature. For example, road traffic noise will be affected by the density and periodicity of the passage of vehicles. The numbers of aircraft using a particular flight path, or the repeated operation of an industrial process may affect greatly the seriousness of a noise disturbance. The nature of noise may change with time, adding further complication. An extreme case is where the sound is totally discontinuous such as gunfire noise or impact noise such as from hammering, stamping, pressing or forging operations. Here the energy of the sound is restricted to a very short interval of time, and the repetition rate may be very variable indeed. Riveting guns or mechanical hammers produce blows at a high rate, but large presses may have a cycle of operations at a comparatively slow rate. The repetition rate of impact noise is of some significance in assessing its potentially harmful effect on hearing.

EFFECTS DUE TO DIRECTION AND DISTANCE OF TRANSMISSION The direction from which sound comes has an important bearing on the reaction of the listener. The judgement of direction is an important factor in nature, and survival may depend upon it for many lower animals or even for man. Judgement of direction depends on the relative strength and time of arrival of the sound at the two ears. As sound spreads from a source, its intensity will diminish with the distance. Due to such factors as local atmospheric temperature variations, or obstructions such as buildings, the sound waves may be deviated from their original direction. In addition, the quality of sounds changes the further it travels, the higher frequencies in particular tending to diminish in strength at a greater rate than the lower.

Velocity of sound waves

We have throughout our discussion of sound emphasised the travelling nature of the waves. The concept of the velocity of sound has, in recent years, been brought to the attention of the public by the increasing speed of modern aircraft, and particularly by the prospect of commercial aircraft possessing the capability of exceeding the local speed of sound. The speed of sound is dictated by the nature of the medium, solid, fluid or gaseous, in which it is travelling. The subject is most conveniently approached by first considering wave motion in solids and the following basic discussion may be supplemented from Stephens and Bate (1966); *see* references to Chapter 2.

TRAVELLING WAVES IN SOLIDS   The propagation of longitudinal waves in solids is best described by taking the case of a long thin rod of circular section and composed of a homogeneous material. Were it possible to have this rod composed of a completely incompressible material, a blow on one end could accelerate the entire rod, so that the impulse would be detectable at the same instant of time at the other end. The speed of transmission of the impulse would thus be infinite. In fact, no such incompressible material exists. If the rod were composed of any real solid material, its elasticity would render it capable of deformation as a result of the application of a force, with subsequent recovery within certain ranges of deformation. Thus in the case of the long thin rod, a blow struck on one end will initially set up a longitudinal wave of compression followed by one of expansion, and these travel down the rod towards the other end. The velocity of such waves is dependent on the elasticity of the material and on its density. To appreciate these factors, let us consider the events in more detail. First, because the rod may be excited to produce a series of waves of compression and expansion, we must consider only a short length at a time. For this purpose the rod can be regarded as being composed of a very large number of very thin discs, or elements, as though it were made up, for example, of a vertical stack of identical coins.

Suppose now that the rod is made to vibrate longitudinally. Each element, as it becomes involved in the compressive phase of the wave, is acted on by a force in the longitudinal direction. Its two faces are thus brought closer together, that is, the element or disc is compressed. The compressive force is known as the *stress* and the degree of deformation or compression of the element the *strain*. If these relations are expressed generally as

$$\frac{\text{force per unit area of cross-section}}{\text{change in length per unit length}}$$

a ratio known as Young's modulus of elasticity (E) is the result. Specifically,

$$E = \frac{\dfrac{F}{A}}{\dfrac{l}{L}}$$

where F = total compressive (or tensile) force

A = cross-sectional area of the rod

l = magnitude of change (either shortening or lengthening) in length of the element

L = original length of the element.

If we assume that Hooke's law is obeyed, that is, that strain is proportional to stress, this ratio will be the same for a range of values of strains and stresses. The compressive force or stress, acting on the mass of the element, will accelerate it. Since acceleration is inversely proportional to mass and mass is proportional to density, then density must also be inversely related to acceleration. The accelerated element will in turn supply the force to accelerate the next element, so that a travelling wave results. From Newton's law of motion, $v^2 = E/\rho$ (where E is Young's modulus of elasticity, ρ is the density of the material, and v is the velocity of the elastic wave in the rod). Consequently

$$v = \sqrt{\frac{E}{\rho}}$$

Thus the velocity of the wave is proportional to the square root of Young's modulus for the material and inversely proportional to the square root of its density.

VELOCITY OF SOUND IN FLUIDS   In the transmission of sound in fluid media, that is liquids and gases, the conditions are somewhat similar to those in solids, but there are certain important differences. In gaseous transmission in air, the waves cause local fluctuations in volume and density which we have noted, but these variations are very small. The essential factors in determining the velocity of sound in air are again its elasticity characteristics and its density.

The concept of a modulus of elasticity is still applicable to gases and liquids, but in this case the nature of the medium requires a modification of the type of strain involved. When a fluid (for example, air) is compressed a reduction in volume is produced without any permanent resistance to change in shape. The compressive force per unit area is effectively a pressure, but may still be termed a stress and it acts uniformly in all directions. The resultant reduction in volume, the strain, is expressed as the change in volume divided by the volume before application of the stress. Thus the version of Young's modulus applicable to gases or liquids is the bulk modulus (K)

$$K = \frac{\text{compressive force per unit area}}{\text{change in volume per unit volume}}$$

This ratio can also apply to tensile (i.e. extending) forces as well as to compressive forces.

When Newton originally considered these relations, he made the assumption that Boyle's law applied. This is that pressure and volume are inversely related, provided the temperature remains constant. Hence if P = initial pressure; V = initial volume; dp = increase in pressure; and dv = decrease in volume:

$$PV = (P + dp)(V - dv) = \text{constant} \qquad (3.1)$$

$$= PV + V . dp - P . dv - dv . dp \qquad (3.2)$$

However, for the magnitudes of sound waves normally encountered dv is a very small (about 10^{-5}) fraction of V, so that $dv . dp$ can be neglected. Thus for such small changes in pressure Equation 3.2 becomes

$$PV = PV + V . dp - P . dv$$

so that

$$P . dv = V . dp$$

or

$$P = \frac{V . dp}{dv} \quad \text{or}$$

$$= \frac{dp}{dv/V}$$

This term is in fact the bulk modulus of elasticity of air

$$\frac{\text{compressive force per unit area}}{\text{decrease in volume per unit volume}} = P$$

where P, it will be recalled, is the initial pressure (Equation 3.1). The bulk modulus of elasticity of air thus becomes equal to P, which in practical terms is the prevailing atmospheric pressure.

We may now attempt to calculate the velocity of sound in air (symbol c) by means of the same formula as for solids, viz:

$$c = \sqrt{\frac{\text{bulk modulus of elasticity}}{\text{density}}}$$

When this was done originally, the results yielded velocities considerably below those found experimentally. The anomaly was clarified by Laplace,[1] who showed that the bulk modulus calculated using Boyle's law, that is, under constant temperature conditions is not usually justified. Anyone who has blown up a bicycle tyre with a hand pump is conscious of the temperature rise in the compressed air. Boyle's law assumes that this rise, which occurs within each situation of increased pressure in sound waves, is prevented by allowing time for the heat to dissipate, i.e. by causing the change of pressure to take place slowly. In the repeated compression of the air in sound waves, there is not enough time for the complete dissipation of the heat, and so there is a greater resistance to compression, due to the expansion of the hot air, than if the compression were slow. The bulk modulus based on Boyle's law, $PV =$ constant, is known as the isothermal bulk modulus, while the equivalent bulk modulus where temperature change occurs, is known as the adiabatic bulk modulus. This is dependent on the adiabatic law $PV^\gamma =$ constant. The adiabatic bulk modulus can be shown to be γP, where γ is the ratio of the principal specific heats of a gas at constant pressure to that at constant volume, instead of P for the isothermal modulus. Thus the velocity of sound in air becomes

$$c = \sqrt{\frac{\gamma P}{\rho}} \qquad (3.3)$$

For air and other gases having 2 atoms per molecule, the adiabatic bulk modulus of elasticity is about 1·4 times the isothermal modulus.

[1] Pierre Simon Marquis de Laplace, mathematician and astronomer (1749–1827).

Using this formula, good agreement is obtained between calculated and experimentally determined velocities.

Certain consequences follow from the above relations.

EFFECT OF PRESSURE ON THE VELOCITY OF SOUND IN AIR   Provided temperature is unaltered, the density of air is proportional to its pressure. Pressure changes will produce corresponding density changes and so the fraction $\sqrt{(\gamma P/\rho)}$ of Equation 3.3 will remain unaltered. The velocity of sound in air at constant temperature is thus independent of pressure.

EFFECT OF AIR TEMPERATURE ON THE VELOCITY OF SOUND   Although temperature does not appear specifically in the velocity equation, its influence on density makes it the main determinant of velocity of sound in air for small amplitudes. Increase of density as a result of lowering the temperature will thus reduce the velocity of the sound wave. The relation simplifies down to the statement that the velocity of sound in air is proportional to the square root of the absolute temperature of the air.

Specific density effects may be seen where air is highly saturated with water vapour. Because of the lower density of the vapour compared to air, the velocity of sound is slightly increased under these conditions.

PRACTICAL IMPLICATIONS   The operation of the velocity equation for different media follows in conformity with their elastic properties and densities. Solids and liquids, although much more dense than air, have in general, much higher relative elastic moduli, so that velocity is higher than in air for most solids and liquids, but conspicuous variations in density occur. For example, at 15°C, steel has a velocity of sound of about 5050 m/s, aluminium about 5200, whereas in lead, the velocity is about 1200 m/s, due to its low value of Young's modulus allied to a high density. In the same circumstances; hydrogen has a sound velocity of about the same value, approximately 1270 m/s, which is high for a gas, mainly due to its low density. The velocity in sea water is about 1500 m/s at 15°C, while in air at this temperature the velocity is much lower, at about 341 m/s. The velocity of sound in air can be determined simply in the following way (Beranek (1971) in references of Chapter 2):

$$c = 20 \cdot 05 \sqrt{T}$$

where c = velocity of sound (m/s) at temperature T

T = absolute temperature ($273 \cdot 2$ + degrees C)
in degrees Kelvin[1] ($^{\circ}$K)

or

$$c = 49 \cdot 03 \sqrt{R}$$

where c = velocity of sound (ft/s)

R = temperature Rankine[2] ($459 \cdot 7$ + degrees F)

Thus the velocity of sound in air, for example at $293 \cdot 2^{\circ}$K (20°C or nearly 68°F), is about 343 m/s (1126 ft/s). Since temperature diminishes with increased altitude, so also does the velocity of sound. Due to this variability in the velocity of sound, and its importance in the operation of high-speed aircraft, it is usual to provide an instrument indication of airspeed on the Mach scale, in which the aeroplane's airspeed is referred to the local speed of sound. For example, Mach 0·9 would equal 0·9 times the local speed of sound. Some confusion can arise in the terminology peculiar to acoustics and aircraft technology. We have seen that frequency is arbitrarily divided into the ranges infrasonic, sonic and ultrasonic; speeds, usually of aircraft, are described as subsonic (less than Mach 1) sonic (about Mach 1) or supersonic (above Mach 1). When aircraft fly at speeds in excess of Mach 1, as we have briefly noted, disturbances in the air known as shock waves occur, resulting in the so-called sonic boom. In certain circumstances small shock waves may be visible where the airflow over the surface of an aircraft locally exceeds Mach 1, as may occur even though the airspeed of the machine as a whole is subsonic.

Standing waves

The reflection of sound, revealed by an echo, is a common experience. In certain conditions sound waves may occur which do not travel. They are known as standing waves, and they are due to

[1] After William Thompson, Lord Kelvin (1824–1907).
[2] After W. J. M. Rankine (1820–1872).

reflection of sound waves from obstructions or surfaces in the path of travelling waves. Since in modern living conditions, sound is heard less frequently in unobstructed space than in built-up areas or the interiors of buildings, reflections of sound are common. In such cases, interaction can occur between the original and the reflected waves so that systems of reinforcement or cancellation can occur, at stationary positions dependent on the wavelength of the sound and the size of the enclosure. Such conditions will obviously influence greatly the distribution of sound pressures within enclosures and demand knowledge and care in making measurements. This brings us to the next step, of appreciating the quantitative aspects of sound.

4

The measurement of sound

Having gained some impressions of the physical nature of sound in a qualitative way, we must now turn to the essential step of expressing sound quantitatively. The measurement and quantitative handling of sound is the essence of the science of acoustics. The practical applications of acoustics in technology and medicine are very numerous: for example, for assessment of the performance of machines; for the assessment of the acoustics of buildings and for judging the habitability of home and office, or of transport vehicles of all kinds; for investigation of conditions for voice communication; or for the avoidance of damage to hearing and for other biological and medical purposes; for all these quantitative description of sound is required.

In considering the way in which sound behaves when radiated in air, let us recall (Chapter 2) the situation of a long tube, at one end of which is a loudspeaker diaphragm (Fig. 2.5); in these circumstances sound energy will flow down the tube. We assume that no losses occur in the tube, which acts simply as a channel for the sound, so that the waves arrive at the end remote from their origin with all the energy imparted to them by the loudspeaker. The type of wave found in these circumstances is called a progressive plane wave.

If we consider what happens when sound is radiated in all directions in an unobstructed volume of air (these conditions are approximately realised where a rocket at a fireworks display ends its flight with the setting off of a charge of explosive material, so making a bang in mid-air), it is obvious that the conditions are different. The sound waves are not now channelled, but are free to spread outwards in all directions so that they fill an imaginary sphere with its centre at the source of the sound. We again assume that no losses occur in transmission through the air. The radius of the sphere is equal to the distance from the source at which we hear, or measure, the sound. Since the intensity of the sound is defined as the energy flow through unit area, it is obvious that the greater the distance from the source, the smaller will be the intensity, because of the increase of the total area of the sphere through which the energy must pass. Since the area of a sphere increases as the square of its radius, the intensity (I) of sound waves of this type, called free progressive spherical waves, must diminish as the square of the distance (r); hence I is proportional to $1/r^2$. It is thus possible to predict the intensity of sound at a given distance if the total acoustic power of the source is known. However, complicating factors exist, and the air itself, as we have already noted, introduces losses during transmission. These losses are small below 1000 Hz but become greater as frequency is increased.

The range of values of sound pressure, and even more so, of sound intensity normally encountered is inconveniently large for normal arithmetical expression. For this and other reasons it has become customary to express the magnitudes in acoustics and in electrical engineering by means of the decibel scale. As we noted briefly in Chapter 2, this scale depends on the use of ratios, so that a given sound intensity, for example, may be twice, 10 times or a million times as great as some defined reference intensity. The ratios are expressed not directly, but as their logarithms, to the base 10. Thus if an intensity is 10 times greater than the reference intensity, the logarithm would be 1, the ratio being designated 1 bel. A ratio of 100-fold would thus be 2 bels; 1000-fold, 3 bels. For convenience, 1 bel has been divided into 10 steps, hence the decibel (dB); and increase of 1 dB means an intensity increase of nearly 1·26. Numerical values in decibels are usually called *levels*, thus:

$$\text{sound intensity level} = 10 \log_{10} \frac{I}{I_{\text{ref}}} \text{ dB re } I_{\text{ref}} \qquad (4.1)$$

where I = intensity

I_{ref} = reference intensity.

Since, for reasons of practical convenience, sound pressure is usually the quantity which is actually measured, it is desirable to be able to use the decibel for pressures also. Assuming that sound intensity is proportional to the square of the sound pressure, an increase of 10-fold in pressure will correspond to an increase of 100-fold in intensity. It is thus clear that, with appropriate modification, dB values for pressure ratios can be derived in the same general way as for intensity ratios. Specifically, all that is needed to accommodate pressure in the formula is to square the pressure ratios in calculating decibels. The intensity equation (4.1) can be rewritten thus

$$\text{sound pressure level} = 10 \log_{10} \left(\frac{p}{p_{ref}}\right)^2 \text{dB re } p_{ref} \qquad (4.2)$$

where p = sound pressure (RMS)

p_{ref} = reference sound pressure

      (p and p_{ref} must be in the same units)

Squaring can, however, be accomplished in logarithmic expressions by multiplying the logarithm by 2, so that Equation 4.2 now becomes

$$\text{sound pressure level} = 20 \log_{10} \frac{p}{p_{ref}} \text{dB re } p_{ref} \qquad (4.3)$$

Equation 4.3 for a particular pressure ratio gives the same dB value as Equation 4.1, using the corresponding intensity ratio.

To summarise, therefore, to obtain the decibel value for a pressure ratio, multiply the logarithm of the ratio by 20; for an intensity ratio, multiply the logarithm of the ratio by 10.

In using decibel values, great convenience and simplicity results, but it must always be remembered that the scale is a logarithmic one. In terms of pressure, a doubling of pressure is an increase of approximately 6 dB, but 12 dB means doubling twice (that is an increase of four-fold) and 18 dB doubling three times, an increase of eight-fold in pressure, in approximate figures. Subjectively, an increase of 10 dB would be judged, on average, to make a sound twice as loud, which is a useful yardstick to remember.

For convenience in use the decibel values for various ratios of pressure and intensity are given in Appendix D (p. 385).

Reference values and sound levels

Having discussed the essentials of the decibel system of denoting ratios, the next step is to consider how it can be used to indicate actual magnitudes of sound. All that is required is a statement of the value, in this case either of pressure or intensity, which is used as the reference. Arbitrary values have been adopted for this purpose. The reference pressure for airborne sound is a root mean square

(RMS) value of 0·00002 N/m^2, alternatively expressed as 0·0002 dyn/cm^2 or 0·0002 microbar. This particular pressure was adopted as being about the smallest audible at the frequency of greatest sensitivity in young people with clinically normal ears. If this reference is used, it is legitimate to refer to decibel values so described as 'sound pressure level' (SPL). Having thus established a reference level for pressure, it is natural to seek also a reference level for intensity. Intensity could then be expressed in dB as 'intensity level' (IL). It would, in addition, be convenient if the intensity reference level could be chosen so that, in using dB values for intensity, the result, for a given sound, would be equal or nearly equal numerically to the SPL values. It is not possible to obtain numerical equality except by confining ourselves to defined conditions of temperature and pressure, for the following reason. It will be recalled that pressure and intensity are related in such a way that intensity is equal to pressure squared, divided by a factor, the characteristic impedance of air (ρc), which is the product of the density of the air and the velocity of sound in air (Equation 2.4). The numerical value of ρc will thus be determined by the prevailing temperature and atmospheric pressure. For combinations of temperature and pressure giving $\rho c = 400$, in conjunction with a reference intensity of 10^{-12} W/m^2, SPL and IL values will be identical. For temperatures and pressures yielding other values of ρc, disparities will occur between SPL and IL values. For example, in temperate conditions and common barometric pressures at some 100 m above sea level (18°C and 0·750 m Hg), quoted on p. 27, the value of ρc is near 409. This value gives only a small difference, about 0·1 dB, between SPL values (referred to 0·00002 N/m^2) and IL values (referred to 10^{-12} W/m^2). The use of 10^{-12} is the recognised convention, but it is always desirable to state explicitly what reference is in use in any particular set of data.

It will be noted that the reference level for intensity, 10^{-12} W/m^2, is expressed in the exponential notation. This use of the powers of 10 confers great convenience and is widely used to express large or small numbers.

For an illustration of the use of simple numerical manipulations involving sound pressures and intensities consult Appendix E (p. 392). These examples are based on the type of presentation used by Beranek (1960, 1971) where more advanced material will be found.

Contribution of individual sound sources to a sound field

In practical situations problems frequently arise in which the part played by separate sound sources in a total effect must be considered. For example, it may be desirable to know the combined sound pressure level to be expected from a number of machines, the SPL values of which are known separately. Or again, if a noise environment is deemed excessive, what effect on the total SPL value would result from removing specific sound sources. For these purposes, decibel values are a convenient language in which to express the results. However, it must be remembered that decibel scales are logarithmic, so that, for example, addition achieves multiplication. We cannot therefore add or subtract sound pressures in dB; the energies, corresponding to the pressures squared, must be found, added or subtracted in arithmetical terms, and the result converted back into sound pressure and expressed in decibels relative to the sound pressure reference. This principle applies to combining any two unrelated signals such as pure tones of different frequency or bands of noise from different parts of a frequency spectrum, and it can also apply to different wide-band sources. In practice decibel values are combined or subtracted with the aid of tables or charts. A chart for this purpose with an explanation of the principles of addition and subtraction of decibel values is given in Appendix F (p. 393).

Measurement of sound: the sound level meter

The basic instrument for the objective measurement of sound is known as the sound level meter. It consists of a microphone, an amplifier and some indicating device, normally a meter which gives visible readings on a scale. The functions of the components are obvious: the microphone transforms sound pressure waves into electrical voltage fluctuations which are then amplified sufficiently to actuate a meter, or alternatively a recorder of some kind. Practical considerations require this basic arrangement to be elaborated in a number of ways. Firstly, if the amplified electrical signal is to be suitable for the actuation of the meter for widely different sound pressures, the amplification must be adjustable. Otherwise, the range of output would far exceed the dynamic range, as it is called, of the meter, which can only show satisfactorily a variation of a few decibels. Thus the meter reads over a small range, and this range is selected to suit the sound being measured, by the manual operation of an attenuator, so varying the effective amplification between microphone and meter. In addition, a rectifier circuit is placed before the meter, which converts alternating currents in the amplifier into

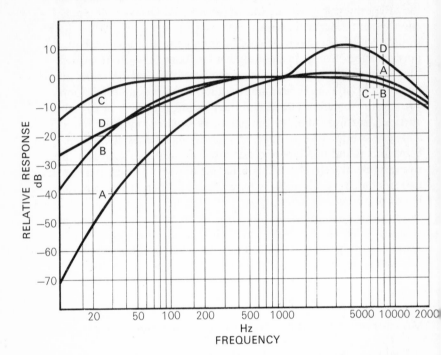

4.1 A-, B-, C- and D-scale characteristics for sound level meters. The relative responses for the four operating conditions are shown as response in dB (vertical scale) for different frequencies (horizontal scale). The A-scale response diminishes markedly as frequency is reduced, so that a tone of 50 Hz would be indicated at about 30 dB less than a tone of 1000 Hz. The C-scale response has only small differences of less than 1 dB, from 63 to 4000 Hz. The D-scale response is enhanced in the higher audible frequencies. (After IEC Publication 179.)

one-directional, or direct current (DC) suitable for actuating the meter, which is calibrated to read sound level over a short interval of time.

Finally, the sound level meter incorporates electrical circuits known as weighting networks. These provide for various sensitivities to sounds of different frequencies, the original object being to simulate the characteristics of the sensitivity of the human ear at different frequencies for different sound levels. These characteristics are known as the A-, B-, or C-scale operating conditions, and can be selected on the sound level meter. The A-weighting has acquired other uses (Chapters 8, 9, 10, 11) and latterly a D-weighting has been proposed (Chapter 13 and Appendix K) to give an indication of perceived noisiness. Although a D-weighting is already tentatively incorporated into sound level meters, in fact the numerical values are still being considered by Working Group 5 of Sub-Committee 29C of the International Electrotechnical Commission (IEC), Geneva. There may also be on the sound level meter a weighting characteristic indicated as 'linear'. This indicates that within defined values the instrument has a similar response to all frequencies. The A, B and C conditions vary mainly in the degree of sensitivity provided at the lower frequencies relative to the sensitivity at 1000 Hz in terms of indicated level for a given sound. Least sensitivity is provided at low frequencies in the A-scale, most in the C-scale, in order to simulate the sensitivity of the human ear for different intensities of sound. These characteristics are published by IEC (1965). It is expected that the D-scale will be added to the above IEC publication; it corresponds to the inverse of the 40 noy contour (Chapter 13 and Appendix K), the insertion loss of the weighting network being (like those of the A, B and C networks) zero at 1 kHz (ISO, 1970a, 1970b). Measurements of the sound pressure level with a weighted response are usually referred to as *sound level* with the appropriate suffix, e.g. sound level A with the actual values expressed as dB(A). As noted previously, any statement of sound pressure level (SPL) without qualification or statement to the contrary signifies a reference value of $0 \cdot 00002$ N/m^2 (2×10^{-5} N/m^2) and it is implied that the measurement was made with an instrument having a linear frequency response. Sound level C readings on a sound level meter approximate to these conditions, and are often used as such. The recommended frequency characteristics for the scales A, B and C, together with a version of the D-weighting, are given in Fig. 4.1 as weighting curves and in Appendix G as numerical values. In the later chapters the significance of various values of sound level on the different scales will become apparent.

Frequency analysis of sounds

OCTAVE BAND ANALYSIS   For many purposes the statement of the total, or overall sound pressure or intensity of a particular sound, by means of a sound level meter operated on a particular scale or number of scales, is not enough. For example, it would be possible to have two sounds which gave the same SPL values in dB on the C scale of the sound level meter, but subjectively one might be highly objectionable, for example the sound of a circular saw, while the other might be entirely acceptable, such as within a moving railway coach or transport aircraft. On the A scale of the sound level meter, the latter type of noise would give a much lower reading, and one could infer, correctly, that its energy was mainly in the lower frequencies, to which the ear is less sensitive. In the case of the circular saw, the energy would be distributed more in the higher frequencies. However, it frequently happens that a noise has to be investigated in detail from the point of view of mechanical considerations, annoyance or potential damage to hearing, and in such cases examination of the distribution of intensity in the frequency spectrum must be undertaken.

TABLE 4.1

Preferred frequencies for acoustic measurements. Geometric centre frequencies and limits of octave filter pass bands

Centre frequencies Hz	Limits of band Hz
63	45–90
125	90–180
250	180–355
500	355–710
1000	710–1400
2000	1400–2800
4000	2800–5600
8000	5600–11200

Note: other bands, in addition to those given above, may also be used. Details are given in Appendix H (p. 397).

The procedure for measurement of sound pressure at different parts of the frequency spectrum is known as frequency analysis. The principle is simple. For the purpose of overall measurement of sound pressure, we have seen how the sound level meter accepts and measures all frequencies throughout an arbitrary band of frequencies on the C scale or linear scale as defined. In frequency analysis we insert a system of filters which will pass only certain bands of frequencies, and again we are helped here by international agreements on quantitative details. The accepted practice for the simplest type of frequency analysis is to use filters which pass a range of frequencies equal to an octave; that is a band in which the lowest frequency is half that of the highest frequency. The bands are contiguous and the centre frequencies by which the octave bands are identified extend upwards and downwards from 1000 Hz. The details of the preferred frequency bands which have been internationally agreed, are shown in Table 4.1 for octave bands (BSI, 1963). The centre frequency of each octave is the geometric centre frequency. Thus the octave with centre frequency 1000 Hz would extend from 710 to 1400 Hz, i.e. have a geometric centre frequency $= \sqrt{(710 \times 1400)} =$ approximately 1000 Hz. 'Octave mid-frequency' has the same meaning.

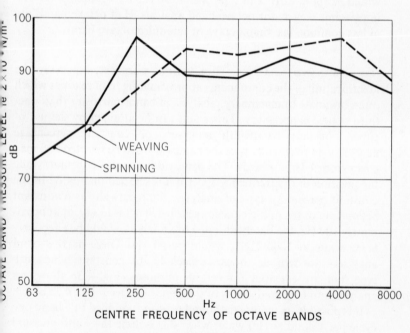

4.2 Octave band analysis of spinning and of weaving machinery.

The SPL of each band is measured and the sound can thus be described in numerical values of SPL in each of the octave bands, or conveniently displayed as a plot of SPL in dB on the vertical axis and the different octave bands spaced out on the horizontal axis. An illustration of the use of this method is given for the noise of spinning and weaving processes (Fig. 4.2). This is a most useful and simple way of describing a given sound and the eye can encompass at a glance the general distribution of energy throughout the frequency spectrum.

ONE-THIRD OCTAVE BAND ANALYSIS However, octave band analysis may be insufficiently selective for some purposes. More detailed information is given by bands of one-half or one-third octave in width; in the latter, the highest frequency is $\sqrt[3]{2}$ times the lowest frequency. The preferred frequencies for these bands are shown in Appendix H. For a uniform spectrum the energy in a one-third octave band is one third of that in one octave; the difference would be about 5 dB. This type of analysis can be recorded or plotted in the same way as for octave band analysis. Half octave bands are in less common use than octave or one-third octave bands.

NARROW-BAND ANALYSIS For various purposes, especially the identification of the components approximating to pure tones which often originate in machinery, analysis in bands narrower than one-third octave is necessary. This result can be attained by the use of either constant bandwidth analysers or constant percentage analysers. In the former type the bandwidth can be selected and may cover from 1 Hz upwards. The bandwidth remains constant, while the position in the frequency spectrum is selected manually. In the constant percentage type of analyser, the bandwidth is a constant proportion of the mid-frequency selected. That is to say, at approximately 100 Hz, the bandwidth of a 6% analyser would be effectively 6 Hz while at 1000 Hz it would be 60 Hz. These narrow-band analysers are tedious to use, especially the constant bandwidth type, but are essential for certain purposes. Fig. 4.3c shows an analysis into 6% bandwidths of the same noise as in Fig. 4.3a and b.

It is possible to express the sound pressure confined to a band one cycle per second (1 Hz) wide; when this is done this band pressure level is known as *spectrum pressure level*.

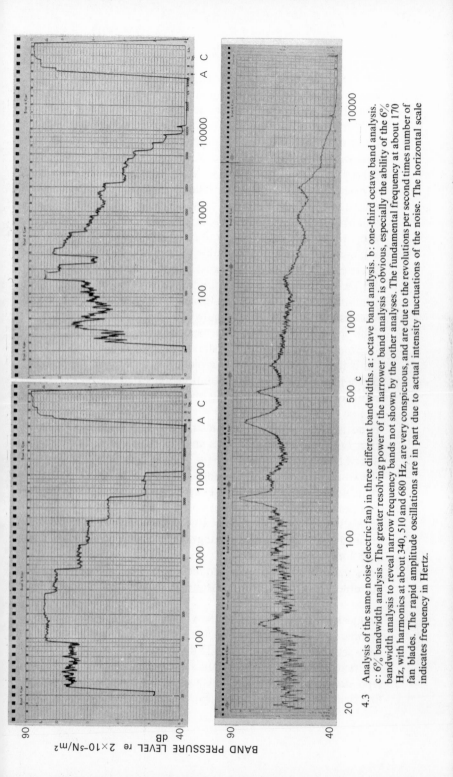

4.3 Analysis of the same noise (electric fan) in three different bandwidths. a: octave band analysis. b: one-third octave band analysis. c: 6% bandwidth analysis. The greater resolving power of the narrower band analysis is obvious, especially the ability of the 6% bandwidth analysis to reveal narrow frequency bands not shown by the other analyses. The fundamental frequency at about 170 Hz, with harmonics at about 340, 510 and 680 Hz, are very conspicuous, and are due to the revolutions per second number of fan blades. The rapid amplitude oscillations are in part due to actual intensity fluctuations of the noise. The horizontal scale indicates frequency in Hertz.

TIME CHARACTERISTICS OF SOUND-MEASURING EQUIPMENT:
IMPACT NOISE   In any sound level meter, the indication obtained
is the RMS value of sound pressure averaged over some definite
short interval. Usually a 'fast' or a 'slow' response characteristic
of the meter is obtainable, the latter being intended to average
fluctuations of sound pressure over a longer period than in the case
of the 'fast'. The 'fast' response, however, is never fast enough, in
the case of a mechanical meter reading on a scale, for the accurate
indication of rapidly occurring noise. Even if it did do so, the needle
would not necessarily be seen adequately by ordinary inspection.
Where rapidity of response is a primary requirement, a cathode ray
oscilloscope and photographic recording is needed. Visually read
meters can still be used to indicate impulse noise, i.e. a single pressure peak or a sequence of impulses, if suitable circuitry is used
including provision for holding the meter indication for long enough
to be easily readable. IEC are known to be considering a supplement
to the precision sound level meter specification, describing impulsive
noise measurement; a German standard already exists (DNA, 1969).

It is thus possible to obtain a picture of a particular sound. For
special purposes, various elaborations may be used. A permanent
record of sound levels covering many hours can be provided by a
graphic level recorder. Where a running frequency analysis of a
sound is needed, the class of instrument known as the frequency
spectrometer is used. This gives an indication of the spectrum as a
plot of SPL against frequency for various bands, such as octaves.
The so-called real-time frequency analyser is a highly elaborated
frequency spectrometer with the capability, in one particular instrument, of continually presenting sound levels, in bands of one-third
octave with the output fed to a computer. The spectrum can be
scanned continually with a readout of the complete spectrum in just
over 2 ms.

In practice, it is convenient to record the sounds on a high grade
portable tape recorder for subsequent analysis in the laboratory.

INTERPRETATION OF FREQUENCY SPECTRA   In interpreting the
graphic records of frequency analyses, the characteristics of the
analyser must be kept in mind. Different types of analyser give very
different results when used on the same noise, since the broad-band
type, such as the octave band analyser, is not capable of showing the
fine structure of noise spectra. For fuller information the reader is

recommended to consult publications on noise measurement (e.g. Beranek, 1960, 1971; Broch, 1969; Peterson & Gross, 1967; BSI, 1967). The noise, analysed into octaves, one-third octaves and narrow (6%) bands, shown in Fig. 4.3 is from an electric fan, and the caption indicates the presence at specific frequencies of a conspicuous component and its harmonics, due to the aerodynamic noise of the blades, at a fundamental frequency corresponding to the blade passage frequency (= revolutions per second of the fan × number of blades).

5

Mechanism of hearing

The sensation of hearing is produced by a train of events which, in principle, is shared by other types of sensory mechanism in the body. The stimulus, in this case sound waves, activates the end organ, a complex mechanism which responds by movement of certain of its parts. These movements, by a series of events, modify activity in the nerve fibres which make up the auditory nerve, so that messages are conveyed to the brain. These various components will be examined in terms of structure and function, and the natural subdivisions of the hearing mechanism provide a convenient framework on which to base the description. We are chiefly concerned here with the peripheral part of the mechanism, that is the structures collectively known as the ear, since it is here that excessive sound can do damage. The central mechanisms, in the brain, are thus of less importance for our immediate purpose, and will not be considered in detail. The subject of the physiology of hearing is only sketched here, and fuller information from which this account has been drawn can be found in the following sources. An exhaustive and highly personal account of the monumental contributions to our knowledge of hearing made by G. von Békésy, will be found in *Experiments in hearing* (1960). A historic work is Stevens and Davis (1938) *Hearing, its psychology and physiology*. Davis (1957) has reviewed the field of auditory physiology more recently and has also reviewed the advances in knowledge of neuroanatomy and neurophysiology of the cochlea up to 1962 (Davis, 1962). He has also stated his present concept of the peripheral mechanism (Davis, 1965). Two valuable recent monographs are available by Littler (1965), *The physics of the ear* and by Whitfield (1967), *The auditory pathway*.

Recognition of the essential unity of structure and function is an essential in attempts to understand cochlear function, and in recent years, advances in knowledge have been conspicuous due to the emergence of new techniques for visualising cochlear structure. The electron microscope and its elaboration, the scanning electron microscope, have opened up new areas of study. Summaries of structural features are given in Davis and Silverman (1970); Smith

(1968); Engström, Ades and Bredberg (1970); an extensive study of the inner ear is in an advanced stage of preparation at the time of writing (Ades, Bredberg & Engström, 1970), and Spoendlin's (1969) description is available. Strikingly beautiful and detailed electron microscope pictures by the scanning technique in the recent work of Engström and his collaborators confer new realism to these features of microanatomy, and complement previous fundamental advances such as that of Smith and Sjöstrand (1961). In seeking compact reviews of cochlear physiology, the sources, in addition to those already cited, include Davis (1970); Kiang, Moxon and Levine (1970) who describe experiments which have contributed notably to knowledge of the activity in the auditory nerve; and the same area is the subject of an extensive review (Kiang, Watanabe, Thomas & Clark, 1965). The latter two publications require prior acquaintance with auditory neurophysiology for a proper appreciation, as do the publications of Eldredge and his colleagues (e.g. Teas, Eldredge & Davis, 1962) on somewhat related fields. An admirable review of the physiology of hearing has been made by Eldredge and Miller (1971); this is a specialist selection of a number of aspects of the subject. Returning to the more general and, as far as possible, non-specialised account of the physiology of hearing, we may begin at the periphery and work inwards.

The study of the peripheral mechanism of hearing has been pursued with great skill and ingenuity over the last two or three decades, but it cannot be said that we clearly understand how the ear performs its function. We must attempt to explain how the physical dimensions of sound, in terms of intensity, frequency and time relations, are continuously received and encoded as nerve impulses in the auditory nerve, subsequently to ascend in the nervous system to be perceived as the sounds which we subjectively experience. In this short account of the peripheral auditory function, it is the coding process, from sound pressure waves in the air, to an organised flow of information in the auditory nerve, which we mainly discuss. At the elementary level, the means for indicating frequency, intensity and time relations of the sound must be sought. The low energies involved, the comparatively large number of variables, and the known performance of the hearing function such as its large dynamic range, its powers of frequency and intensity discrimination, and its time-resolving abilities, must be looked at against the background of its small dimensions, physical complexity, inaccessibility and remarkably beautiful and intricate ultrastructure. We are impeded in many directions by deficiencies of understanding of function and even of structure, but enough of a picture emerges to clarify to some extent the clinical aspects, and to provide enough basic physiological understanding for the construction of hypotheses, and for their experimental exploration, in an attempt to improve the present picture.

The peripheral mechanism of hearing

GENERAL   The general structure of the peripheral hearing mechanism is shown in Fig. 5.1, in a diagrammatic but partially realistic manner. The events to be described start with the occurrence of sound pressure fluctuations in air; these in turn set into vibration the ear drum or tympanic membrane, which in turn actuates a lever system of three very small bones, the auditory ossicles, situated in the air-filled cavity of the middle ear. The ossicles are grouped in

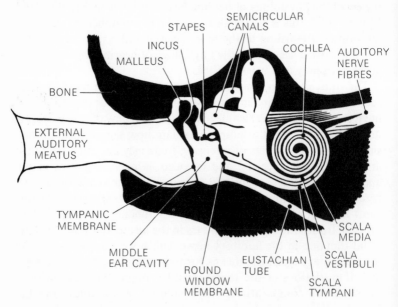

5.1   Diagram of the main components of the peripheral hearing mechanism, showing the external ear canal, middle ear and inner ear. The coils of the cochlea would not, due to overlapping, be visible complete as shown here. (After Davis & Silverman, 1966.)

such a way that the vibrations of the tympanic membrane drive the ossicle known as the malleus, which drives the second ossicle, the incus; and finally the third ossicle, the stapes, sets up vibrations in fluid and certain structures contained in a complex system of coiled canals in the bone of the skull, known as the inner ear. These canals include the semicircular canals, concerned with balance, and the cochlea, in which is situated the end organ of hearing or organ of Corti where mechanical vibrations set up nerve impulses in the fibres of the auditory nerve. The information travels in this coded manner to the appropriate parts of the brain, there to be perceived as a sensation, probably stored as a memory, capable of arousing pleasure or annoyance, as well as having other secondary effects. We now consider the peripheral parts of the hearing mechanism in more detail. The reader who is more concerned with the general picture of the effects of noise may prefer to omit the ensuing more detailed description of the peripheral mechanism of hearing.

External ear canal and middle ear

Referring to the diagram of the parts of the ear in Figs. 5.1 and 5.2, and pursuing the same course as would the sound waves, we first encounter the external ear. In man this is less important than the large movable external ears of many animals, but the position on each side of the head with the mass of the head in between makes possible the directional identification of sound by differences of intensity, and of the time at which the sound arrives at each ear. The sound is conducted down the external auditory meatus or external ear canal which is about 25 mm or 1 in long, and next impinges on the tympanic membrane or ear drum, a stiff conical diaphragm with flexible margins; this is the boundary between the external ear canal and the cavity, filled with air, known as the middle ear. The tympanic membrane is made to vibrate in synchrony with the sound pressure changes, with an amplitude dependent on their magnitude. The mechanical properties of the tympanic membrane and its attachments are such that a large proportion of the sound energy is utilised in setting the drum in motion. The middle ear cavity is in communication with the back of the nose through the Eustachian tube (Figs. 5.1 and 5.2), which is normally closed, but opens on swallowing, in order to maintain equal air pressure on the two sides of the tympanic membrane. The cavity of the middle ear is also continuous with air-filled cells in the mastoid bone of the skull, behind the external ear, and bacterial infection may involve both middle ear and mastoid cells.

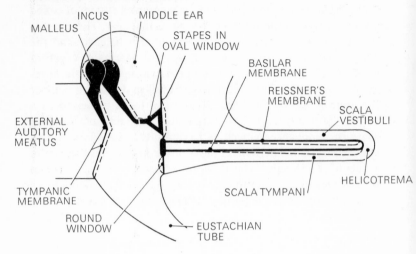

5.2 Diagram of the main components of the middle ear and cochlea. The cochlea is not specifically labelled but is shown 'unrolled' as a straight structure extending to the right, for clarity. The solid outlines show the movable parts at rest; the interrupted lines show in an exaggerated manner the deflection caused by the positive (i.e. above atmospheric pressure) phase of a low-frequency sound wave. Note that both in structure and function, this diagram is an over-simplification.

The sound vibrations are imparted to the chain of ossicles in a manner dictated by their linkages and supports. The malleus and incus are normally locked together; they oscillate through a small arc about an axis which is approximately a tangent to the upper rim of the tympanic membrane, driven by the handle of the malleus, which is attached to the inner surface of the membrane. The incus in turn drives the stapes, which forms a small partition, intervening between the air of the middle ear cavity and the fluid filled spaces of the internal ear and filling the orifice in the bone known as the oval window. The oscillations of the stapes are made possible by a flexible rim, the annular ligament, which seals the fluid in the internal ear, and the little bone moves inwards and outwards, like a piston or a diaphragm pump, in the oval window, provided the amplitude of movement is small. For large amplitudes, the movement is more like a door on a hinge. This is associated with the shape of the annular ligament which is wider and more flexible at the anterior part, so that the stapes hinges at the posterior end for the higher ranges of amplitude, as noted below. A third mode of vibration, involving rotation on the long axis of the stapes, thus transmitting less energy, operates at the highest amplitudes. The inner ear fluid is thus also made to vibrate, and the necessary elasticity is given to the closed fluid-filled cavity by a second aperture in the cavity, the round window, which is sealed by an elastic membrane. The area of the stapes in the oval window is only about one-thirtieth that of the tympanic membrane, and the vibrational displacements of tympanic membrane and stapes foot-plate as well as the forces at each, are about equal. The force per unit area on the fluid of the internal ear is thus much greater than that on the tympanic membrane and the transfer of energy to the fluid of the internal ear is rendered more efficient. This is probably the main function of the middle ear. It also protects the internal ear to some extent against excessive noise.

MIDDLE EAR MUSCLES: AURAL REFLEX   Two small muscles (tensor tympani and stapedius) are attached to the malleus and the stapes, respectively, and when intense sounds occur, or objects touch the external ear canal, these muscles contract, tending to pull the stapes and the tympanic membrane towards the middle ear cavity, their actions being thus antagonistic. In consequence they introduce increased resistance to vibration into the ossicular chain; the movement may be very small and variable. Sounds, particularly of low and of high frequency, are then transmitted with increased loss, so partially protecting the internal ear from damage. This action is analogous to that of the iris of the eye, which reduces the pupil diameter to admit less light to the eye in bright conditions, and vice versa. Stapedius responds best to sound; tensor tympani to touch.

Jepsen (1963) has reviewed the action of the middle ear muscle reflexes in man. This reflex is usually referred to as the aural reflex, and it appears to be able to offer significant protection against sound of lower frequencies. The reflex is capable of being set up by a wide range of frequencies and is activated most readily by sound in the frequency range 1000–2000 Hz. In order that the reflex shall be activated the sound must be fairly intense, the threshold for activation being just under 80 dB above the threshold of hearing at 1000 and 2000 Hz and about 85 dB above the threshold of hearing at 250 and 4000 Hz. The protection, according to Reger (1960), during maximum voluntary contraction of the tympanic muscles, which some people can achieve, is considerable. The reduction of sound reaching the inner ear, estimated by the elevation of threshold during maximum voluntary contraction, exceeds 30 dB in the range 125–500 Hz, but this finding is open to different interpretations; with rising frequency this protective action diminishes until, at 2000 Hz, Reger found no alteration in the threshold. Møller (1961) assumed that not only did the aural reflex limit the movement of the ossicles, but also altered the mode of the vibration of the stapes. Instead of a hinge-like movement on a vertical axis through the posterior end of the footplate the movement became (Békésy & Rosenblith, 1951), more in the nature of a rotation about the long axis of the footplate. The large reductions in sound transmission found by Reger on voluntary contraction of the middle ear muscles should perhaps not be assumed necessarily to occur in normal reflex action due to sound.

Inner ear

This system of canals filled with fluid is deep in the temporal bone of the skull, and so protected mechanically and acoustically. The fluid which is set into vibration by the movements of the stapes is contained in the coiled tubular canal in the temporal bone, called the cochlea from its resemblance to a snail's shell. The reason for the coiled structure has never been satisfactorily explained. Starting at the base, the tubular cavity of the cochlea ascends for about $2\frac{3}{4}$ turns to the apex in the human ear. In Fig. 5.1 the cochlea is shown as a flat spiral. Actually its conical shape would result in the apical turns partially hiding the basal turns in the view shown. In Fig. 5.2 for simplicity the canal of the cochlea is shown as a straight tube. Fig. 5.3 shows, in a simplified geometrical form, approximately how the coiled canals would appear if the cochlea were cut along its axis. The cochlear canal is split longitudinally by a partition which ends below the apex, so forming a separate canal on each side of the partition, with a communication known as the helicotrema at the apex. These two canals are called the scala vestibuli and scala tympani and are filled with a fluid, perilymph, similar to the cerebro-spinal fluid surrounding the brain and spinal cord. The oval window is situated at the basal part of the scala vestibuli, while the round

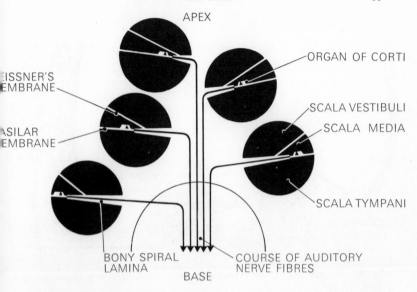

APEX

ORGAN OF CORTI

REISSNER'S MEMBRANE

SCALA VESTIBULI

SCALA MEDIA

BASILAR MEMBRANE

SCALA TYMPANI

BONY SPIRAL LAMINA

COURSE OF AUDITORY NERVE FIBRES

BASE

5.3 Diagrammatic representation of a 'vertical' section across the canals of the cochlea. They have all been shown as circular in section (which they are not). *Black:* fluid-filled canals.

window is at the corresponding part of the scala tympani (Figs. 5.1 and 5.2). Vibration of the stapes will set the fluid into motion and with it, the round window and the partition between the scalae. This partition is all important, since it carries the end organs which transform movements of the partition into patterns of nerve impulses in the auditory nerve. The partition is actually another tube, the scala media, containing a different fluid, the endolymph, which appears to be secreted by cells of a long ribbon of tissue rich in blood vessels and thus called the stria vascularis, covering the inner surface of the outer wall, Fig. 5.4. The other boundaries of the cochlear partition separating the scala media from the scala vestibuli are the two long ribbon-like membranes; on the side towards the scala vestibuli is Reissner's membrane, a thin flexible wall, while on the side towards the scala tympani is the basilar membrane, on which rests the receptor organ, the organ of Corti. The human basilar membrane is about 35 mm long from base to apex, narrow (0·04 mm) at the base and broad (0·5 mm) at the apex, with a gradual change in width. The cochlear partition and particularly the basilar membrane, which are structures possessing elasticity, have a graded

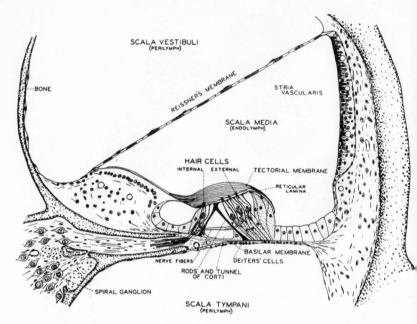

5.4   The microscopic appearance of the guinea-pig cochlear partition sectioned in the 2nd turn. The scala vestibuli and scala tympani are incompletely shown. (Davis and associates, 1953.)

variation of stiffness and mass along their length, so that the different parts of the membrane resonate at different frequencies; high frequencies being located at the basal end. The whole partition possesses other physical characteristics which also determine its motion in response to sound vibrations conveyed through the tympanic membrane and ossicles.

ORGAN OF CORTI   This is a complicated system of cells and associated tissues extending along the entire length of the cochlear partition. Except for the different width of the basilar membrane and consequent gradual variations of dimensions along the length of the organ of Corti, the structure remains unchanged in general architecture throughout the whole of the cochlear partition. Basically it consists in man of 3, 4 or 5 rows of external hair cells, the number increasing from base to apex (Fig. 5.4) with certain supporting cells and structures, which rest on the basilar membrane. Conspicuous elements in this supporting category are Deiters' cells (Fig. 5.4). The bases of these elongated cells rest on the basilar membrane, the upper end of the cell bodies support the external hair cells, and there is a long process which reaches up to the reticular lamina. If the organ of Corti were viewed from the top right of Fig. 5.4, the appearance, from a scanning electron microscope photographed by Engström, Ades and Bredberg (1970) seen in Fig. 5.5 would be the result. This striking three-dimensional view shows

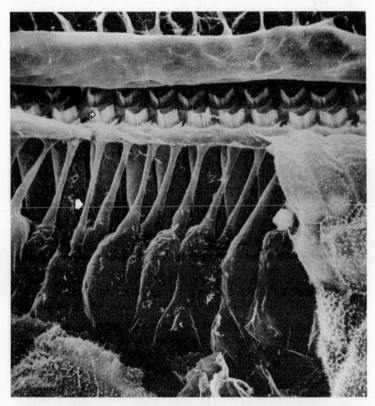

5.5   Scanning electron micrograph of the guinea-pig organ of Corti, viewed as from the top right of the section in Fig. 5.4. The outer wall is partially removed, together with part of the tectorial membrane. The Deiters' cells form a row along the bottom of the picture, with their long thin phalangeal processes (large arrow) reaching upwards to the reticular lamina. The outer hair cell bodies are seen behind the Deiters' cells as narrow columns terminating in the hairs (small arrow) projecting above the level of the reticular lamina. (Engström, Ades & Bredberg, 1970.)

the surface of the organ of Corti, Deiters' cells, the hairs from the outer two rows of hair cells, the outer row of hair cells under the reticular lamina, and the tectorial membrane lifted off the hairs. The hair cells are elongated, with the small hair-like projections which give the cells their name projecting upward into the scala media, and probably anchored by their tips to the underside of the tectorial membrane, which runs as a spiral shelf over the organ of Corti. Separated from the external hair cell complex by the row of A-shaped struts known as the rods of Corti, are the internal hair cells, with supporting cells. The internal hair cells are on the side

5.6   Scanning electron micrograph of the upper surface of the organ of Corti of the guinea-pig, viewed towards the modiolus without the tectorial membrane, and showing the W-formation of hairs on each cell, magnification × 3520. (Engström, Ades & Bredberg, 1970.)

nearer the axis of the cochlea and are arranged as a single row, again for the entire length of the cochlea, and they lie over the inner margin of the basilar membrane. Their hair-like projections are also probably attached to the underside of the tectorial membrane. The upper extremity of the hair cells is anchored by a thin strong sheet, the reticular lamina, with openings into which the ends of the hair cells fit. The other ends of the cells rest in sockets formed by the cells known as Deiters' cells. The outer hair cells are cylindrical (diameter about 8 microns[1]) and elongated. The rounded lower end contrasts with the flat upper end, from which the sensory hairs project (Fig. 5.5). These hairs are arranged, on each cell, in the form of a W (Figs. 5.6, 5.7). The numbers of hairs approach 80 per cell. The hairs vary in size and disposition somewhat, but are, on average, about 1 micron in diameter and about 10 microns in length or less. Each human ear contains about 9000 outer hair cells (Davis, 1970).

The internal hair cells number about 3000, are more rounded than the outer hair cells, and their diameter is about 10 microns. The hairs are arranged somewhat differently, they are longer and coarser (Engström, Ades & Bredberg, 1970) than those of the outer hair cells, but the general arrangement, with the hair-bearing surface of the cells fitted into the reticular lamina (Fig. 5.4), and the nerves in

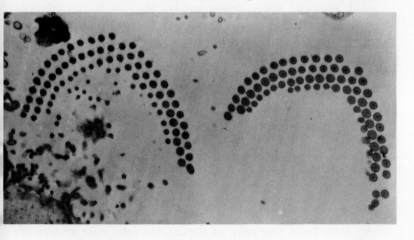

5.7 Electron micrograph of a cross-section of the hairs, from two outer hair cells of the guinea-pig. Each hair can be seen to have a dense central core, magnification × 8260. (Engström, Ades & Bredberg, 1970.)

[1] 1 micron = 0·001 mm.

contact with the other end, is the same for both inner and outer types. The shape of the long shelf-like tectorial membrane, as seen in section (Fig. 5.4) is significant. It becomes thin and flexible at the inner edge, and in fact hinges along this edge, as may be seen when we consider the mechanical movements of the organ of Corti. The nerve fibres which make up the auditory nerves start at the hair cells. They are applied to the exterior of the hair cells, the cell bodies of these nerve fibres being grouped together as the spiral ganglion as they converge towards the modiolus (Fig. 5.4) or central axial cavity of the cochlea, on their way to the brain. A feature of the structure significant from a functional viewpoint is that only the ends of the hair cells carrying the hair processes are in contact with endolymph; the space between the rods of Corti, known as the tunnel of Corti, contains perilymph from the scala tympani, or a fluid similar to it.

NERVE FIBRES   The nerve connections of the hair cells are both afferent (to the brain), and efferent (from the brain). Considerable uncertainty still surrounds their actual arrangement. Each auditory nerve is composed of some 25,000 afferent fibres and 500 efferent fibres. The afferent fibres run from the bases of the hair cells to the spiral ganglion which accommodates the cell bodies of these neurons, and then to the cochlear nucleus, where they establish connections with other neurons in the auditory pathways. The efferent fibres originate in the superior olivary nucleus and run peripherally in the olivo-cochlear bundle, to the cochlea, where they end on either the hair cells or on the afferent endings. This is seen in proximity to the inner hair cells (Engström, Ades & Andersson, 1966).

The exact manner in which the nerve fibres are distributed to the hair cells is still under investigation. On the basal portion of both inner and outer hair cells nerve endings of both afferent and efferent type have been identified by electron microscopy and physiological experiment (Smith, 1968). The arrangement of the afferent nerve fibres is at present not satisfactorily resolved. Spoendlin (1966, 1969) has shown that apparently as many as 90% of afferent fibres originate only from inner hair cells, in the cat, the remainder presumably serving outer hair cells. If this finding is substantiated, and if it obtains in other species than the cat, it must have important implications.

In all these areas of contiguity of nerve endings and hair cells (or nerve endings and nerve endings in the case of some of the efferent fibres) there is evidence that these sites are synapses (Smith & Sjöstrand, 1961). The activity here is believed to be transmitted by the release of a chemical substance, or substances (see below) so far not identified.

The role of these structures in generating nerve impulses and the effects of mechanical movement must now be examined.

The cochlea as the end-organ of hearing

This aspect of cochlear function may be considered from two viewpoints. The first concerns the nature of the movements of the cochlear structures, by which the cochlea is able to act as a frequency analyser of sound, and so to contribute to the appreciation of its pitch and loudness; the second viewpoint concerns the means whereby the movements are translated into patterns of impulses in the auditory nerve.

Movements of the cochlear structures

Let us assume that, due to a sound wave, the tympanic membrane is pushed inwards towards the middle ear, so actuating the chain of ossicles, and causing the footplate of the stapes to move inwards into the oval window. The resultant rise in pressure of the fluid in scala vestibuli sets up a complicated pattern of movement in the structures of the cochlear partition. As a permissible simplification, these structures can be considered to move together in response to the pressure changes in scala vestibuli, but the exact nature of the movements depends on the fact that the partition has different values of stiffness and mass along its length, and that mechanical coupling between adjacent areas, and damping, are also present. Some of these physical properties are particularly dictated by the characteristics of the basilar membrane. In such a system, with stiffness maximal and mass minimal at the basal end, a rise of pressure in scala vestibuli will set up a travelling wave of transverse and longitudinal displacements in the cochlear partition. This wave, as originally shown by Békésy by direct microscopic observation and stroboscopic illumination, starts with a high velocity at the basal end of the partition, and travels upwards towards the apex, increasing in amplitude as it goes. At a position which is different for different frequencies, the wave reaches a maximum amplitude, and then suddenly diminishes in velocity and amplitude and is soon extinguished. The tuning curves of the cochlear partition have subsequently been derived (Johnstone, Taylor & Boyle, 1970) in which the frequency response of a point on the membrane was found to agree with Békésy's findings, by the use of a Mössbauer technique in which a small radioactive source is placed on the basilar membrane. With sounds of appreciable duration, trains of waves of displacement will thus pass up the cochlea from base to apex, while the round window membrane vibrates in the opposite phase to the vibrations of the oval window, so contributing the essential elasticity to an otherwise almost rigid fluid-filled cavity. As a result of this complex behaviour of the cochlear partition, it can act as a frequency analyser for sound. High-frequency sounds will excite trains of waves which will reach their maximum amplitude at the basal part of the partition, and thereafter become extinguished. The velocity of travel here is sufficiently rapid to regard the cochlear partition as vibrating almost as one unit in the basal turn. For lower frequencies the wave motion becomes more obvious and travelling waves ascend towards the more apical regions, before suddenly diminishing and dying away; while for the lowest frequencies the waves involve the entire length of the cochlea and reach their maximum in the apical regions.

In summary, the movement of the partition is almost simultaneous at the basal regions, but higher up takes the form of obvious travelling waves, with different velocities at different parts of the cochlea and for different sound frequencies but reaching a maximum amplitude at a specific position along the cochlea, dictated by frequency in an orderly manner: highest frequencies nearest the base, lowest nearest the apex. Following the peak is a rapid reduction of amplitude and velocity and therefore wavelength, followed by extinction.

Generation of nerve impulses

The cochlea belongs to the category of mechano-receptors, in which the stimulus is mechanical displacement. We must now consider how the movements of the cochlear partition are translated into nerve impulses, with particular regard to the means used to attain the remarkable sensitivity, and the type of coding employed in the auditory nerve to carry the information on frequency and intensity of the sound stimulus.

The auditory nerve shows spontaneous activity modified by mechanical displacement of the cochlear structures; this is achieved, we believe, either in conjunction with or as a result of variations in electric potentials. These electrical changes are of various kinds and are anatomically identified with different structures. To describe these phenomena we must look in more detail at the whole electro-mechanical situation in the cochlea.

In general, biological activity in tissues is associated with the production of measurable potential differences in certain parts with respect to other parts, attributable to particular distributions of ions. Thus, electrical potentials can be established and ionic currents can flow. These potential differences may be fairly stable and permanent, and in this category we find that in general the interior of living cells is negative to the exterior of the cells. Sudden and dramatic re-groupings of ions across membranes may give characteristic electrical changes, often very rapid and transient, such as those accompanying the passage of nerve impulses along nerve fibres or the contractile changes in active muscle.

ENDOCOCHLEAR POTENTIAL   When the interior of cells of the organ of Corti are explored to determine their electrical potential by means of a microelectrode,[1] we find the interior electrically negative with respect to the exterior, as with other types of cell. In addition, there is a unique situation in that the endolymph in the scala media is at a potential of nearly a tenth of a volt positive to the perilymph in scalae vestibuli and tympani, and remains so if no sound stimulation occurs. This potential difference is actually about 80 millivolts (mV) positive with respect to the scala tympani and other extra-cellular situations in the body, so that the potential difference with respect to the negative interior of the hair cells is thus some 140 mV. The electrical positivity of scala media is called the endocochlear potential and it is maintained by the metabolic activity of the cells of the long ribbon of tissue on the inside of the outer wall of the scala media, known as the stria vascularis; the internal negativity of the hair cells is a result of their own metabolic activity. The

[1] Microelectrode recording systems consist normally of a very fine wire, or glass tube of diameter 1 micron or less containing a conducting fluid, coupled to a suitable electrical amplifier and recording system.

endocochlear potential is acutely dependent on adequate blood oxygen and largely fails in its absence. We do not yet know with certainty the physiological basis of this potential.

If the relative positions of the structures in the cochlear partition are altered by 'upward' and 'downward' movement (i.e. towards scala vestibuli or scala tympani respectively) the value of the endo-

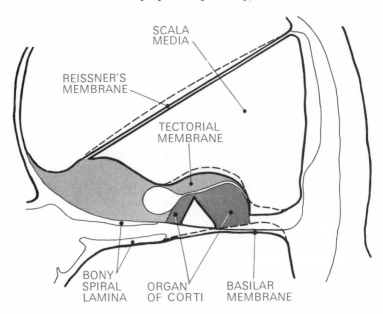

SCALA
MEDIA

REISSNER'S
MEMBRANE

TECTORIAL
MEMBRANE

BONY
SPIRAL
LAMINA

ORGAN
OF CORTI

BASILAR
MEMBRANE

5.8   Block diagram of a cross-section of a cochlear turn to show the deflection of the structures of the cochlear partition due to an outward movement of the stapes. The extent of the organ of Corti is indicated in dark grey. This diagram should be studied in conjunction with Fig. 5.4.

cochlear potential alters. If the cochlear partition is moved towards scala vestibuli (Fig. 5.8) by an outward movement of the stapes (and so also of the tympanic membrane) the endocochlear potential is diminished. The opposite movement of the organ of Corti, i.e. downward movement of the basilar membrane and organ of Corti, caused by an inward movement of the stapes, will increase the endocochlear potential. These potential changes are not influenced by movement of Reissner's membrane, but can be evoked by appropriate isolated movements of the tectorial membrane by a micromanipulator. The essential movement is relative displacement of the hair cells and tectorial membrane, so that the hairs are bent one way or the other. Bending the hairs outwards, i.e. away from the axis of the cochlea, reduces the endocochlear potential, and vice versa. Because of the different pivoting axes of the basilar and tectorial membrane, any simultaneous 'rise' (or 'fall') of these two membranes, causes a shearing or wiping action across the free ends of the hair cells by the tectorial membrane, so bending the hairs in a radial plane with respect to the cochlea (Fig. 5.9). The maintained

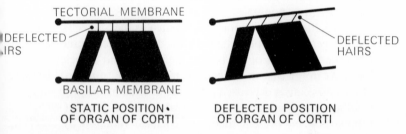

5.9 Diagram to show the movements of the hairs of the hair cells in response to an 'upward' movement of the cochlear partition.

changes in the endocochlear potential are a consequence of the bending of the hairs due to this displacement of the cochlear structures. The endocochlear potential will remain at a fixed value as long as the mechanical stimulus of modified position of the structures is maintained. Thus endocochlear potential is determined by the position of the hairs. Davis (1965) has attributed this variation in endocochlear potential to change in electrical resistance causing change in the voltage drop across the hair-bearing ends of the hair cells, in presence of the leakage current driven by the negative interior polarisation of the hair cells, aided by the positive endocochlear potential. The variation in electrical resistance of the hair cells, he suggests, is determined in turn by the position of the hairs.

COCHLEAR MICROPHONIC   From the endocochlear potential and its variations it is but a short step to the consideration of the alternating potentials, evoked by sound stimulation, known as the cochlear microphonic potentials. The movements of the cochlear partition, in response to sound stimulation, set up voltage fluctuations across the endocochlear partition, which are known as the cochlear microphonic. The potential changes may be picked up by electrodes in the anaesthetised guinea-pig ear in various situations such as in scala media and scala tympani. The cochlear microphonic follows the sound pressure variations with great accuracy, being linearly related in voltage to the sound pressure values, up to a certain sound pressure level, after which further increases of sound pressure produce diminishing increases of cochlear microphonic potential. The response finally reaches a maximum and thereafter diminishes. High intensities, particularly at low frequencies, will cause distortions of the electrical response.

The cochlear microphonic is attributed to the hair cells, responding to 'vertical' displacement of the basilar membrane resulting from the vibrations of the basilar membrane and associated structures. As in the case of endocochlear potential, 'outward' bending of the hairs causes decreased positivity of scala media, and 'inward' bending increased positivity of scala media. Davis (1965) has summarised his present theory of cochlear function by ascribing the cochlear microphonic to changes in resistance of the hair-bearing face of the hair cell, which produce fluctuations of the voltage drop across this part of the cell, as noted above. The great importance of this concept is the fact that the extremely small energies of the faintest sounds, which are only able to move the basilar membrane through very small amplitudes, perhaps about 10^{-8} cm (Whitfield, 1967), are increased by the energy stores represented by the potential across the hair-bearing face of the hair cells. Minute sound energies can thus be used to trigger off the energy stored across the face of each cell as the resting potential difference, so helping to explain the extreme sensitivity of the ear. The further events following the cochlear microphonic oscillations are open to some speculation. However, taking the response to a low-frequency tone, auditory nerve impulses tend to occur during the part of the cycle when current is flowing from scala vestibuli towards scala tympani. As postulated by Davis, this could be the consequence of the release by the hair cells of a chemical transmitter, yet unidentified, which would activate the nerve endings. Relations between endocochlear potential, cochlear microphonics or the role of the latter are not clear, nor does the above represent the complete picture as seen by Davis and other workers. Further implications will be noted in the context of nerve activity, and in the summary at the end of this chapter.

SUMMATING POTENTIALS   Associated with the records of cochlear microphonics are changes of the baseline which are due to longer duration, steady shifts in mean endocochlear potential. These are evoked by sound pressure waves in the higher range of intensity. These electrical responses are not yet fully understood; and Davis, Deatherage, Eldredge and Smith (1958) have made certain proposals in their earlier studies. One more recent hypothesis has been advanced by Whitfield and Ross (1965). This, without prejudice to other possible mechanisms, ascribes the summating potential to non-linearity of the response of the basilar membrane to vibrations, in conjunction with the assumption that the shapes of the cochlear microphonic and summating potentials are due to the presence of out-of-phase components in the outputs of numerous hair cells.

ACTION POTENTIALS IN THE AUDITORY NERVE   The usual response of a sensory end organ to graded stimuli is that increased intensity of stimulus is signalled by increased frequency of discharge of nerve impulses in the associated nerve fibre. Without elaborating on the physiology of nerve transmission, the principle is as follows. Nerves are composed of numbers of individual nerve fibres. These are about 8 microns in diameter in the case of the auditory nerve, and this nerve contains many thousands of such fibres carrying information from the cochlea to the brain, as well as some conveying information to the cochlea, which seem to be able to control the sensitivity of the ear to some extent, each operating as a separate channel of communication. The messages are coded in a very simple way. They are carried by nerve impulses, which are short pulses of activity, all identical, which have been likened to a morse code composed only of dots. Each impulse is a complex physicochemical change occurring very rapidly on the outer membrane of the nerve fibre as a distinct entity. The impulse travels at a definite velocity depending on the type of nerve, and is physically detectable as a short length of electrical negativity relative to quiescent areas, on the surface of the nerve fibre. The velocity may be over 100 m/sec (about 220 mile/h) according to the diameter of the fibre; the larger the diameter the faster the conduction rate. Nerve impulses signal events by activity in specific fibres. The stronger the stimulus on the end organ, the higher the frequency of occurrence. Each auditory nerve fibre in the cat, for example, can respond to sustained sound stimulation at a frequency of about 150 impulses per second although for very short periods of a few milliseconds, higher rates are possible, at the onset of a sound stimulus. The frequency of discharge is determined, for any given nerve fibre and sound frequency, by the intensity of the sound. Thus the greater the sound pressure, the more will the basilar membrane be displaced, associated with a higher frequency of discharge of action potentials in any appropriate nerve fibre. The discharge of a single impulse in each of a number of nerve fibres can theoretically result from each appropriate movement of the basilar membrane. This may happen at the lowest frequencies, but as the frequency is increased, the nerve fibres cannot respond to these frequencies, and they respond at a lesser rate than the frequency of the sound. The actual rate of discharge in these circumstances is still determined by the magnitude of the sound pressure and its frequency relative to the sound frequency characteristic of the particular part of the cochlea.

Place principle

The hair cells and associated nerve fibres thus act as detectors of the vibration of the basilar membrane and other structures of the cochlear partition. As we noted, the cochlea acts as a mechanical analyser of frequency. High frequencies cause vibration at the base only, low frequencies in the whole cochlea, but each frequency sets up vibration of maximum amplitude at a point dictated by frequency. It is the function of the thousands of hair cells and associated nerve fibres to signal to the brain the pattern of vibration. The location of this pattern on the basilar membrane, including its vibrational maximum, is determined by the frequency of the sound. This relation is recognised in the term place principle or theory. The actual location on the membrane is signalled to the brain by impulses in the nerve fibres serving the particular area along the basilar membrane.

To recapitulate, the study of the pattern of activity of the cochlea has been approached in a number of ways. There is, first, the direct visual inspection of Békésy (1960). The second possible approach is by the study of the patterns of cochlear microphonics, particularly by Davis and his collaborators (see references of Davis, 1965). Thirdly, we have the investigation of the activity in the auditory nerve, either from the action potential patterns of the whole nerve as used by Eldredge (Teas, Eldredge & Davis, 1962), or by the study of the activity of single fibres of the auditory nerve (Tasaki, 1954; Kiang, Watanabe, Thomas & Clark, 1965). All three methods corroborate the pattern of mechanical vibration described by Békésy referred to above and in this context we may logically conclude with a brief reference to the activity in the individual fibres of the auditory nerve.

The extensive studies of Kiang (Kiang, Watanabe, Thomas & Clark, 1962; Kiang, Moxon & Levine, 1970) give the most complete description of the activity in the nerve fibres carrying the primary information to the central nervous system. Obviously a full description of this work is out of place in this text, but certain aspects of fundamental importance can be readily summarised.

If the activity of a single fibre in the auditory nerve is recorded by a microelectrode, the nerve impulses can be studied in relation to sound stimulation (as well as in its absence). Such nerve fibres may be assumed to be in functional contact with a hair cell or cells, the identity and number of which is not precisely known, but the complex can be called an *auditory sensory unit*. It was found that most individual nerve fibres showed normally a spontaneous discharge, even with any source of sound excluded from the animal's ears. The frequency of the spontaneous discharge appears to be characteristic of each unit, and may be from 5 or 6 spikes/min up to 100/s. In order to determine whether a given unit (from which a record was being obtained) was activated by a sound stimulus, an arbitrary percentage increase in discharge rate was selected as the criterion of threshold activation. In this way the intensity of a pure tone (or of some physical derivative) at any particular frequency, which was necessary for threshold activation of the unit, was established. Establishment of this threshold activation, for different frequencies, of a given unit yields the quantitative information known physiologically as the *response area* of the unit. This is defined as the range of thresholds as a function of frequency for each unit. Fig. 5.10 shows the response areas of a number of units of the cat, in this way, appropriately termed their *tuning curves* by Kiang; the threshold in this case is in terms of stapes displacement. The *threshold* and *characteristic frequency* of each unit are defined respectively as the level and frequency co-ordinates of the lowest point on a tuning curve, plotted on a graph of stapes displacement against frequency (Kiang, Moxon & Levine, 1970). These curves show the frequency range, at various sound pressures (as indicated on the graph by stapes displacement), over which each unit responds (as indicated by audio-visual criteria of an arbitrary increase in frequency of discharge above the spontaneous discharge rate). The characteristic frequency, i.e. the apex of each curve, is well marked, and the selectivity is quite high, but the unit will respond to other frequencies provided the sound level is increased. It can be clearly seen that a moderately intense pure tone, at just under 20 kHz for example, will activate four of the units, but at a low level, only one of those illustrated might be above threshold.

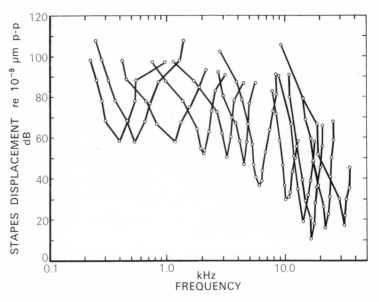

5.10   Tuning curves for twelve auditory sensory units of the cat. The criterion of response in each particular auditory nerve fibre is an arbitrary increase in impulse discharge rate; and the range of sound level and frequency over which the unit responded is plotted as stapes displacement against frequency; p–p, peak to peak. (After Kiang, Moxon & Levine, 1970.)

The same information is provided to a higher degree of frequency and level resolution by an automated method, shown on Fig. 5.11. The curves, as previously, indicate the lower limit of the response area for pure tones and thus are equal-response contours, using the threshold criterion of a specific rate increase of the spike potentials of the unit. The equipment, under computer control, sets the frequency and sound level to achieve the arbitrary response condition. Correction from a computer store gives the stapes displacement scale of the figure. The shape of the curves is consistent with Békésy's findings on mechanical movement, and shows the abrupt cut-off at the high-frequency end indicating that the envelope of vibrational amplitude of the cochlear partition is moving towards the stapes, leaving the particular unit eventually in a quiescent part of the basilar membrane, as the frequency is raised. Other significant features (not shown on Fig. 5.11) include the relation between threshold response level and the frequency of the spontaneous discharge. Units with

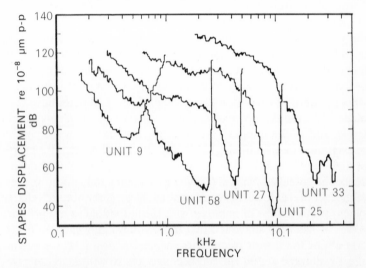

5.11 Tuning curves for five auditory sensory units of the cat. The derivation is basically similar to that used in Fig. 5.10, but here an automated method giving a high degree of resolution of level and frequency, as described in text, is used; p–p, peak to peak. (After Kiang, Moxon & Levine, 1970.)

lower spontaneous discharge rates tend to have higher thresholds, and their maximum rates of discharge elicited by intense tone bursts is lower than with units with higher spontaneous discharge rates. This resting frequency is variable, but most fibres, under prolonged acoustic stimulation, cannot be driven much over 150 impulses per second, even at the highest sound intensity (Kiang, Moxon & Levine, 1970). At the onset of a sound stimulus, a few impulses at a much higher rate can be elicited, but only briefly. A change of SPL of only about 25 dB may be enough to reach the maximum rate of discharge. It is clear from this small dynamic range that the extent of the pattern of units must be important in indicating the intensity of the stimulus.

The question of variability of threshold among different hair cells is of much theoretical and clinical interest. It has been suggested that the inner and outer hair cells may have different thresholds, but no evidence for such separation appears. For units in the cat's auditory nerve, the range of 'thresholds' for units of nearly the same characteristic frequency is only 20–30 dB. Contrary to some previous belief, large spreads of 'threshold' will occur, however, as noted above, by the inclusion of units of different characteristic frequency, at higher sound levels, so that the pattern of activity will be spread more widely over the membrane. Thus, the frequency of a tone must be indicated by the pattern of activity in large numbers of fibres. The higher the sound stimulus level, the greater the number of fibres will become involved. Change of frequency involves a movement of the whole pattern of excitation along the cochlea. Estimates have been made of the numbers of active fibres for given frequencies and sound levels, and of the implications of frequency change on the pattern of excitation (see Whitfield, 1967); suffice to say here that frequency of the sound is indicated by the position of an array of active units, hence the so-called *place principle* of frequency indication. In view of the fact that at low frequencies, the cochlear nerve fibres discharge synchronous volleys of impulses with the individual sound pressure waves, it has been suggested that these volleys may carry information about the lowest frequencies, hence the *volley principle*. This effect is marked at 500 Hz, but becomes less so with increasing frequency. It is not possible to assess its quantitative significance at the present time. Whatever its significance, it is clearly subsidiary to the place principle. Interactions between units such as inhibition of the response to a tone by the addition of another tone of a nearby frequency, occurred in Kiang's records. The explanation is not clear.

We can now try to summarise the general principles on which the cochlea acts as the end organ of hearing.

Summary

The sound waves enter the external ear, set in motion the tympanic membrane and middle ear ossicles, culminating in the movement of the fluid in scala vestibuli, media and tympani. The characteristics of different sensitivity to different frequencies are determined by the physical properties of the external ear canal, middle ear and cochlea, and the transfer of energy to the fluids of the inner ear is notably efficient.

The vibrations of the fluids set up complex waves in the cochlear partition, travelling from the base to apex with both transverse and longitudinal bending of the basilar membrane, and the other parts of the cochlear partition. The membrane at the basal end follows the waveform of the vibrations, but each frequency component in the wave reaches a maximum, slows down and disappears at a position on the cochlear partition characteristic of that frequency. High frequencies die out first, low last; with the lowest frequencies, the whole cochlea vibrates.

These vibrational patterns modify the normal spontaneous activity in auditory nerve fibres, possibly as follows (Eldredge & Miller, 1971). 1. Basilar membrane movement, in one direction, excites the nerve endings possibly by the release of a chemical transmitter manufactured by the hair cells and acting on the associated nerve endings. The decay of this transmitting substance in time is constant. The excitatory process might last longer than one cycle of a pure tone. 2. The electrical currents, detectable as the cochlear microphonics, are postulated to be capable of enhancing or suppressing activity in auditory nerve fibres, whether spontaneous or that induced by the transmitter substance. The timing of the cochlear microphonic is such that the increased or diminished probability of discharge operates alternately to produce the periodic discharge patterns seen in the auditory nerve; for example during a pure tone, impulses tend to occur when the current flow is from scala vestibuli towards scala tympani. Either enhancement or suppression may predominate, the latter presumably must be effective in suppressing nerve discharge cyclically when the sound frequency is high enough to involve persistence of the activity of the transmitter substance over a number of cycles of a tone. 3. The extreme sharpness of the cut-off above the characteristic frequency of a unit (Fig. 5.11), is not explained by basilar membrane dynamics. The effect thus must depend on some form of active suppression of nerve responses at frequencies above the characteristic frequency of the unit. The summating potential may have some function in this connection, but no direct evidence is available. Two-tone inhibition at cochlear level, however, is described (Sachs & Kiang, 1968). In this, the nerve activity in response to a tone is diminished by the second tone, in appropriate stimulus conditions.

On the basis of these facts and concepts, the auditory nerve shows characteristic patterns of activity, peculiar to any particular stimulus. For a pure tone, the block of active fibres may be comparatively narrow (in terms of extent along the basilar membrane) if the intensity is low. For increasing intensity, a moderate increase in frequency of impulses occurs, together with the involvement of additional active fibres, so that the active block widens. Thus fibres whose characteristic frequency is considerably removed from that of the tone, also become active. The block of active fibres is, however, still in a position indicative of some particular frequency. This frequency is determined by the position of the pattern of activity: intensity by its extent, and presumably by the discharge rate of the active fibres.

Time relations are probably effectively indicated by activity in the basal turn of the cochlea, due to the minimal effects in this region of the dispersal in time at the higher reaches of the cochlea due to the travelling wave.

Another means of indicating frequency has been postulated for the lowest frequencies. During stimulation with a pure tone of, say, 400 Hz, direct frequency following by the nerve impulses occurs, i.e. the nerve fibres tend to discharge in volleys, at some instant, within the position of the sound wave cycle corresponding to 'downward' flow of current through the hair cells, although not necessarily at each cycle. On the basis of this synchronised activity or 'volley theory' information might be conveyed for frequencies at the lower range of audible frequencies.

This completes the peripheral part of the hearing process. The central aspects are not discussed here.

EFFECTS OF INTENSE SOUND   The moving parts of the hearing mechanism can be damaged by intense sound. Prolonged exposure affects the hair cells, e.g. of the guinea-pig (Davis & associates, 1953; Engström & Ades, 1960; Beagley, 1965a, 1965b; Engström, Ades & Bredberg, 1970). In man, damage in the inner ear is detectable as a reduction of sensitivity of hearing. The measurement of hearing is treated in the next chapter.

6

Normal hearing and its measurement

The essence of most scientific investigation is measurement of some kind. In attempting to examine the effects of noise on man, attention has been paid to the nature of the stimulus, so that sound can be appreciated in a quantitative way. The next step is to appreciate how hearing can be measured, and to examine some features of the 'normal' human sense of hearing. With these two fundamental factors as a basis, the relations between noise and the psychological and physiological functions of the body can be examined quantitatively.

Measurement of hearing

The technique of measuring hearing is known as audiometry, and various forms are used for different purposes. The simplest basic technique is pure-tone air-conduction audiometry in which the audiometer is controlled manually and the sound is presented to the ear by means of an earphone. In this method the minimum intensity of sound, in the form of a pure tone, which is audible (the auditory threshold) is determined. Since the sensitivity of the ear is not the same for all frequencies in the audible spectrum, a threshold determination is necessary at a number of different arbitrary frequencies. Each ear is tested separately. The threshold values obtained could be expressed, for example, as actual values of the sound pressure presented to the ear, and measured preferably at the ear drum, but this would be inconvenient and difficult to interpret. Instead, as will be explained later, the threshold values at the different frequencies are expressed in dB relative to a standard set of threshold values, representing the average auditory thresholds at each frequency, for normal young people, measured at the entrance to the external auditory meatus; this series of 'normal' values is known as a standard for normal hearing. Other diagnostic forms of audiometry are used for special purposes. These forms include bone-conduction audiometry and speech audiometry. Comprehensive accounts of the techniques of audiometry are given by Hirsh (1952) and Watson and Tolan (1949). Hinchcliffe and Littler (1958) and Littler (1962) deal with technique; Robinson (1971) has reviewed, very informatively, certain basic aspects of the subject.

Basis of pure-tone air-conduction audiometry

In the conventional form of audiometry the essential requirement is that the person being tested should be presented with pure tones of known and controllable intensity and frequency. While it is

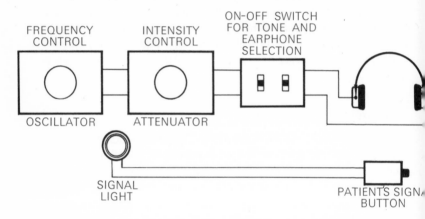

6.1   Block diagram of a conventional audiometer for air-conduction audiometry.

possible to perform these threshold determinations by setting up a sound field of known intensity in which the subject is placed (free-field measurement), this is not normally done; earphone listening is the normal procedure, high grade moving coil type telephones being essential. In the conventional audiometer (Fig. 6.1) the tones are generated by an electrical oscillator circuit capable of providing sine waves at a number of frequencies from which the operator selects one at a time for each threshold determination. The electrical output to the telephones is controllable by an attenuator, also in the control of the operator, and the resultant sound generated by the earphone can therefore be set at any desired value, within the capacity of the instrument. Switching arrangements to give an on–off control of the tone in a gradual manner, free from clicks or other unwanted sounds, together with earphone selection for right and left ears, provide the basic facilities for audiometry. The test must be done in a sufficiently quiet environment to avoid masking of the audiometric tones by other sounds. This type of audiometry is a procedure based on the judgement of the subject. The audiometrician can only form his conclusions on the opinion of the subject as to when he can or cannot hear the tones. These are applied by the audiometrician by operating the appropriate switch on the audiometer, and should be about 1 second in duration in this type of audiometry. Usually, all that the subject has to do is to raise a finger, or press a button, which operates a light signal, as long as he hears a tone, and refrain from making any signal when he hears nothing.

It is important that clear and uniform instructions on what is being done, and what his or her share in the test is, should be given to the subject in an audiometric test. Hinchcliffe and Littler (1958) have found the following instructions to be satisfactory:

> 'You will now be tested by asking you to listen through these earphones to a number of simple musical tones. Each ear will be tested separately, beginning with the right (or left) one, and every time you hear a tone no matter how loud or how quiet it may be, you are to press the button. Keep the button pressed as long as you hear the sound, and release it as soon as you think that the sound has disappeared.'

Simple illustration of what the tones sound like completes the demonstration, and the earphones are correctly positioned on the subject's head by the audiometrician.

The auditory threshold

The decision on what is the threshold value for any particular frequency is affected by the performance of the ear near threshold. Suppose that, in the absence of appreciable other sound, a tone of some particular frequency, lasting about 1 second, is presented by means of an earphone, and the subject is asked to indicate when he hears it by pressing a signal button. It is found that if the tone is repeated at different intensities the response will vary in a particular way with the intensity of the tone. There is an intensity at and above which every test tone elicits a response; that is, the tone is heard 100% of the times it is applied. There is also a somewhat lower intensity at, and below which, no response is elicited no matter how often the test is repeated. Between these levels there is a range of intensity where there is uncertainty whether the tone has been heard or not. The size of this range of uncertainty varies between individuals, but may be as small as 2 dB in some persons and as large as 6 dB or more in others (Chocholle, 1954). For intensities intermediate between those eliciting a 100% response and no response at all there is a gradation of the percentage of responses to applied tones, and about the middle of the range 50% of tones will elicit responses.[1] This point can be taken as representing a good practical indication of the auditory threshold, and it has the advantage that it is within the range of greatest sensitivity to change in intensity. In practice, it is neither practicable nor necessary to make many trials at each intensity for each frequency, and if two responses are elicited in response to four applied tones, we accept the particular intensity as the auditory threshold. If we proceed to find the auditory threshold in this way, certain practical points present themselves. We start at an easily audible intensity and work downwards to the threshold represented by the 50% response level. This is called the descending threshold. Alternately we can start at an inaudible intensity and work upwards to the 50% response level, so eliciting the ascending threshold. However, these two thresholds are not identical, for the descending threshold is normally lower. In other words, if the tone is first heard it apparently is followed down lower in intensity than if it is being identified in the rising direction. However, the descending and ascending thresholds are not usually very far apart and practical audiometry is not greatly complicated by this effect. The other factor is that in its usual form, the audiometer cannot have an infinite number of steps of intensity, and in fact it has become standard practice for manual audiometers of the type discussed to have steps of 5 dB on the intensity control.

The pattern of manual audiometry follows from these various factors. Taking one particular frequency, we apply to the earphone a tone which we expect to be about 40 dB above the threshold for the frequency. At this level a single tone should elicit a response, and we then proceed to present single tones, each 10 dB less intense than its predecessor, until we reach a level which elicits no response. This gives a quick and approximate indication of the threshold.

Recalling the criterion for threshold of two responses out of four

[1] It should be noted, that in experimental audiometry of the highest precision, signal detection theory is applied, taking into account the occurrence of false positive responses.

presentations of the tone, we return to the lowest intensity which was heard and present four tones. The test now consists of testing systematically at progressively less intensity, in 5 dB steps, to find the lowest intensity at which two, three or four responses are elicited from four tone presentations, provided that at the next less intense 5 dB step, either no response or one response is elicited by four presentations. This establishes the *descending threshold*. The test is now repeated, starting at an intensity at which we had ascertained that no response was elicited irrespective of the number of tone presentations, and progressively increasing the intensity, using the same criterion for threshold, in order to establish the *ascending threshold*. Frequently the ascending and descending thresholds are the same, but if they are not the arithmetical average of the two thresholds in decibel values is taken as the true threshold.

It is thus obvious that the size of the 5 dB step, which is quite suitable for most forms of audiometry, nevertheless may obscure the difference between the ascending and descending thresholds and frequently the actual intensity which would give two responses out of four presentations cannot be precisely selected because it falls between the 5 dB steps.

Having thus established the threshold for one frequency in one ear the other frequencies are tested in turn on the same ear as described later.

Quantitative aspects of audiometry

These basic principles and procedures of audiometry are, however, only steps towards the actual measurement of hearing. The attainment of a satisfactory quantitative basis for the measurement of hearing has entailed an immense amount of work in many countries, and international agreement by the relevant committees of the International Organization for Standardization (ISO) has only recently been reached (ISO, 1964).

The problems of pure-tone audiometry have been very clearly stated by Robinson (1960) in three phases. The first phase is the determination of the threshold of hearing for normal listeners (this is the 'normal' hearing for young people to which reference has already been made). The second phase is the 'transfer of this information to the dials of practical audiometers'. The third is the actual technique of hearing measurement.

These will be considered in the same order.

Normal threshold of hearing

The requirement to decide what is to be regarded as the normal or average value in a series frequently occurs in biology and medicine. The pitfalls are many, and arise from such factors as the variability of the individual values, and the problems of securing a suitable sample, both in quantity and quality, to make up the individual values. In the case of hearing, the need is to secure data on the auditory thresholds of a sufficient number of persons of suitable type, e.g. without disease or injury of the hearing mechanism and of appropriate ages, from which to derive a representative average value for each of the arbitrary frequencies. This type of investigation has been done, over the years, by a number of different investigators in a variety of ways. Since the important contribution of Sivian and White who reviewed the data available in 1933, numbers of surveys of hearing have given somewhat conflicting results. On the basis of American work subsequent to the Sivian and White publication, an American standard for normal hearing was published (ANSI, 1951). This standard has been increasingly regarded as an under-estimate of the sensitivity of young normal ears. This view has been confirmed on a number of occasions, and in Britain in particular by two independent investigations of the auditory thresholds of young people by Dadson and King (1952) at the National Physical Laboratory (NPL), and Wheeler and Dickson (1952) at the Central Medical Establishment of the Royal Air Force. These two investigations showed excellent agreement and formed the basis of the original British standard, BS 2497, *The normal threshold of hearing for pure tones by earphone listening* (BSI, 1954). The thresholds regarded as normal for young people, expressed as modal[1] values, were on average approximately 10 dB lower than the corresponding values of the 1951 American standard. This curious disparity persisted for some years, and it is now accepted that this American standard, due in part to the nature of data on which it was based, understated the hearing acuity of normal young people. An international standard (ISO, 1964), which is quantitatively somewhat similar to the original British standard is now in use, and will be discussed later. Although the original British standard of 1954 is now superseded, it was a significant contribution, and it is instructive to note how it was derived from the audiometric data, and the developments which

---

[1] The modal value is the one which most often appears in a series of values.

have contributed to the present status of audiometric standardisation.

In all, the auditory thresholds of some 1200 otologically normal ears of men and women of age 18–25 years were determined at various arbitrary frequencies, using the Standard Telephones and Cables type 4026A earphone. The sensitivity of the ear is such that the sound pressures at threshold, although of course different at different frequencies, are extremely small. They are in fact so small that attempts to measure them directly, for example under the earphone earcap at the entrance to the external ear canal, are practical only in laboratory conditions, and even so, are only marginally feasible with the most elaborate apparatus. We can thus exclude this direct measurement as a routine procedure, and in its place we must substitute other means, of various degrees of indirectness, admittedly with some added uncertainty in the results. Fortunately, in audiometry, the voltage applied to the earphone is easier to measure at threshold; in the investigations leading to the 1954 standard, the measure of the auditory threshold of the subject was the RMS voltage across the earphone terminals when the tone was just audible according to the defined criteria for the auditory threshold. This electrical index, or threshold voltage excitation, once established, may then be transformed into acoustical terms, specifically sound pressure level. The locations for which these SPL values could be derived are various. In the original British standard these values were specified for two conditions.

The more obvious of these two approaches is a statement of SPL values actually experienced by the ear at threshold, i.e. corresponding to threshold voltage excitation of the earphone. Various possible anatomical situations suggest themselves, such as in the immediate vicinity of the eardrum, or more accessibly, the entrance to the external ear canal. The latter site was in fact quoted in the original NPL data and subsequently in the standard. Access to this situation is possible by means of a probe-tube microphone. This consists of a tube of about 1 mm bore, which passes through a hole drilled diagonally through the earphone cap. The open end of the tube, when the earphone is applied, is situated just at the entrance of the ear canal. The outer end terminates in a microphone. The system must possess suitable frequency characteristics, and it must not disturb the acoustic conditions under the earphone. The sensitivity of such an arrangement is limited, but this disadvantage can be overcome in the following way. The threshold voltage excitation of the earphone is determined, then increased by, say, 50 dB so bringing the sound pressure within the ability of the system to measure satisfactorily; the recorded sound pressure is then reduced by 50 dB, the result being the threshold sound pressure. Some small element of uncertainty must persist, since the validity of this elevated level derivation must rest on the assumption, basically justifiable, that a linear relation exists between the electrical input to the earphone and its sound output. In the original NPL data, and subsequently in the 1954 standard, modal values of threshold sound pressure level, at the entrance to the external ear canal, were given for the various frequencies (see Appendix I, Table I, Column 2) and were designated *normal threshold of hearing*.

Turning to the second method, while such a statement of *normal threshold of hearing* is occasionally of direct use, the quantity that we manipulate in normal audiometry to arrive at the auditory threshold is the voltage applied to the earphone. It is still necessary, however, somehow to calibrate the earphone in terms of sound output for voltage input. The method now routinely used is solely electro-acoustic, so avoiding the complications of actual measurements of sound pressure experienced by the ear, and having great advantages in simplicity and reproducibility; it utilises a type of device known alternatively as an artificial ear, or coupler. An artificial ear attempts to simulate the acoustic conditions of the human ear; the earphone sets up a sound pressure in a specially shaped and dimensioned cavity, or a series of cavities, while a calibrated microphone and amplifier system measures the sound pressure. A coupler uses the same principle, but without any very rigorous effort in its design to simulate the acoustic properties of the human ear. By means of either device, the sound pressures corresponding to threshold voltage excitation can be conveniently derived from sound pressure measurements made at an elevated level, in the same way as described for the measurement of the ear canal pressures, but with the convenience and reproducibility of solely electroacoustic physical measurements. It must be remembered, however, that the pressure generated by a given voltage excitation of an earphone, on any artificial ear or coupler formerly in use, does not bear a close relation to the pressure which would result in the average human ear for the same voltage excitation of the earphone. This restriction does not apply to a new artificial ear (IEC, 1970b), to be described below.

To return to the evolution of the 1954 British standard, both of the contributory investigations expressed the modal threshold, at the various frequencies, as RMS sound pressure levels in an artificial ear described in BS 2042 (BSI, 1952), in conjunction with the Stabdard Telephones and Cables Type 4026A earphone. In the standard a unified set of sound pressure values derived from the two investigations was described as the *normal equivalent threshold sound pressure*. The 1954 British standard remained in use in Britain until the advent of international standardisation in 1964, which in any case did not entail any drastic change. The subsequent history and present position can thus best be followed from the time of publication of an international standard by ISO (ISO, 1964), whereby a great rationalisation of audiometric practice, at an international level, was achieved.

International standardisation

For more than a decade prior to 1964, great difficulties in the field of hearing measurement resulted from the divergence of national standards and practice. The American standard of 1951 (ANSI, 1951) recommended a threshold markedly above (i.e. implying less sensitive hearing) the British standard of 1954, while the German practice was based on a threshold slightly lower than the British. This unsatisfactory situation came under the consideration of an international body of technical experts in 1955, under the auspices of ISO, set up for the purpose of arriving at a best value for an international threshold (or reference zero) for the measurement of hearing. For this purpose all available published data were considered, and extensive additional work carried out on a co-operative basis between the five participating countries, in France, Germany, United Kingdom, USA and USSR. The details of the work have been described by Weissler (1968), and its outcome was the publication of ISO Recommendation R389, *Standard reference zero for the calibration of pure-tone audiometers* (ISO, 1964). This was a notable international scientific achievement and has received widespread acceptance. The R389 recommendations are in the form of equivalent threshold values of SPL to be taken as representative of the modal values for the thresholds of hearing of persons 18–30 years of age, in a normal state of health and free from all signs or symptoms of ear disease and from wax in the ear canal, and with no history of undue exposure to noise. The values are stated for specified artificial ears or couplers, in conjunction with specified earphones. No real-ear sound pressures are given. The artificial ear or coupler pressures are termed *reference equivalent threshold sound pressure level (monaural listening)* usually abbreviated to RETSPL. The standard thus consists of RETSPL values at each of the arbitrary audiometric frequencies (125, 250, 500, 1000, 1500, 2000, 3000, 4000, 6000 and 8000 Hz). A complication arises, however, inasmuch as these values are not identical for the different combinations of earphone and artificial ear (or coupler) adopted as instrumental reference standards in the different countries. Specifically, the situation arises from the fact that the various earphones and artificial ears or couplers exhibit different values of the property known as acoustic impedance (Appendix A, p. 370). The *characteristic impedance (ρc) of air* (p. 27) has already been met; it is analogous to resistance of electrical circuits, and is the ratio of sound pressure to the particle velocity in a sound field. *Acoustic impedance* could refer to a hypothetic surface in a sound field (e.g. the mouth of a loudspeaker horn) or an actual surface (e.g. a loudspeaker diaphragm); it is, in general terms, the ratio of sound pressure to volume velocity of the sound waves. Volume velocity is defined as the product of linear particle velocity and area at the surface concerned. In the use of earphones the acoustic impedances of the source (the earphone) and of the ear (or artificial equivalent) interact so influencing the value of sound pressure developed for a given electrical excitation of the earphone; and for any earphone of practical design, the sound pressure resulting from a given electrical excitation is thus markedly influenced by the acoustic impedance of the artificial ear, coupler or human ear on which it is acting. The various national standard types of artificial ear or coupler have tended to evolve separately, and it is hardly surprising that their impedance characteristics differ, and in addition, fail to reproduce those of the average human ear. National standard earphone types are also different, so that the RETSPL values in R389 had to be specified for each national combination of earphone and artificial ear or coupler, making, in effect, five national specifications to describe the same standards of hearing acuity. This is highly inconvenient since comparisons between the different systems are tantamount to a repetition of the work on which the standard was based (Robinson, 1971).

Further progress in standardisation subsequent to R389 has,

however, occurred (Delany, 1971; Robinson, 1971). In the first place, the British standard 2497 of 1954 has been revised; it is issued in two parts (BSI, 1968 and 1969, respectively). Part 1 is technically identical with R389. Part 2 recognises the fact that many more earphone types are in use than those specified in R389 or BS 2497: Part 1, and specifies RETSPL values for eight commonly used earphone types, in conjunction with the 9-A coupler, and an equivalent document is being issued by ISO. Similar data are also included in the American audiometer standard (ANSI, 1969). This helpful simplification in the technique of international audiometric comparisons has been recognised formally by ISO, by specifying the 9-A coupler (ANSI, 1969) as the basis for an Interim Reference Coupler, and the IEC have issued its specification (IEC, 1970a). The reader may well enquire at this stage why this long and devious process of standardisation could not have been expedited by international agreement to use a single earphone and artificial ear combination. The IEC did, in fact, explore the possibility of a standard earphone specification, but difficulties proved prohibitive. However, the pursuit of simplification is best served at present by attention to the artificial ear or coupler component of the system, by devising an artificial ear which reproduces very accurately the impedance, as a function of frequency, of the average human ear. In this way, calibrations of individual earphones could be performed against a single set of RETSPL values, this unique threshold standard being independent of earphone type and so valid for any earphone. This step has in fact now been achieved as a result of work over a period at the National Physical Laboratory (Delany, 1964; Delany & Whittle, 1966; Delany & Whittle, 1967; Delany, Whittle, Cook & Scott, 1967). Since this solution depends on the closest possible simulation of the impedance of the average human ear, the NPL work logically started with a new determination of the impedance, as a function of frequency, of human ears. For this purpose, an impedance measuring unit was produced in a form closely resembling a normal earphone, was worn as such, and could be fitted with any desired standard form of earcap. The static force exerted by the earphone caps on the ears was adjustable by a pneumatic headband. The impedance unit consisted of two electroacoustic transducers, the diaphragms of which faced the ear through the aperture of the earcap. One acted as a sound source generating a known volume velocity, the other measured the sound pressure developed across the acoustic impedance of the ear, as viewed through the aperture of the earcap. From the relative magnitudes of sound pressure and volume velocity, the acoustic impedance was derived. The impedance of the ear varies with frequency, and in Delany's (1964) data, at 1 kHz, a volume velocity of about 2×10^{-8} m^3/s produced a sound pressure of about 70 dB SPL in the average ear. The impedances viewed through various earcap types in use for audiometry varied somewhat, but not so much as to preclude the achievement of a satisfactory compromise in the impedance as a function of frequency in the NPL artificial ear (Delany & Whittle, 1966). This artificial ear has proved to be capable of use with any conventional earphone, it is superior to any previous artificial ear or coupler generally used for calibration of earphones, and reproduces closely the real ear threshold (Delany, Whittle, Cook & Scott, 1967). It has now been formally adopted in a recommendation by IEC (IEC, 1970b). The only step still left, at the time of writing, is its general adoption, which would entail less discrepancy than do the present practices. It is to be hoped that international agreement can be achieved for the reference threshold appropriate to the new artificial ear. Such a step would provide an opportunity to rectify some particulars in which R389 appears (Delany & Whittle, 1967) to be capable of improvement. It should be emphasised that these improvements in specification of auditory threshold for the calibration of audiometers must co-exist with some imperfections. These include the impossibility of allowing for such variations as the different volumes under different earcaps, and the uncertainty of the acoustic conditions due to lack of uniformity of placement and fit of earphones, in conjunction with the flexible and complicated shape of the external ear. Finally, the inherent variability of the human auditory threshold (either real, or an artifact of the technique of earphone audiometry) is recognised by specifying the auditory threshold by some form of average, in this case the modal value. As an example of the consequence of the use of earphones, the variability due to different placement and pressure has been mentioned already; in addition, a person with a large external ear canal and associated space, or with greater than normal compliance, will have a less than normal impedance. Lower sound pressures will be developed, with the probable result that his hearing acuity may appear to be less than normal.

Thus, standards can provide the necessary quantitative basis for the calibration of audiometers, so that the next step in Robinson's

(1960) phrase the 'transfer . . . to the dials of practical audiometers' is clear. The audiometer is provided, as has been described, with two main controls, one for frequency and one for the intensity, of the tone produced by the earphones. Also, the audiometer is constructed to produce, at any frequency selected, a sound pressure in the earphone corresponding to the normal threshold of hearing, when the intensity control is appropriately set. In practice, this is achieved by graduating the intensity control on the audiometer in dB, on a scale where 0 dB indicates that the normal threshold pressure is being set up by the earphone. Originally described as a scale of *hearing loss*, this is now called *hearing level* (Davis, Hoople & Parrack, 1958) or *hearing threshold level*; its meaning is as follows. If a subject's threshold at any frequency, by the criterion already described, is found, and the intensity control reads 0 dB, the hearing is designated 0 dB hearing level. If the control has to be set at 10 dB hearing level, the sound pressure in the earphone is greater by 10 dB than the normal threshold, i.e. the hearing is less sensitive than 'normal'. This may or may not indicate any abnormality, as is discussed later. On the other hand, should the hearing be more acute than that specified in the standard, the threshold sound pressure in the earphone will be less, and the hearing level will then be indicated by a negative sign on the dial, e.g. − 10 dB hearing level.

Audiometric frequencies

Pure-tone audiometers are normally designed to provide a number of tones of standard frequency. The British standard for pure-tone audiometers (BSI, 1958) specifies eight frequencies; these are 250, 500, 1000, 2000, 3000, 4000, 6000 and 8000 Hz. Sometimes higher or lower frequencies may be employed, or some of these frequencies may be omitted. The tones are applied in a systematic order; starting with one ear a convenient way is to find the hearing level at 1000 Hz, and work upwards through the other frequencies until the highest; after which the lowest is examined, and the others in ascending order upwards until 1000 Hz is reached, which is then repeated (Watson & Tolan, 1949). If the second threshold at 1000 Hz produces a lower hearing level than the first, it is preferable to use this as a truer indication of the threshold. The other ear is then examined in the same way.

Finally, it should be emphasised that a clinical examination by the usual methods of otology is a fundamental part of any hearing examination, since the measurement of hearing is only a part of the examination, and can be grossly misleading without the full examination; as an example the ear may be obstructed by wax, or the person suffering from a cold, both of which would throw doubt on the reliability of the audiogram, quite apart from the risk of missing a serious condition if audiometry alone were used.

In summary therefore, in audiometry a scale, in dB, of hearing level is used, the zero of which corresponds to the normal threshold as defined by the standard in use. This zero level is automatically set by the audiometer for each frequency used. The hearing level value is defined as the deviation in decibels of an individual's threshold of hearing from the normal threshold value to which the audiometer is adjusted.

The data on hearing are plotted in the form of an audiogram in which frequency is indicated on the horizontal axis and hearing level on the vertical axis. The two ears are designated by different symbols. An audiogram form showing what would be regarded as normal hearing in a young person is shown in Fig. 11.3A (p. 223).

Maintenance of calibration of audiometers

Audiometers must be kept in a proper condition in order to produce reliable results. They must be maintained and checked for acoustic output and adjusted or calibrated at regular intervals (Knox & Lenihan, 1958; Hinchcliffe & Littler, 1958).

For the purpose of calibration or adjustment of the output of audiometers, when the procedure outlined above, by means of an artificial ear, is not possible, a biological calibration may be used.

BIOLOGICAL CALIBRATION   The audiometer can be checked by recording audiograms from a group of young people aged 18 to 30 years, conforming to the criteria of otological normality specified above, from ISO R389. The numbers should be normally not less than 20, but if the group is of individuals known to possess unimpaired hearing, perhaps half of this number is permissible. Having obtained audiograms from the group, the hearing levels at the various frequencies are reduced to one value by some averaging procedure. An arithmetical average, or mean, value in a normal group will be less useful than the modal or median values. The object is to obtain an expression of the central tendency of all the values which is not influenced by occasional extreme values, and the median value is a convenient way of doing this. The median value is the value which occurs halfway along a series of observations; that is, if there were 21 observations, and they were all grouped in ascending order of size from the lowest, which we will call No. 1, to the highest which we will call No. 21, the median value would be that of No. 11 in the series. When there are even numbers, or other complicating factors, these can be overcome by simple arithmetical methods (Bradford Hill, 1966; Moroney, 1962). The median values for the hearing levels of the group at the various frequencies can then be taken as representative of normality, and the audiometric readings of hearing level provided by the audiometer corrected accordingly.

Permissible noise in audiometric rooms

If the prevailing noise in audiometric rooms is excessive the test tones will be masked, that is, their thresholds will be raised, so producing higher hearing levels than the true hearing levels. This aspect of audiometry is discussed in Appendix I (p. 398).

The definition of normality of hearing

When we examine the hearing thresholds of numbers of people, such as have been the basis for national or international standards for normal thresholds of hearing, it is obvious that 'normal' hearing is a statistical concept. What is found is, of course, a range of hearing thresholds. The distribution of values of auditory threshold when numbers of persons are measured has been found to conform reasonably closely to the concept of a statistical normal distribution, in persons who have no evidence suggesting any kind of defect or abnormality of their hearing. Distribution curves state how values in a series are grouped. For example, in a group of persons, their various heights could be measured and averaged. This average would only give a limited idea of the stature of the group as a whole. More information would be required to describe the range and variation of heights. If the numbers, or percentages of the total number, of persons are plotted on a vertical scale (ordinate) and the actual values of height on a horizontal scale (abscissa) a frequency distribution curve is obtained. Such curves may be symmetrical about the average, or they may be asymmetrical. In the case of symmetrical curves, one particular form is the 'normal' distribution. This type of distribution curve is a fundamental concept in statistics. It can be derived mathematically, and is applicable to a great variety of situations. Many, but not all, of the measurable quantities in the world of animals and plants, show an essentially normal distribution of values. Height in human beings is an often-quoted illustration of a dimension in biology which for practical purposes, shows a normal distribution. Fig. 6.2 from Bradford Hill (1966), illustrates a hypothetical distribution of the heights of 1000 men, in the form of a histogram. Superimposed upon the histogram is a normal distribution curve. This diagram also illustrates the concept of the standard deviation. This is a fundamental measure of the scatter of values round an average and is widely used in statistical analysis. As an example of the use of the standard deviation, it is possible to have some groups of values with a small scatter, so that they are all closely spaced about the average value. Other groups may, however, be widely scattered at large intervals. The standard deviation indicates the average of the deviations from the mean of all the values in the series. In statistical calculations, refinements are used in the calculation of the standard deviation. As an example, some values will be greater than the mean and others less, so that some differences from the mean will be positive and others negative. In order to avoid the problem of positive and negative values, each difference is squared, so eliminating the negative signs. After averaging these values, the square root of the average is taken. In addition, we are usually working with samples from a larger group of values, and it has been found that a more accurate indication of the standard deviation in the larger group is obtained by dividing the squared deviations, not by their number, but by one less than their number, e.g. if n = number, divide by $n - 1$. When we relate the standard deviation to the curve of normal distribution, we find (Fig. 6.2) that the shape of the curve can be described in terms of standard deviation. The peak of the curve corresponds to the arithmetical mean or the median or the modal values of the observations; symmetrically spread on either side, the curve falls away with a slope such that a spread of one standard deviation on each side of the mean includes 68·27% of the observations; a spread of two standard deviations on each side of the mean includes 95·45% of the observations; and a spread of three standard deviations on each side of the mean includes 99·73% of the observations.

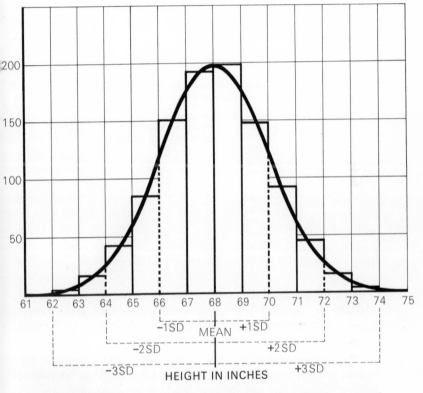

6.2   Histogram of statures of 1000 men (hypothetical figures) compared to a normal distribution curve. The spread of values of height, in inches, corresponding to one, two and three times the standard deviation on each side of the mean, is shown. (Bradford Hill, 1966; metric quantities, Bradford Hill, 1971.)

Thus, in a normal distribution, we can quickly picture what the spread is like if we know the standard deviation. For example, values differing from the mean by more than twice the standard deviation (plus or minus) will only amount to about 5%, or 1 in 20 of the total.

Range of hearing levels

In the apparently normal ears such as those used to define the British standard for the normal threshold of hearing, the scatter of threshold values, by earphone listening, found by Dadson and King (1952) at the National Physical Laboratory, for the various frequencies, yielded the standard deviations shown in Table 6.1. Similar data from a number of different sources have been collected by Knight and Coles (1960).

TABLE 6.1

Standard deviations of auditory thresholds
(Dadson and King, 1952)

Frequency Hz	Standard deviation of threshold sound pressure dB
125	6·8
250	7·3
500	6·5
1000	5·7
2000	6·1
4000	6·9
6000	9·1
8000	8·7

These figures mean that, for example, at an audiometric frequency of 1000 Hz, about 5% of persons with apparently quite normal hearing will differ in hearing level from the standard value by more than plus or minus 5·7 × 2 = 11·4 dB; similarly at 6000 Hz the deviation could exceed 9·1 × 2 = 18·2 dB. In Dadson and King's subjects, if the frequency range 80 Hz to 4000 Hz is considered, 90% of the subjects had thresholds which were included in a total range of 25 dB. The range of hearing levels is in fact due to a combination of true biological variation and variation of an instrumental nature inseparable from the use of earphones. The origins and significance of these variations have been discussed by Robinson (1960) and will be referred to later.

These degrees of scatter, in normally hearing young people demand the utmost caution in the interpreting of audiograms. Decisions on the basis of audiograms should only be entrusted to those qualified to do such work.

CUMULATIVE DISTRIBUTIONS   Distributions may be displayed as a cumulative series. For this purpose, the observations are ranked in order of magnitude on the vertical scale at equal intervals. On the horizontal scale is the magnitude of each observation. For normal distributions such as the curve of Fig. 6.2, the cumulative curve is S-shaped and symmetrical. The vertical scale would cover 1000 intervals (conveniently reduced to a percentage scale 0–100); the horizontal scale would be the observed heights of the men, The curve would thus show the relation of height to progressively vary-ing proportions of the population. For example, in this ideal normal distribution, about 25 men (strictly 2·275%) would equal or exceed a height of 72 inches, and other associated percentages and heights would be graphically displayed. From distributions other than normal, different curves result. The cumulative concept can be expressed by quantiles, which denote position in the ranked series, and not a proportion, as do percentages. If the ranking is on a 100-point scale, the individual positions are called centiles. A quartering of the range yields quartiles, commonly used to denote the 25th and 75th centiles. The median is the 50th centile. In use, since it is a position in a series which is denoted, not a proportion of the series, quantile usage is less cumbersome. For example, 'a hearing level of 45 dB was equalled or exceeded by 5% of the population', would become 'the hearing level for the 5th centile was 45 dB'. Both usages will be found below.

The effect of age on hearing

Ageing processes are apparently an inevitable feature in the life of the animal body. They are manifest in a variety of ways, and per-haps one of the most conspicuous is the change which occurs in vision. Here, due to the progressive reduction of the elasticity of the lens of the eye, the near-point, or minimum distance for distinct vision, recedes as age advances. The effect is known as presbyopia. A less obvious change occurs in hearing, known as presbycusis or presbyacusis. This takes the form of a progressive deterioration of hearing for high tones. At any given time the degree of elevation of hearing level, by comparison with young people, is directly related to frequency; in audiometric terms, the higher the frequency of the test tone, the greater the hearing level; and the older the person the greater the deterioration.

A knowledge of the normal trends in presbycusis is highly important, since any audiogram must be viewed with the age of the person in mind, for it will be recalled that the international (ISO, 1964) standard audiometric zero, as well as national standards conforming to it, refer to the hearing of persons 18–30 years of age inclusive. The effect of age has been assessed in various surveys of hearing of people of different ages, using persons who have not suffered, as far as can be ascertained, from ear disease or any other condition likely to affect hearing. The question of previous exposure to noise is highly important in attempting to decide what constitutes normal hearing at various ages. The hearing of people in considerable numbers has been measured, on a volunteer basis, where opportunity has offered, such as at State Fairs in the USA (Glorig, Wheeler, Quiggle, Grings & Summerfield, 1957). Various surveys of hearing of random samples of different types of population have been conducted. In addition to Hinchcliffe's (1959a) study on a rural population, Hinchcliffe and Littler (1960) have examined a coal-mining population. The hearing of a primitive people has been examined (Rosen, 1962), and Hinchcliffe (1965) has made a survey of hearing in a tropical rural community. These studies have not, however, produced definitive data of the effect of age on hearing. The situation is complicated by the fact that it has proved difficult to obtain populations free from the effect of noise. Where noise exposure has occurred, there are certain well-marked tendencies in the pattern of hearing deterioration, as will be seen in Chapter 10 and subsequently. For industrial noise, with a wide variety of different frequency spectra, which in general are highest in the middle audio frequencies, the noticeable feature is a localised dip in the audiogram, due to raised thresholds in the frequency range from 3000 to 6000 Hz. This feature is the normal consequence of noise exposure, and is seen in Hinchcliffe's male rural population. Thus in the data of Dadson and King (1952) which contributed to the 1954 British standard for hearing threshold, there was no significant difference, in the age group 18–25 years, between the hearing of males and females, but even in the younger age groups, in the surveys of Glorig, Wheeler, Quiggle, Grings and Summerfield (1957), and of Hinchcliffe (1959a) significant sex differences in hearing level began to appear at certain frequencies. For example, in Hinchcliffe's (1959a) hearing survey of a random sample of otologically normal people from a rural population in Scotland the standard audiometric frequencies from 125 Hz to 12000 Hz were employed to determine the hearing levels of groups ranging in age from 18 to 74 years. In all groups covering the age range 18–54 years, the hearing levels of women were significantly lower than those of men at 3000, 4000 and 6000 Hz, while in older groups significant differences were also found at 2000 and 8000 Hz. This consistent superiority of women over men in hearing sensitivity at these particular frequencies is attributed not to an intrinsic difference, but to the greater noise exposure sustained by male ears in the course of work, military service or recreation. In this, small arms fire can be an important source of noise-induced hearing loss (Hinchcliffe, 1959b). Evidence from temporary threshold shift also tends to suggest an essential similarity between male and female hearing (Ward, Glorig & Sklar, 1959).

An assessment (Hinchcliffe, 1958) of the effect of age on hearing is shown in Fig. 6.3. These are smoothed median curves from data collected under very well controlled conditions. The relationships shown apply to women, but the data for men are significantly different only at the frequencies and ages mentioned above. If, as seems highly probable, these sex differences can be attributed to greater noise exposure sustained by the men, the curves would then represent the effect of age on the hearing of men or women who have been minimally exposed to noise. It is suggested that they can be used as such.

While uncomplicated cases of hearing deterioration due to noise or to age can be clearly distinguished by audiometry and the normal methods of otology, it is still necessary to recognise the possible difficulties of demarcating normal hearing from hearing impaired to a moderate degree by noise, especially in older people. The noise environments of work or even of some recreations may be such that, especially in the case of men, sufficient noise exposure may be sustained to produce degrees of hearing loss in excess of that which would accrue from age alone at frequencies in the region of 4000 Hz. When we recall that normal hearing lies within an appreciable range of hearing levels at each frequency, and that susceptibility to noise-induced hearing loss has a wide individual variability, together with the variety of characteristics of the audiogram, it is not surprising that inspection of a single audiogram may leave considerable uncertainty as to the condition and previous history of the person's hearing, as opposed to the practical statement of hearing levels.

Nevertheless, it is necessary to try to allow for the effects of age in assessing the effects of other factors, particularly exposure to noise. Up to the present, it has been customary to assume that the effects of age and of noise are additive in a simple manner, the implication being that the mechanisms are different to a significant degree. Thus a person who has worked in noise for a number of years has a hearing level for example, of 40 dB at 4000 Hz, and it is necessary to give an opinion on the possible effect the noise exposure may have

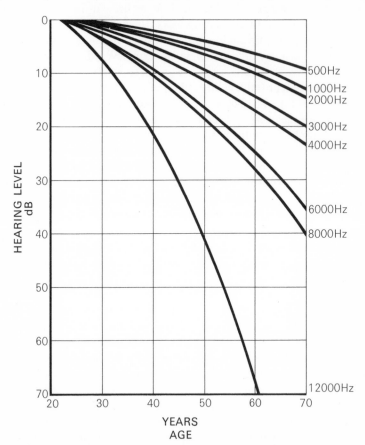

6.3   The relation between age and median hearing level. The values apply equally to men and women up to age 54 years, and for frequencies up to 2000 Hz inclusive. At greater ages and higher frequencies the curves apply to women, but in absence of noise-induced hearing loss it is probable that they would also apply to men. (After Hinchcliffe, 1958.)

had. If the age is 65 years and there is no indication of any present or past aural disease, it could be assumed that the hearing level, due to age alone (Fig. 6.3), would have been 20 dB. Simple subtraction of dB values would indicate that some other cause, in this case presumably noise, had contributed to the difference of 20 dB between the estimated and the recorded hearing levels. This procedure appears to have been justified by the findings of Glorig and Davis (1961), in which hearing level, as a function of age, has been investigated by air-conduction and by bone-conduction audiometry, which is described below. They conclude that a part of the presbycusis deterioration is in the middle ear, and therefore distinct from the deterioration due to the effects of noise on the inner ear. Threshold shifts due to noise and to age· would therefore be to this extent separate and so simply additive, in decibel values. Other components of presbycusis are believed to exist: deterioration of hearing from physical changes in the cochlea, from neural or end organ changes in the cochlea, and from ageing processes in the brain. Nevertheless a significant component appears to be associated with the middle ear at ages when senile changes are not conspicuous in the hearing mechanism. This being so, subtraction of the presbycusis allowance from the measured hearing level appears to be permissible in attempting to estimate the effect of noise. One word of caution is desirable, however, due to the various unassessable factors involved, this process may give erroneous results in the case of individuals. In the case of average figures of groups, it is widely used and has given useful information of the relations between hearing loss and noise exposure. The applications of presbycusis recur in Chapter 12.

Special types of audiometry

BONE-CONDUCTION AUDIOMETRY Our previous considerations of audiometry have been concerned solely with air-conduction audiometry. This obviously tests the entire chain of events, starting at the outside of the external ear canal. In certain circumstances, however, for diagnosis or in the study of Glorig and Davis (1961) just described it may be necessary to exclude the conductive mechanism of hearing and to test separately the part from the cochlea onwards. This can be done by supplementing the results of air-conduction audiometry with bone-conduction audiometry. The latter technique consists of applying vibrations to the bone of the head by means of a bone-conduction transducer, which takes the place of the telephone earpiece. The transducer is applied to the bone of the head and the latter set into vibration. The vibration in the bone excites the cochlea to vibrate, so that a sensation of sound results, without normal intervention of the conductive mechanism. This method thus tests the function of the inner ear. Also, an abnormality in the difference between the hearing levels by air and by bone conduction may indicate a defect in the conductive mechanism.

The technique is not so developed as for air conduction. Békésy (1960), Naunton (1963) and Hood (1962) have discussed various aspects of the theory and application of bone conduction, and Delany and Whittle (1966) have described a new artificial mastoid, a device simulating the physical properties of the superficial tissue of the skull and the mass of the head, with a vibration-sensitive transducer to record the vibrations set up by the bone-conduction transducer under test. The whole is analogous to the artificial ear for air conduction, and such artificial mastoids are required for the calibration of the transducers used in bone-conduction audiometry. A standard (BSI, 1966) for an artificial mastoid has been issued.

SPEECH AUDIOMETRY   In the types of audiometry we have so far discussed, the aim has been to ascertain the auditory threshold. This may not give the kind of information desired, since we may wish to know what practical handicap, in terms of difficulty in understanding speech, is being imposed on everyday life by deafness. In such cases some type of speech audiometry may be used. These are techniques designed to present recorded speech sounds at controllable intensities so that some quantitative indication is obtained of the ability of the person to perceive words or sentences correctly. It is found that meaningful curves expressing the relation between percentage of words correctly identified and the SPL values of the speech sounds are obtained, and that these curves are of value in differentiating different forms of deafness. The subject is highly specialised and an essential for the assessment of social adequacy of hearing. It is, however, outside the more basic field of routine air-conduction pure-tone audiometry as normally used in the monitoring of hearing in industry. For this reason it is not discussed in detail here, and the reader is referred, for further information on the essentials of the method, to an informative early paper by Knight and Littler (1953). With the development of this technique as a normal clinical procedure commercial speech audiometers are now freely available using recorded speech sounds in the form of tapes or disc material.

BÉKÉSY AUDIOMETRY Various modifications to the standard audiometric technique, such as we have noted, have been devised for specific purposes. The most important of these is the type of audiometry due to Békésy (1947) which gives an automatic record of the audiogram. In the most commonly used form of Békésy audiometry, the only instructions which the subject needs to receive are that a switch should be pressed as long as a tone is heard, and if no tone is heard the switch should be released. The audiometer attenuator is controlled by the switch so that, when the latter is pressed, the attenuator is moved by an electric motor and the tone gradually becomes weaker. If the switch is released the motor reverses and gradually increases the intensity of the tone. The rate of variation of the tone may be in the region of 2 dB to 5 dB per second. The movement of the attenuator is recorded by a pen or stylus on a rotating drum of paper or moving card. In the form originally described by Békésy, the frequency of the tone is changed continuously, so that the horizontal scale is one of time and thus of audiometric frequency, while the vertical scale is of hearing level. The trace has upwards and downwards excursions, showing how the subject is regulating the intensity of the tone, so that it rises and falls above and below his threshold. The mean of the excursions is usually taken as representing the auditory threshold. Experimental evidence, such as that of Burns and Hinchcliffe (1957) indicates that the results obtained in this way are, for practical purposes, the same as those obtained by the conventional pure-tone manual method. Another form of Békésy audiometry (McMurray & Rudmose, 1956) provides a programmed sequence in which the tone at each of six frequencies is maintained for approximately 30 seconds, and the two ears are tested separately, in the usual way, the results being obtained on a single card. These cards must then be read individually to decide exactly where on the trace the true threshold line should lie. Normally, a line through the middle of the excursions is acceptable, but this is not always the best index, and certain precautions must be observed; in some traces a considerable amount of experience and discretion is required to decide where the true threshold probably lies. This is a thoroughly practical and convenient method for performing numbers of audiograms for screening or individual purposes. It is particularly suitable for routine audiometry, for example in industry; only occasionally is it found that a particular person is unable to master the simple task involved. In this case resort would have to be made to manual audiometry. A tracing from the Rudmose type ARJ 4 audiometer is shown in Fig. 6.4. Rudmose (1963) describes the practical aspect of the technique in detail. The accuracy to be expected of this, and other types of audiometry, is of great importance; it is discussed in Appendix J (p. 400).

RUDMOSE ASSOCIATES, INC.
RICHARDSON, TEXAS

Johnson & Quin, Inc., Chic

6.4   Audiogram obtained with Rudmose ARJ4 audiometer. (Rudmose, 1963.)

Audiometry in children

The care and education of deaf children is a special service of great importance, and much effort in many parts of the world has been given to this work. As a part of this, and of the examination of the hearing of children in general, audiometry of children has been much studied, and the development of techniques suitable for young children has produced ingenious and successful procedures. The use of Békésy (1947) audiometry, particularly in the form developed by McMurray and Rudmose (1956) has many advantages, as we have already noted. These include a standardised procedure, rapidity, the elimination of changes due to fatigue of the audiometrician, greater ease in taking large number of audiograms and a permanent record of the result. These factors have prompted Delany, Whittle and Knox (1966) to investigate the use of this type of self-recording audiometry for children. They conclude that the lower age limit for self-recording audiometry with Rudmose audiometry is approximately 6 years, and that children above this age give results quite equivalent to those from conventional manual audiometry. The same slightly lower thresholds in the case of self-recording audiometry compared with manual audiometry were found in children as have previously been observed in adults (Burns & Hinchcliffe, 1957).

Evoked response audiometry

In any of the conventional forms of audiometry in which the criterion is a subjective judgement of the presence or absence of a sensation of sound, clear limitations exist. For example, in young children or in cases of non-organic hearing loss, the method may be impossible or the results misleading, respectively. An objective method of audiometry for some purposes would thus have peculiar advantages. The evoked response technique fills this requirement. It depends upon the production of characteristic electrical potential waves by the brain in response to sound stimuli. First described by Pauline Davis (1939) the response has only been examined specifically as a form of audiometry in recent years. These potential waves can be recorded through the scalp by surface electrodes, in the manner used for many years in the technique of electroencephalography. The subject need pay no attention to the sound stimuli, which are usually short tone-pips repeatedly presented at intervals of a few seconds. The resultant electrical potentials are amplified, with averaging and storage by a digital computer.

The technique, interpretation and uses of this type of audiometry are described by Davis and Silverman (1970) and by Beagley and Knight (1967). It is still an elaborate and slow process compared to subjective pure-tone audiometry and it offers no advantage in terms of precision and repeatability, over the latter method. But for certain purposes it is of unique value, despite the comparatively early stage of its development as an audiometric technique.

Other procedures related to audiometry

Many elaborations of the basic procedures of audiometry have been developed for special diagnostic purposes. It is not the aim of this book to venture into these fields of clinical audiology. Should the reader wish to gain a general impression, however, of the diagnostic contribution to clinical otology by specific techniques, he may refer (preferably after reading the next chapter) to Harrison (1969), who summarises the present practice in audiological diagnostic methods in a specialist department of neuro-otology. An interesting insight into the realities of clinical diagnosis is provided; a variety of audiological techniques are briefly described, involving determinations of threshold, loudness function, speech perception and other criteria. Coles (1972) also discusses these aspects.

Sources of error in audiometry

It is of the greatest importance that a clear appreciation of the accuracy to be expected of pure-tone air-conduction audiometry should be possessed by anyone who attempts to consider this technique as a tool for the assessment of hearing. The actual judgement of audiometric findings cannot be divorced from an otological examination of the patient, and is thus a medical responsibility. However the full understanding of the effects of noise on hearing as disclosed by audiometry is impossible without an appreciation of the limitations of audiometry. A consideration of the sources of error in audiometry has been given in Appendix J, which the reader is strongly advised to make the effort to assimilate, for a proper understanding of the subsequent discussion of noise-induced hearing loss.

Measures of sensation

The ability accurately to measure sound has for many years stimulated efforts to produce sound-measuring instruments which would give a realistic indication of loudness or some other aspect of the sensations which sound can arouse. This objective has proved highly elusive, and in fact the purely instrumental measurement of quantities such as loudness has not been achieved satisfactorily in any direct and simple manner such as by the reading of a practical portable instrument. Measurement of sensation is often a necessity in assessing the loudness, annoyance or other subjective attribute of sounds, and several methods are in use. These will be encountered in later chapters in connection with a variety of practical applications. The basis of some of these measures, including the sone and the phon for loudness, and the PNdB for perceived noisiness is discussed in Appendix K. It should be emphasised that for many practical purposes an understanding of these scales of sensation is essential.

7

Deafness

This book is intended to be a guide to the effects of noise, rather than a description of hearing, normal and abnormal, as a whole. Since intense sound is capable in various ways of damaging hearing, this is our main concern. However, to look at the effect of noise in complete isolation would result in an imbalance, and would leave untouched the great field of clinical otology. Such an omission would be the more undesirable, since this branch of medicine has seen great advances in recent years. The ear has in fact provided opportunity for some of the most elegant and skilful surgery ever to be accomplished. In addition, those with hearing defects are an appreciable proportion of the population and much effort is continually being expended on education, care and treatment of the deaf, and in improvement of the methods used. The importance and scope of this branch of medicine is indicated by the existence of many books on deafness (Ballantyne, 1970; Davis & Silverman, 1970), which are intended to be of assistance to the wide range of interests converging on the subject. Mention of the categories of people whom these authors can assist will indicate the breadth of involvement in the care of the deaf: for example, the deaf; the parents of deaf children; the families, teachers and friends of deaf persons; social workers; audiologists; family doctors; paediatricians; otologists; educational psychologists; public health and industrial medical officers.

In considering deafness the first problem we encounter is one of definition. No general agreement exists on the terminology, and efforts to define different degrees of hearing impairment by different terms have not yet been generally accepted. All this is unsatisfactory for the specialist and confusing to the non-specialist. Deafness is not a specific term. It usually means that the patient and his relatives recognise that his hearing is impaired; but it is deeply entrenched in clinical terminology in terms such as 'conductive deafness'. But some have felt, like Davis (Davis & Silverman, 1970), that it should be restricted to severe conditions in which 'everyday auditory communication is impossible or very nearly so'. Where the impairment is not as bad as this, the term 'hearing loss' is preferred by Davis. We shall use 'deafness' in this text in the sense of a hearing defect which is clinically recognisable as a hearing impairment. Such a condition would usually be appreciated as such by the patient or his relatives. Where hearing defects are of a less obvious nature, perhaps only detectable by audiometer or other means, we shall use the term 'hearing loss'. Either 'deafness' or 'hearing loss' may be used in descriptive terms, e.g. 'conductive deafness', 'occupational hearing loss'.

Varieties of hearing defects

From the point of view of structure and function of the ear we can distinguish three clearly separable types of defect. This is not to say that these, in one way or another, cannot co-exist, for they may do so in various combinations. The three types of defect are as follows:

1) Those in the pathway from the exterior to the cochlea.
2) Those in the cochlea or the auditory nerve.
3) Those in the hearing pathways and perceptive mechanism in the brain.

The main features of these types we may now consider.

Defects of the sound-conducting pathway: conductive deafness

The conducting mechanism consists of the pinna, the external ear canal, the tympanic membrane and the ossicles and their attachments, including the orifices, the oval and round windows, which are the boundaries between the middle ear and the inner ear cavities. The defects which may involve the conduction mechanism may therefore be easily anticipated. Obviously, since the passage of sound from the exterior to the oval window is a physical process involving the movement of air molecules and of anatomical structures, any interference with these events will tend to reduce the transfer of sound energy into the cochlea. Taking the sound path in the natural sense, from outside inwards, the most obvious situation would be an obstruction to the entry of the sound waves into the external ear canal. Covering the ears with the hands or closing the meatus with a finger tip are both almost instinctive actions which achieve this. Ear muffs or ear plugs are merely an elaboration of this natural action, and the effect is to produce a conductive hearing loss. The obstruction may be in the meatus itself, from a variety of causes, the simplest of these being a foreign body, water or ear wax.

The conductive defect may be in the middle ear itself. The communication, through the Eustachian tube, between the middle ear and the part of the back of the nasal cavity known as the nasopharynx allows infective processes in the latter to involve the middle ear. Infection may even extend to the mastoid air cells. These communicate with the middle ear and occupy the mastoid portion of the temporal bone, behind the external ear. Infection of the middle ear can occur with the common cold and with diseases such as measles and scarlet fever. Fluid collects in the middle ear, and the infective process may proceed to the formation of pus, and the tympanic membrane itself may rupture, the discharge then running out of the ear canal. The condition may develop suddenly and may be very painful. This is the condition known as acute suppurative otitis media. The condition may be arrested by treatment or by the reaction of the body itself at any stage, or it may proceed to a chronic stage, in various forms. In these the perforation of the tympanic membrane may persist. Chronic types of middle ear infection may appear without a preceding acute stage. The result of these conditions is that interference to varying degrees with the mobility or even of the integrity of the ossicles, as well as the damage to the drum, reduces the energy transfer to the oval window and hearing is affected.

Another most important category of disease where conductive defects occur is the condition known as otosclerosis. Here growth of bone in the neighbourhood of the oval window gradually interferes with the movement of the stapes. This is an inherited condition which usually appears in young adults. A conductive deafness is the result, and over the years aural surgeons have worked, with success, to evolve various methods to improve the hearing of patients with this condition. Work of outstanding ingenuity and skill continues to be done in the surgery of otosclerosis.

Defects of the cochlea and auditory nerve: sensory-neural deafness

In this category the defects are either in the cochlea itself or in the fibres in the auditory nerve; they can collectively be described as sensory-neural (or sensorineural) deafness. The possible causes are numerous. This form of deafness or hearing loss is one of which the majority of adults in later life become aware, for deterioration of hearing especially for the high frequencies, as we have seen in Chapter 6, is apparently a characteristic of growing older. High frequencies are particularly affected, and the name presbyacusis or presbycusis denotes the connection with age. The changes in this condition are progressive and may be detected from young adult life onwards. In later life a well-recognised aspect consists of changes in the cochlea near the base, where the hair cells degenerate to varying degrees, and the nerve fibres which would have carried their messages to the brain degenerate also. The reason, despite the fact that the condition affects most people to a lesser or greater degree, and is almost physiological, is not known.

From our point of view in this book, we are concerned particularly with the acquired form of sensory-neural hearing loss due to noise. At this stage we may note that changes attributable to noise also occur in the sensory cells of the cochlea (Beagley, 1965), but the distribution of the degeneration is not identical with that due to age. We shall discuss the effects of noise in detail in subsequent chapters.

A large number of other conditions resulting in sensory-neural hearing defects can be acquired. Certain drugs are now known to be capable of damaging the cochlea, as are diseases not specifically otological in nature, such as meningitis resulting from any one of a number of different kinds of infective agents; in addition certain virus diseases, such as mumps and measles, appear to be able to damage the organ of Corti, in addition to the liability to middle ear infection associated with measles and some other diseases to which we referred in connection with conductive deafness.

Children may be born with a variety of comparatively rare inner ear disorders which result in deafness. There is a hereditary category due to inherited characteristics. Another type of inner ear deafness occurs where some factor has interfered with the proper development of the embryo. The first group are true hereditary defects and knowledge is steadily accumulating about the various types and causes. The pre-natal group is likewise becoming better understood, and one of the best known and now largely avoidable dangers to the unborn child is of the mother becoming infected with the virus of rubella, or german measles.

A condition which includes a sensory-neural deafness is known after Ménière, the French physician of the last century, as Ménière's disease. This is characterised (Williams, 1952; Cawthorne & Hewlett, 1954) by attacks of dizziness, with perhaps nausea and vomiting, some deafness and noises in the ear (tinnitus) and is usually confined to one side, at least initially. It is associated with appearances of the endolymphatic system of the labyrinth, suggesting a distension of these spaces, but this has not been demonstrated as the specific cause in all cases. The characteristic of the condition is its tendency to come and go, and the attacks may be sudden and violent, the sense of rotation making standing, or perhaps even lying still in bed, impossible, due to the involuntary movements which are set up by the nerve messages from the organs of balance.

Among sensory-neural deafnesses is a condition characterised by the presence of a tumour of the auditory nerve. The ill-effects are due to the growth of the tumour, which presses on the fibres of the auditory nerve, causing characteristically a high tone deafness, and other symptoms. This is a dangerous cause of deafness of which the physician or otologist will always be aware.

Central deafness

By central deafness is meant a hearing impairment that cannot be explained by abnormality of the cochlear sense organ or auditory nerve. The cause is to be sought in any condition of the brain which may affect the nerve pathways from the auditory nerve to the temporal lobe of the brain, where auditory sensations are represented. As Davis and Fowler have pointed out, in the diagnosis and prognosis of these conditions, it is the neurologist, the neurosurgeon, the psychiatrist, the psychologist and the educator, rather than the otologist, who must be the main sources of guidance. This category of deafness is sometimes described as of intracranial origin, that is inside the space in the skull occupied by the brain and brain stem. Intracranial conditions which may have deafness as a symptom include various types of tumours or abcess formation in the brain, interference with the blood supply to parts of the brain such as from cerebral thrombosis or haemorrhage, or from birth injuries. In these conditions, measurement of hearing threshold for pure tones, which is so valuable in defects of the conductive or peripheral sensory mechanisms, is not necessarily helpful. As an example, injury to the temporal lobe of the brain has for many years been associated with a condition known as aphasia, where the sounds are heard in a normal way, but due to the failure at the higher levels of the brain, their meaning is not understood. A head injury may thus render speech incomprehensible.

Enough has been said to illustrate the kinds of problem which face the otologist in his daily practice. The methods of diagnosis and of treatment are not our concern here. This brief reference to the variety of hearing impairment, and the place occupied by the effects of age and of noise, may however, enable us to view the latter two causes in a better perspective.

8

Disturbance

In this chapter are grouped certain effects of noise which are distinct
from the potentially damaging effects on hearing and other body
systems. These effects are sufficiently varied and dissimilar to render
a satisfactory single title difficult to find. In this context, we use the
term disturbance to imply interference on the broadest basis, with
sleep, rest or activity. To escape the error of classifying all sounds as
either potentially disturbing or not, the subjective nature of this
description must be remembered. With the exception of the most
intense noises, no single noise is likely to evoke exactly similar
responses from all of the individuals in a population exposed to it.
A given sound is pleasurable or objectionable in the opinion of the
listener, and these opinions may be exceedingly divergent. Thus
certain sounds, especially those unusual or uncommon in the particular environment, be they even very faint, may cause significant
personal or community disturbance. On the other hand very loud
sounds may be regarded by a minority with indifference. The
quality of the same sound may evoke great loyalties or antagonisms.
A moment's reflection will provide examples. The symphony orchestra may have little emotional appeal to the habitual listener to pop-
music, while the latter may sometimes receive less than a sympathetic
reception from those of classical musical tastes. Again, a motor race
conducted in near silence, if such were possible, might lose some of
its appeal to the hardened spectator, and yet be regarded as a wel-
come relief by the community in general.

It is from an infinitely wide variety of sound stimuli, each eliciting
a range of individual subjective effects and responses, that we must
try to extract some unifying pattern of relations. In seeking to
discern such a pattern, we must also be prepared to study in some
detail the characteristics of particularly important sources of disturbance, such as road traffic noise and aircraft noise. These categories will be discussed in subsequent chapters.

In this chapter the various effects will be considered, followed in
Chapter 9 by discussion of possible measures for their restriction or
elimination.

Classification of disturbing effects

In order to confer some degree of order to this varied and compli-cated field, systematic subdivision is required. We will discuss a number of different aspects, including disturbance of sleep or rest; annoyance; disturbance of any form of activity (including speech communication); possible effects on health, mental or physical. Temporary and permanent impairment of hearing are excluded at this stage.

Following the subdivision used by Robinson (1970a) these effects can conveniently be accommodated within two categories: direct effects on the individual, and indirect effects on the individual. The direct category constitutes aspects of the perception of the noise itself, or of the immediate subjective consequences. It includes the phenomenon of loudness sensation; a similar but psychologically distinguishable aspect of the sensory experience of noise, perceived noisiness; and thirdly, interference with the perception of speech (or difficulty in hearing sound signals) resulting from the masking of speech or other meaningful sound by the prevailing noise. The indirect effects include the disturbance of sleep or rest; annoyance; disturbance of activities involved in work or leisure; possible effects on health. The direct effects are therefore, in a sense, the primary phenomena; the indirect effects are the consequences. The essential problem is to find the relation, if any, between physical aspects of the sound and the effect it produces. The physical aspects of sound can be related fairly closely to the direct effects; the indirect effects are less clearly dependent on the nature of the sound stimulus. Vary-ing degrees of success have attended attempts to correlate subjective judgements with physical parameters of the sound (including measures of some of the direct effects). We must now attempt to give greater reality to these concepts by considering them in more detail.

Direct effects

Robinson's (1970a) review of criteria for limitation of urban noise provides an illuminating discussion which has been liberally drawn upon here, and may be consulted for further information on many of the following sections.

Loudness

This is defined (BSI, 1969) as an observer's auditory impression of the strength of a sound. It is thus a purely subjective concept, determined, for any particular sound, by the strength of the sensation which the person experiences. Thus a scale of loudness would logically originate at the auditory threshold, and for increases of intensity above threshold for any specific sound, loudness would increase. This in fact occurs, and in an orderly way, but nevertheless subject to individual variation. In psychophysical experiments, a doubling of loudness occurs, to a good approximation, with an increase in sound level of 10 dB. This relationship can be expanded to form a scale, so that a 20 dB increase in sound level results in a four-fold increase in loudness, and so on. These relations are average findings for normally-hearing subjects, and are regarded as a close enough approximation to the laboratory data to form a practical method of expressing loudness. The details are given more fully in Appendix K, but for the immediate purpose the essentials are these. Loudness can be expressed in two ways, in the *sone* scale of loudness or the *phon* scale of loudness level (BSI, 1958). The sone (Appendix A) is the unit of loudness on a scale designed to give scale numbers approximately proportional to the loudness. It is defined by its relation to the phon scale. This relation is such that a loudness of 1 sone is equal to a loudness level of 40 phons. The loudness level of a sound, in phons (Appendix A), is numerically equal to the SPL in dB (re 2×10^{-5} N/m^2) of a pure tone of frequency 1000 Hz, consisting of a plane progressive sound wave, coming from directly in front of the observer, when the sound and the pure tone, when presented alternately, are judged to be equally loud. The sone scale extends upwards and downwards in conformity with the approximation noted above, i.e. that doubling or halving loudness (in sones) is caused by changes of plus or minus 10 phons, respectively. This relationship is described numerically (BSI, 1958), based on the work of Stevens (1955) and Robinson (1957). It will have been noted that the loudness level, as defined above, is a numerical value in phons (symbol P), based entirely on a subjective comparison between the sound being assessed for loudness, and a pure tone of frequency 1000 Hz. It is also possible, and much more practical, to calculate the loudness level from a knowledge of the frequency spectrum of the sound, as described by Stevens (1956, 1957, 1961) and Zwicker (1960, 1961); where such calculations are made, specific designations for the resulting phon values are used. The basis of these calculations are explained in more detail in Appendix K. This is a procedure of considerable practical importance, since from it has evolved the method of calculating perceived noisiness, which is extensively used in the measurement of aircraft noise.

It is well to emphasise again that the relationships of the phon and sone scales are engineering approximations to the way that the human ear behaves. Thus the precise relations of phons and sones given in the relevant British standard and ISO recommendation (BSI, 1958; ISO, 1959) are not strictly observed for individuals.

A particular aspect of loudness, that of subjective discomfort, has been studied by Hood and Poole (1966) and by Hood (1968). They found that, in persons with normal hearing, the mean SPL at which a tone of 1000 Hz is judged to be unpleasantly loud is near 100 dB. This level they designate the *loudness discomfort level* (LDL), and in a group of 200 subjects with unilateral deafness, the normal ears gave a mean LDL of 98·2 dB SPL at 1000 Hz, and similar values at other frequencies. The scatter of individual values of LDL was such that 90% of subjects selected levels between 90 and 105 dB. It was subsequently shown (Hood, 1968) that LDL is not related to the pure-tone threshold for the particular frequency. Thus persons with very acute hearing have LDL values little different from those with markedly less acute hearing. This can be expressed in the statement that LDL is independent of sensation level (Appendix A). Further examination of the data shows that the more acute the hearing the higher the SPL judged to be unpleasantly loud, and vice versa. This variability is within a narrow range, but the trend is clear. Hood speculates on the significance of these findings, and shows different LDL values for different kinds of hearing disorders (Hood & Poole, 1966; Hood, 1968). The test is now used clinically for diagnostic purposes (Harrison, 1969).

This digression is included to emphasise the finer structure of the relation between sound intensity and loudness, but does not affect the broad working principles used for the practical measurement of loudness, as exemplified in the relevant standards (BSI, 1958, 1967; ISO, 1959, 1966). These relations have important practical applications. The subjective judgement of 'half as loud', which in practical noise reduction would be regarded as a modest improvement, implies a reduction of 10 dB, which diminishes the energy to one tenth of its previous value. If loudness reduction is to be achieved by increasing distance from source to listener, say in the case of an aircraft flying overhead, a theoretical reduction of 6 dB is achieved each time the distance is doubled. This is due to the fact that, radiating into free space, the sound energy is spread over an imaginary spherical surface, the area of which increases as the square of its radius. Thus the sound energy flows through a surface which increases in area as the square of the length of the sound path; doubling the distance will thus give a four-fold increase in area, or one quarter of the intensity, which is a reduction of very nearly 6 dB.

In addition to this fundamental relation based simply on the solid geometry of the system, the air itself possesses the power of absorption of the sound. This is very small below about 1000 Hz, but increases with frequency until, approaching 10 kHz, the attenuation is about 0·07 dB/m, but is a strong function of humidity and temperature. Other factors also operate (Parkin, 1962) and we may take 7 dB per doubling of distance, for attenuation in air (Robinson, 1970a). Thus, if we wish to decrease the loudness of the noise of an aeroplane overhead by increasing the height, the following example will serve. Let us say that we require a reduction of appreciable amount, subjectively judged to be between one half and one quarter in loudness; for this we must achieve a reduction in the region of 15 dB, i.e. about two doublings of distance. Thus the height of the aircraft would need to be increased four-fold. This rather discouraging result is especially unhelpful when the sound, as from a large aircraft, is already loud at a considerable distance.

These considerations have been confined to relations between intensity and loudness without including the effect of frequency. This is a fundamental factor in determining the quality of subjective sensations of sounds. It has been under investigation for nearly half a century, and much valuable work has been devoted to it. The subject is dealt with in some detail in Appendix K, where the important points referred to in this chapter are briefly re-stated, together with the quantitative details of the physical basis of loudness, and its calculation. Here we will concern ourselves with qualitative generalities. The essential relations between loudness, intensity and frequency have been demonstrated by the establishment of *equal-loudness contours* (Robinson & Dadson, 1956), shown on Fig. K1. These curves state the SPL of pure tones, of different frequencies, necessary to evoke the same sensation of loudness as that of a reference tone of 1000 Hz. The data are given for various arbitrary sound pressure levels of the reference tone and by formulae for interpolation. Study of the shape of the curves will show that, for a given loudness, the smallest values of SPL are required in the region of 3 to 4 kHz, and that higher SPL is needed at low frequencies. The greatest disparity between the low- and high-frequency tones in their effectiveness in evoking loudness sensation is when the reference tone is least intense, i.e. as threshold is approached. These relations are the basis of an international standard (ISO, 1961). Extensions to lower frequencies are noted in Appendix K.

In practice, the consequence of these relations is that the loudness of sounds of the same overall SPL may vary markedly depending on the distribution of sound energy over the frequency spectrum. Thus sounds predominantly of low frequency will tend to be less loud than sounds of predominantly high frequency, despite similar overall SPL values. This statement must be qualified, of course, in the light of the effect mentioned above, that the absorption in air of high frequencies is greater than for low frequencies, so that at considerable distances, the spectral characteristics of a sound will change as a result of the relatively greater reduction of the high frequencies. The sound will then be said to be more 'mellow' or less 'harsh', so that distance may lend enchantment to the sound, as well as to the view. Equal loudness contours for octave bands of random noise have also been determined by Robinson & Whittle (1964).

In summary, the subjective attributes of a sound are thus seen to be dependent on these, and other characteristics, and the resulting sensation is not measurable in any direct manner. The subjective judgements based on the phon scale of loudness level are supplemented by methods of calculating loudness described in Appendix K. These methods do not give identical results, nor yet results exactly in agreement with phon values even for broad-band steady noises, but they serve a useful, and indeed indispensable function, in enabling calculations of loudness level (if need be by computer) to be made in conditions which would not be compatible with subjective determinations of loudness. The calculation of loudness for non-steady or impulsive sounds is on less secure foundations. Different national outlooks are evident here, but ISO is known to be considering the question of loudness of impulsive sounds on a multi-laboratory basis. We now proceed to a modification of the loudness concept, to include the element of disturbance or noisiness.

Perceived noise

On the basis of the background described in the immediately pre-ceding section it is possible to calculate loudness level in terms of phons, using various versions of the technique evolved particularly by Stevens, which is described in more detail in Appendix K. In the development of heavy piston-engined transports after World War II aircraft noise was becoming a nuisance and the phon was used as a measure, calculated in the manner then current, of the loudness of such aircraft noise. By mid-1949, the first jet transport flew, and a new source of community disturbance thus arrived, and with it problems of quantifying the disturbance caused. The use of the phon scale proved to underestimate the disturbance of the jet noise, and led to the work of Kryter and his colleagues (cited in Appendix K), which produced a new scale of measurement, in which the judgement of 'loudness' was replaced by one of 'noisiness' and the resultant values were designated '*perceived noise level*', the unit being the perceived noise decibel (PNdB). This basic concept has been retained, but modified repeatedly to meet changes in the type of engine noise and refinements of measurement. For example, if an allowance for spectral irregularities and for the time history of the noise of a single flyover of an aircraft is made, the resultant value is known as the *effective perceived noise level*, expressed in effective perceived noise decibels (EPNdB). At the time of writing agreement has been reached (ISO, 1970), and the current position is given in Appendix K. In summary, however, there is no fundamental difference be-tween the calculation of PNdB or its variants, and the long-estab-lished approach of Stevens to the calculation of loudness; the differ-ence lies in the weight given to the different parts of the frequency spectrum in assessing noisiness as opposed to loudness. The PNdB, despite its habit of acquiring modifications or accretions, has virtu-ally universal use in the expression of aircraft noise for engineering purposes, but for quantifying community disturbance dB(A) also has a considerable following and further reference will be made to this scale later. The other principal standard measures of the magni-tude of sounds (A, B, C and D weightings) achieved by incorporat-ing a frequency-weighting network in the amplifier, have been mentioned in Chapter 4. Since these measures are derived by simple circuitry, and give direct readings on a sound level meter without any need to perform calculations either manually or by computer the data are immediately available. Sound level A has now received widespread acceptance for a variety of purposes, as will be seen in subsequent sections of this book. It bears some relation to the subjective quality of annoyance, and has been found to be a valid basis for indices of other effects, such as the ability of the sound to cause impairment of hearing (Robinson, 1970b; Burns & Robinson, 1970). Sound level D is related to noisiness to a useful degree, as will be discussed later.

The two direct subjective effects, loudness and noisiness, are both associated with a general result of noise, the interference with the perception of speech and other communication by sound; this will now be discussed.

Interference with communication

Noise impairing the perception of other sounds is an important and clear effect, quite separate from annoyance and disturbance, although it may be associated with them. Speech, directly or by telephone or loudspeaker, or any other meaningful sound such as a warning signal, may be interfered with by noise. The potential dangers, as well as the inconvenience and disruption of work or leisure of these effects, are self-evident.

In Chapter 6, the problem of ensuring that in audiometry the auditory threshold is not artificially elevated by the presence of noise was encountered, and some aspects of the masking of one sound by another were considered. Wegel and Lane (1924), Fletcher (1940), French and Steinberg (1947) studied the way in which the threshold of the masked sound is elevated in presence of a masking sound. At low intensities, a masking tone elevates thresholds of masked tones at and around its own frequency, but at higher intensities, masking effects are more pronounced at frequencies above that of the masking tone, than below. This result is consistent with the pattern of vibration of the cochlea described in Chapter 5. In experiments on the masking of one tone by another, uncertainty arises when the masked and masking tones are close to one another in frequency, due to the phenomenon of beats. The effect is less pronounced if noise is used to mask tones, and this was the situation seen in Chapter 6. The quantitative aspects of the masking of pure tones are complex, but certain fundamentals can be noted. The critical band concept has proved to be very valuable; it implies that, in the presence of a noise, a pure tone is masked mainly by the components of the noise in a band of frequencies centred round the frequency of the pure tone. According to Fletcher, when a tone is just audible against a noise background the intensity of the tone is the same as the total intensity in the critical band. Further (Hawkins & Stevens, 1950), this applies very precisely for a wide range of intensities. To avoid masking of the tone, the intensity per critical band of the noise should be about 10 dB less than that of the tone, and this is the criterion which was adopted in the calculation of the maximum permissible noise environment for audiometry in Appendix I, p. 398. Some differences in the apparent widths of critical bands have been found (Zwicker, Flottorp & Stevens, 1957) according to the method employed, and it appears that for the summation of loudness, or even for masking measured in a particular way, the critical band may be some $2\frac{1}{2}$ times as wide as previously found. For this reason, the wider critical bands were used in the calculation of the permissible noise in audiometric environments in Appendix I, which is a more rigorous interpretation of the data, so as to yield lower permissible noise levels.

The situation of speech in noise is a particular case of masking; whether speech is intelligible in a given noise may be discovered by a practical trial or by calculation, given certain facts about the acoustic properties of the speech and of the noise, and the degree of interference with the perception of the speech which it is possible to accept. The significant factors are many. Obviously, there is a gradation of interference between quiet and noisy surroundings. If the background sound is continuous, interference will be greater than if it were intermittent. The special characteristics of the speech or signal and the noise, as well as their relative intensity, are also relevant. Intelligibility in speech is not merely present or absent; there is a range of intelligibility as defined by some particular criterion, and the level of intelligibility of a communication system can be estimated. The intelligibility of speech will also depend on the type of spoken material used, and the latter may be adapted to suit the acoustic conditions. Thus satisfactory communication may be attained by special speech procedures and vocabularies, in conditions where normal conversational habits would be inadequate.

There is an extensive literature on communication by speech, which should be consulted to supplement the brief summary of the quantitative aspects of communication by speech which follows in

Chapter 9. In particular, Morgan, Cook, Chapanis and Lund (1963) give detailed information on communication problems; Fletcher (1953) gives a revision of his historic earlier writings on speech and hearing in communication, and Licklider (1951) surveys much material relevant to speech communication as does Kryter (1970).

In assessing the effectiveness of speech communication in a given situation, a direct practical test with a proper experimental plan and adequate numbers of subjects is the most effective method. This is frequently inconvenient or impracticable and in any case it deals with an established situation. It is usually preferable to calculate the effects of a given noise, or it may be that an assessment of a hypothetical situation in the course of design is needed. The methods available are also indicated in Chapter 9.

Indirect effects

We now address ourselves to the less concrete concepts of the various aspects of disturbance which were mentioned at the beginning of the chapter. These may be described with reasonable clarity in a qualitative way, but the real problem, as we noted above, is to relate quantities, such as various noise measures, to the disturbing effects, so as to produce meaningful scales, usable in the control of these indirect effects.

In such a complex and diverse subject as this, attempts to reduce the large body of information are difficult, and the danger of misplaced emphasis is considerable. The reader who wishes to pursue the subject should consult the writings of Broadbent (1954, 1955, 1957a, 1958), Broadbent & Little (1960), Broadbent & Gregory (1963), Carpenter (1959, 1962), Committee on the Problem of Noise (1963), Kryter (1970) and particularly the recent major contribution by Broadbent (1971) *Decision and Stress*. This work collects and assesses a mass of material, and should be consulted for detailed reference.

Disturbance of sleep

Of the various effects being considered in these psychological fields, the interference with rest or sleep is perhaps the most serious. It is an undisputed fact that adequate sleep is a physiological necessity. Thus health is likely to be prejudiced by insufficient sleep. It is true that the powers of adaptation to the environment possessed by the human body may go a long way to solving problems of interference with sleep, and people can become accustomed to noise, as they can to other environmental factors, so that satisfactory sleep can be obtained in conditions which would at first render sleep impossible. Individual peculiarities in the reaction to noise in this respect are legion. It is common experience that some households can sleep oblivious to a selection of chiming clocks which wake the guest unaccustomed to these nightly noises, each time the chimes occur. Those accustomed to certain noises may even find that their absence is an impediment to falling asleep. This may occur even with aircraft noise. On the other hand, some persons exposed to moderate noise at night never become accustomed to it, and may adopt every conceivable means of securing the quiet conditions which they find necessary for adequate sleep. Despite these varied reactions, there is a limit to the intensity of noise and to its characteristics, such as frequency and intermittency, which are compatible with adequate sleep. Noise may make falling asleep impossible or may awaken the sleeper, who may then find difficulty in falling asleep again. If, as is likely, there is resentment against the cause of the noise, the frame of mind suitable for sleep is even less likely to be achieved. On the evidence of normal experience, people appear to vary markedly in the ease with which they can be awakened by noise. It is also known that sleep varies in depth in the same person at different times. When sleep is deep, which occurs from time to time, or during dreams, awakening by noise is less likely. On the other hand, there are times when sleep is light and awakening is easy; much fainter noises are sufficient to awaken, compared to the deep sleep condition. Apart from the intensity of the sound, the type or significance of the sound has a powerful influence on its awakening effect. Sounds which are familiar and which do not require anything to be done about them, are less liable to awaken. Examples are ventilating fans and air conditioning equipment, which in the past have not been particularly quiet, but are apparently compatible with satisfactory sleep. When a sound is especially significant for a particular person, awakening is possible by comparatively faint sounds; the baby's cry will arouse the mother, because it is highly meaningful and may demand instant action. The mechanisms in the nervous system which make possible this type of response are of great interest, but do not concern us specifically in this particular context.

In face of these varied and complicated relations, many of them of a highly individual and personal nature, it is not surprising that it is virtually impossible to lay down rigid rules of a practicable nature for preventing disturbance of sleep by noise. The general tendency is to suggest maximum permissible noise levels for sleeping accommodation (Beranek, 1971; Kosten & van Os, 1962) but an additional factor is that of intermittent noise, such as that from passing road or air traffic, and attempts must be made to account for the consequent individual disturbances on the basis of their frequency of occurrence. This factor is of particular importance in the case of aircraft noise.

All that has been said about actual wakening from sleep will, in general, obtain in the case of rest, which might or might not eventually merge into sleep. Presumably the benefit of rest, short of actual sleep, would be diminished if disturbance from noise occurred. The quantitative aspects of these effects will be considered later.

Annoyance

Annoyance may be regarded for the present purpose as the displeasure or resentment caused by sound, either by its physical presence, or because of the implications arising out of its presence. There is no objective method of measuring annoyance as such, but it is possible to obtain some indication of the annoyance caused by sound by asking a sufficient number of people about their reactions. On the basis of these replies to specific questions, together with a knowledge of the relevant sound environment, some quantitative indication of the way in which noise interferes with people's lives can be obtained.

An investigation on these lines was carried out in collaboration by the Building Research Station, the then London County Council and the Central Office of Information in 1961–2. The area investigated was 36 square miles of Central London (Committee on the Problem of Noise, 1963; McKennell & Hunt, 1966). The noise survey on which the Committee reported has subsequently been published in full (Parkin, Purkis, Stephenson & Schlaffenberg, 1968). This full version extends the coverage but does not materially alter the conclusions of the Committee on the Problem of Noise (1963). Some 1400 people living in this area were questioned about noise, and the actual noise was measured at 540 points throughout the area. The relative importance of various factors which might give rise to dissatisfaction was investigated by asking the question, 'If you could change just one of the things you don't like about living round here, which would you choose?' The answers given are summarised in Table 8.1. Of those questioned 11% wished to change the

TABLE 8.1

Relation of noise to other factors

The one thing that people most wanted to change	The percentage of people who wanted to change it
Noise	11
Slums/dirt/smoke	10
Type of people	11
Public facilities/transport/council	14
Amount of traffic	11
Other facilities/shopping/entertainment	7
Other answers	1
No answer, or vague reply	5
Would change nothing	30

noise conditions, but it is possible that the 11% who objected to traffic were not uninfluenced by the noise associated with it. Noise is more liable to cause disturbance while people are at home than when they are outdoors or at work. Table 8.2 gives the results of questioning each 100 people about noise disturbance at home, outdoors and at work. The columns 2, 3 and 4 show the number of times that noises were mentioned as being disturbing in these situations, respectively. Noises found disturbing in the home were mentioned 99 times per 100 people questioned, outdoors 35 times and at work 26 times. The striking preponderance of traffic noise as a source of disturbance is to be expected in an area such as central London. The nature of the disturbance, whether it be annoyance, disturbance with activities or disturbance of sleep, will be dependent on the circumstances, but annoyance may reasonably be regarded as being present in all these circumstances. In Table 8.2 the most obvious result is the leading position of traffic noise as a source of disturbance. Aircraft noise in this area, on average, was of much less importance, but of course areas near airports or flight paths show intense disturbance by aircraft noise, as will be discussed in Chapter 13.

It is of some interest, in view of the marked antipathy to noise shown by some persons, to enquire what proportion do not notice noise, and if they do, what proportion are disturbed by it. The percentages of those interviewed occurring in these categories, again at home, outdoors or at work, are shown in Table 8.3, columns 2, 3 and 4. Apparently a large majority, on the basis of this survey, notice noise, but of those, not all professed to be disturbed. The

TABLE 8.2

Noises which disturb people at home, outdoors and at work

| Description of noise | Number of people disturbed, per 100 questioned | | |
Column 1	when at home Column 2	when outdoors Column 3	when at work Column 4
Road traffic	36	20	7
Aircraft	9	4	1
Trains	5	1	—
Industry/Construction works .	7	3	10
Domestic/Light appliances .	4	—	4
Neighbours' impact noise (knocking, walking, etc.) . . .	6	—	—
Children	9	3	—
Adult voices	10	2	2
Wireless/T.V.	7	1	1
Bells/Alarms	3	1	1
Pets	3	—	—
Other noise	—	—	—

TABLE 8.3

Percentage of people who were ever disturbed by noise at home, outdoors and at work

| Individuals' reaction to noise | Noise source | | |
Column 1	at home Column 2	outdoors Column 3	at work Column 4
Those who are disturbed by noise . . .	56	27	20
Those who notice but are not disturbed . .	41	64	70
Total of people who notice noise . . .	97	91	90
Those who do not notice noise . . .	3	9	10
	100	100	100

most disturbance was at home, where a clear majority suffered disturbance; outdoors, about one third were disturbed, but at work only 2 out of every 9 suffered disturbance. The importance attached to a quiet home suggests further enquiry about the origins of noises which disturb people in these conditions. Table 8.4 gives a similar

TABLE 8.4
Origins of noises which disturb people when they are at home

Description of noise	Origin of noise (per 100 people questioned)		
Column 1	External noise Column 2	Internal noise Column 3	Noise from own home Column 4
Road Traffic	36	—	—
Aircraft	9	—	—
Trains	5	—	—
Industry/Construction Works . . .	7	—	—
Domestic/Light appliances	1	2	1
Neighbours' impact noise (knocking, walking, etc.)	1	5	—
Children	8	1	—
Adult voices	7	3	—
Wireless/T.V.	2	5	—
Bells/Alarms	3	—	—
Pets	3	—	—
Other noise	—	—	—

presentation to that of Table 8.2, but restricts the noise sources to those giving disturbance at home. Column 2 (external noise) refers to noises originating outside the dwellings; column 3 (internal noise) is that caused by neighbours in, for example, adjacent rooms in flats. The points of interest are that road traffic remains the greatest source of disturbance, while aircraft and children are almost equally disturbing and of much less importance than road traffic in the category of external noise. Internal noise, the traditional disturbance by neighbours, seems to consist principally of miscellaneous impact noises due to walking, etc., with the expected disturbance from loudspeakers of radio, television and presumably record players of equal importance.

The importance of road traffic noise invites a study of noise levels. At the time of the publication of the Report of the Committee on the Problem of Noise (1963), the Central London Noise Survey had available an analysis of the noise at some 400 of the total of 540 points at which noise measurements were made. Of these 400 locations, traffic noise predominated in 84%. The levels in dB(A) and

TABLE 8.5
Range of noise levels at locations in which traffic noise predominates

Group	Location	Noise climate in dB(A)*		Percentage of the total numbers of points measured falling in each group
		Day 08.00–18.00	Night 01.00–06.00	
	Column 1	Column 2	Column 3	Column 4
A	Arterial roads with many heavy vehicles and buses (kerbside)	80–68	70–50	4
B	(i) Major roads with heavy traffic and buses (ii) Side roads within 15–20 yds of roads in groups A or B(i) above	75–63	61–49	12
C†	(i) Main residential roads (ii) Side roads within 25–50 yds of heavy traffic routes (iii) Courtyards of blocks of flats, screened from direct view of heavy traffic	70–60	55–44	17
D	Residential roads with local traffic only	65–56	53–45	18
E	(i) Minor roads (ii) Gardens of houses with traffic routes more than 100 yds distant	60–51	49–43	23
F	Parks, courtyards, gardens in residential areas well away from traffic routes	55–50	46–41	9
G	Places of few local noises and only very distant traffic noise	50–47	43–40	1
			Total	84%

* By noise climate is meant the range of noise level recorded for 80% of the time. For 10% of the time the noise was louder than the upper figure of the range and in the case of Group A attained peak levels of about 90 dB(A): for 10% of the time the noise was less than the lower figure in the range.

† In Groups C to F, noise from other sources, such as trains or children's voices, predominated over road traffic noise at particular times, but traffic was the most frequent noise source.

details of types of location, are shown in Table 8.5, and in columns 2 and 3 are noted the effect on the noise of the time at which it was measured. Of the remaining 16% of points, the predominant noises were: industrial noise 7%; railway noise 4%; building operations noise 4%; unclassified noise 1%. Again, it must be remembered that the locations concerned did not include those subject to significant aircraft noise. Aircraft noise was only noticed in a certain area, and then only when other noises were less than 60 dB(A).

In the definitive London Noise Survey report (Parkin, Purkis, Stephenson & Schlaffenberg, 1968), much additional information is provided. The measured noise levels may be studied in terms of day or night, and as statistical distributions of level as a function of time, throughout the period. This is illustrated in principle, in Fig. 8.1. Two sound levels are shown, plotted as cumulative distributions from a statistical distribution level recorder. The latter in this case produced an output in 5 dB steps, but the curves obtained were reasonably smooth and are shown as smooth curves of level versus

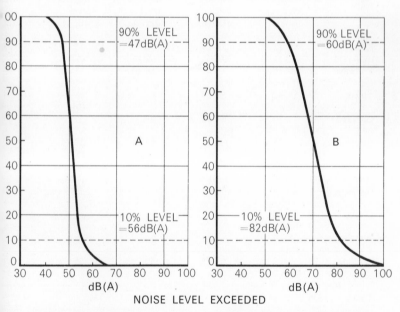

NOISE LEVEL EXCEEDED

8.1 Statistical distribution of sound level as a function of time, showing the levels exceeded for 10% of the time, and for 90% of the time. A: shows a comparatively steady noise, B: one with greater variation. (After Parkin, Purkis, Stephenson & Schlaffenberg, 1968.)

percentage of total time, which was a period of 24 hours. A small proportion of each hour was sampled. Having plotted the curves in this way, two arbitrary statistical levels were read off. These were (1) the value, in dB(A), which was exceeded for 10% of the time, designated L_{10}; and (2) the value, in dB(A), exceeded for 90% of the time, designated L_{90}. In Fig. 8.1A it can be seen that in this particular sound environment, with a peak value of 65 dB(A), L_{90}

TABI

Kind of noise heard by day. Note: Figures in parentheses are average values

Subsidiary noise heard	Principal noise heard by day—levels and number of positions in each cat									
	Road traffic		Railways		Voices		Children		Bird	
	No.	Range and average level dB(A)	No.	Range and average level dB(A)	No.	Range and average level dB(A)	No.	Range and average level dB(A)	No.	Ran and ave leve dB(
Road traffic	70	55–82 (72·5)	22	55–79 (65·5)	15	55–76 (63)	15	55–73 (62)	3	55
Railways	22	58–76 (67)	—	—	2	58–61	2	58 and 67	—	—
Voices	139	55–82 (68)	4	58–70	—	—	8	58–64 (61)	1	58
Children	41	52–76 (65)	7	61–79 (67)	1	52	—	—	3	55
Birds	17	52–73 (62)	—	—	2	55 and 61	6	52–61 (57)	—	—
Building operations	13	58–82 (70)	6	61–82 (70)	1	58	2	64 and 67	—	—
Industry	6	61–73 (66)	1	64	—	—	—	—	—	—
Docks	5	61–70 (66)	—	—	—	—	—	—	—	—
Weather	3	58–64 (61)	—	—	—	—	—	—	—	—
Aircraft	20	58–79 (64)	8	61–76 (67)	1	55	1	55	—	—
Totals	336		48		22		34		7	
Percentage of each kind	62·5		9		4		6·5		1·5	

was 47 dB(A) and L_{10} 56 dB(A). A somewhat more variable noise is shown in Fig. 8.1B. These two levels are in fact the numerical basis of the sound level ranges quoted in Table 8.5, and designated noise climate. We shall have occasion to return to this method of stating levels later. This survey presents data for day and night separately. The noise day is defined as the period between the times at which the smoothed curve of L_{10} plotted at each hour of the 24, falls

Principal noise heard by day—levels and number of positions in each category

Building operations		Industry		Docks		Weather		Unspecified	
No.	Range and average level dB(A)	No.	Range and average level dB(A)	No.	Range and average level dB(A)	No.	Range and average level dB(A)	No.	Range and average level dB(A)
13	61–79 (68)	15	58–79 (67)	10	55–67 (61)	1	55	4	52–64 (58)
1	61	1	67	—	—	—	—	—	—
1	67	5	55–73 (64)	4	58–70	1	55	3	58–79
2	61	2	61 and 64	—	—	—	—	1	61
—	—	—	—	1	61	—	—	1	52
1	76	1	61	1	70	—	—	—	—
—	—	4	64–79 (70)	—	—	—	—	—	—
1	67	—	—	6	55–64 (61)	—	—	—	—
—	—	—	—	—	—	—	—	1	55
—	—	5	64–73 (67)	—	—	1	58	2	64 and 67
19		33		22		3		12	
3·5		6		4		0·5		2	

3 dB(A) below the steady level obtained over most of the day-time period. Defined in this way the day-time period may last for 8 to 16 hours, at different noise-measuring positions, but average values of about 12 are found. The range of noise levels and sources by day and night are shown in Tables 8.6 and 8.7, and in Figs. 8.2 and 8.3.

URBAN NOISE LEVELS   The kind of noise heard by day as defined above (Table 8.6), divided into 10 categories according to sources, of principal and subsidiary origin, is shown in terms of prevalence and level. The levels in L_{10}, for each principal source of noise, are given as a range and an average value for the group of measurements of each category of noise, together with the numbers of measuring points at which the particular source predominated. Subsidiary noise is noted if present. In Table 8.7 are given the corre-

TABLE 8.7

Kind of noise heard by night

Subsidiary noise heard	Principal noise heard by night and number of positions in each category									
	Road traffic	Rail-ways	Voices	Chil-dren	Birds	Building opera-tions	Indus-try	Docks	Weather	Un-classifie
Road traffic	159	38	1	—	7	—	2	6	9	6
Railways	48	15	—	—	2	—	3	1	—	6
Voices	46	1	—	—	2	—	—	1	—	1
Children	—	—	1	—	—	—	—	—	—	—
Birds	47	5	—	—	—	—	—	—	—	2
Building operations	1	—	—	—	—	—	—	—	—	—
Industry	1	1	—	—	—	—	5	—	—	—
Docks	7	1	—	—	1	—	—	6	1	1
Weather	18	1	—	—	2	—	—	1	1	3
Aircraft	36	16	1	—	—	—	—	2	1	—
Totals	363	78	3	—	14	—	11	17	12	19
Percentage of each kind	70	15	1	—	2·5	—	2	3	2·5	3·5

Note: Because night-time range of levels is generally large and average values are r particularly relevant, range and average values corresponding to those in the day-time tab (Table 8.6) are not quoted.

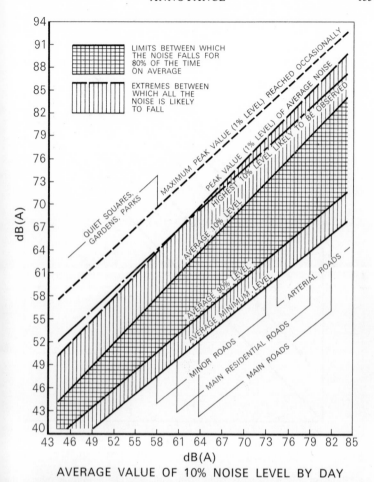

8.2 Range of noise levels observed by day in various locations, with respect to the average value of the day-time 10% noise level. The 10% and 90% levels are seen to differ by about 6 dB in the quietest situations, and this difference increases to about 12 dB in the noisiest situations. The average minimum level is about 3 dB below the average 90% level irrespective of the values of the 10% level. The principal source is road traffic. (After Parkin, Purkis, Stephenson & Schlaffenberg, 1968.)

sponding data for noise heard by night, except that the sound levels are not quoted in view of the greater variability of night-time noise. In each of these tables, it can be seen that the principal source of noise is that of road traffic. In the day-time this applied to 62·5% of

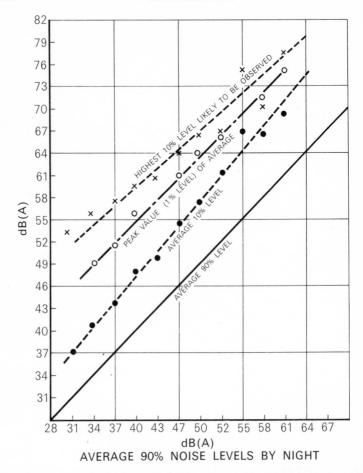

8.3 Range of noise levels observed by night, with respect to the average value of the
     night-time 90% level. The principal source is road traffic. (After Parkin, Purkis,
     Stephenson & Schlaffenberg, 1968.)

positions, at night to 70%. If the situations where road traffic noise
was a subsidiary source are also included, these percentages are
increased to the region of 80%. It is also notable that, in terms of
principal source in this urban area, the other 9 categories of noise,
viz. railways, voices, children's activities, birds, building work,
industry, docks, weather and a miscellaneous category, were found
at a relatively small number of measuring positions.

The distribution of the noise levels as a function of time is worthy of further study. In addition to L_{10} and L_{90}, other parameters may be added. The 'maximum peak value' (arbitrarily defined as the 1% level) reached occasionally, or the 1% level ('peak value') relative to the average value for the various measuring points, can be derived; as can the 'average minimum level' (i.e. theoretically 100% level). These values are shown, with some indication of the sources, in Fig. 8.2, for day-time noise. This figure, it should be noted, shows how these values varied with respect to the L_{10} average value. Entering the graph vertically from a point on the L_{10} average scale will indicate on the vertical scale the values of the other parameters. The value $L_{10}-L_{90}$ is seen to increase with increase of L_{10}, while the average minimum is about 3 dB below L_{90} everywhere; peak value (1%) is fairly constantly about 6 dB above the average value of L_{10}. Some corresponding values for night noise, in this case referred to L_{90}, are given in Fig. 8.3. The same tendencies to some extent persist, but the sound levels are, as might be expected, appreciably lower.

In view of the unequivocal position of road traffic noise as a source of disturbance in urban areas, a recent publication by the British Road Research Laboratory is very opportune (Road Research Laboratory, 1970). The problem of road traffic noise is reviewed under these headings; subjective reactions; reduction of vehicle noise; traffic noise and methods of reduction; choice of the form and level of a criterion for limitation of noise. In this chapter we are concerned with the nature of the disturbance.

Factors in traffic noise

Although the measurement of road traffic noise can conveniently be made in sound level A, differences may be caused in the spectrum by a number of factors, such as distance from source, road surface and presumably the type of vehicles in the traffic flow. The spectra, in terms of L_{10}, of a motorway and a city street, are shown in Fig. 8.4, from the Road Research Laboratory report. The city street kerbside noise is such that shouted conversation would just be possible at a distance of about 1·2 m (4 ft), on the basis of *speech interference level* (see Chapter 9). This kind of condition, and the high noise levels of motorway noise thus demand attention as a source of community disturbance. The factors of importance in

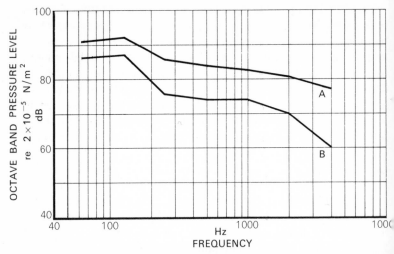

8.4   Traffic noise spectra for levels exceeded for 10% of the time (L_{10}). A: motorway, 4 m from the edge of the nearest traffic lane (3000 vehicles/h). B: the Strand, London (rush hour). (After Road Research Laboratory, 1970.)

their causation are summarised from the Road Research Laboratory report.

TRAFFIC FLOW   The relations here are complicated, but noise increases with increase of flow rate in terms of vehicles per hour. Urban road noise increases with flow rate up to about 1200 vehicles/h, after which little increase in noise occurs. This is believed to be due to congestion reducing speed. The same phenomenon occurs at about 2500 vehicles/h on motorways. Relations between traffic density and noise are discussed by Meister (1964); Stephenson and Vulkan (1968); Rucker and Glück (1964); Johnson and Saunders (1968). The last-named publication is a valuable contribution and includes the development of an empirical equation which predicts well the noise to be expected from given traffic conditions. Thus, for freely-flowing continuous traffic containing 20% of heavy vehicles, the noise measured at a point distant laterally more than four times the interval between vehicles, the relation is

$$L_{50} = 46{\cdot}5 + 10 \log_{10} Q/d + 30 \log_{10} \overline{V}/65$$

where L_{50} = median sound level in dB(A)
  Q = traffic flow in vehicles/hour
  d = distance from edge of road in metres
  $\overline{V}$ = mean traffic speed in km/h.

TRAFFIC COMPOSITION   The tendency is for heavy vehicles to be more noisy, so that increase in their proportion to lighter vehicles increases the noise. This difference is more marked in conditions of higher power, e.g. on hills (Galloway & Clark, 1962). At higher speeds the disparity between cars and heavy vehicles is less, so that the composition of traffic on motorways influences noise less than on urban roads (Johnson & Saunders, 1968).

TRAFFIC SPEED   A convenient relation between mean traffic speed has been shown by Johnson and Saunders (1968); doubling of speed almost doubles subjective loudness (actually an increase of 9 dB(A)).

OTHER FACTORS   The effects of road conditions, such as the presence of intersections, pedestrian crossings, gradients, road width and the presence of buildings and road surface are noted by the Road Research Laboratory report (1970), which should be consulted for further detail.

PROPAGATION OF TRAFFIC NOISE   This highly important factor is discussed in the Road Research Laboratory report under the headings of distance; screening by acoustic barriers, sunk roadways and other devices; ground absorption; sound shielding and reflection effects, such as by buildings intervening between the road and the receiving area. The Urban Design Bulletin No. 1, Traffic noise, of the Greater London Council (1970) gives data on the noise fields of roadways, and is specifically orientated towards practical guidance. It is mentioned later in this connection.

Basic acoustical considerations on radiation of sound are discussed by Parkin (1962). The effect of distance from a road is complicated by the multiple, moving sources of noise. As Johnson and Saunders (1968) point out, at small distances from a motorway every vehicle that passes is heard more or less individually, so that level L_{10} (the highest component) is clearly distinguishable from level L_{90}, the general level of traffic noise. At greater distances the individual vehicle noise, which can be considered to be a point source, will be attenuated at about 6 dB per doubling of distance, in the normal way. The whole stream of traffic, on the other hand, is in effect a distributed, or line, source and as such the noise is attenuated at only 3 dB per doubling of distance. The levels L_{10} and L_{90} thus tend to converge with increasing distance. These relations and the additional effects of traffic volume are discussed in detail by Johnson and Saunders.

The effects of screening, of elevated and sunk motorways are all of practical importance, and are dealt with by the Greater London Council (1970) report. Enough has been presented to establish traffic noise as the main urban source of disturbance. It may be assumed that this conclusion is representative of urban conditions in other British cities as well as in London, and probably has general validity. The inference is that though the disturbance may not be so acute as with aircraft, the numbers of people disturbed are probably much greater.

The contribution of different classes of vehicle, their particular characteristics as sound sources, and the possible palliatives will be discussed in the next chapter.

In addition to the general findings on disturbance, there are innumerable sources capable of giving much disturbance, annoyance and distress which result principally from lack of consideration and the exercise of imagination. Nocturnal social activities; the noise of slamming of car doors, which often cannot be shut quietly; motor vehicles driven without adequate thought in residential areas and near hospitals, at night especially; moving of goods at night; and many other sources can all contribute to disturbance, most of which could be avoided with little cost or none at all, if the will existed.

Aircraft noise is a special topic which has assumed great importance since the introduction of the heavy jet transport aircraft; with the evolution and possible introduction of the supersonic transport aircraft (SST) into scheduled commercial operations the problem becomes more serious, and the whole topic will be considered separately in Chapter 13.

Disturbance of work

Experimental psychologists have for many years been interested in the effects of noise on work, and laboratory investigations designed to elucidate these problems have been pursued for more than forty years. Bartlett (1934) published a review of the position at that time and discussed experiments in which various sounds such as speech, music, or bells were used. These sounds were meaningful rather than a loud and continuous background noise. The type of tests which the subjects were asked to perform were of comparatively short duration and either did not show any effects attributable to the noise, or effects which were judged to be due to irritation or distraction, and consciously recognised as such. Such effects are, of course, experienced by probably everyone in one circumstance or another, and the degree of irritation or distraction from tasks is presumably dependent on the frame of mind, on the nature of the task and the nature of the sound, including the amount of meaningful distraction it contains, such as speech. This effect we can thus accept as a normal response, and in a sense, an aspect of annoyance.

However, the problem which must be examined is whether detrimental effects on work in the sense of reduced rate of performance, liability to error, or some other criterion of reduced efficiency, may be caused by noise which is not consciously irritating and distracting to the performer. This aspect can be investigated by studying real work in the field, or by setting up experimental situations in the laboratory. The two approaches can be complementary to one another, but the difficulty of controlling field studies is great, and care must continually be taken in avoiding anomalies in the investigation or the interpretation of the results.

A classic early investigation of a real work situation with noise as a factor was that of Weston and Adams (1932 and 1935). This study concerned the textile industry, the occupation cotton weaving. The type of work involved the supervision of machines in which the operation was automatic, but in the event of any stoppage due to such events as thread breakage, the weaver had to intervene to rectify the fault and start the machine again. The work involved vigilance and attention to keep the equipment working satisfactorily. Weaving is a fairly noisy occupation (Burns, Hinchcliffe & Littler, 1964; Taylor, Pearson, Mair & Burns, 1965) and the weavers were asked to wear earplugs which provided a modest reduction in the level of noise reaching the ears. These earplugs were not liked and, significantly, the weavers did not believe that their use would improve their work. Despite this, to their credit, they co-operated in the experiment and the results showed a significant improvement in the efficiency of work, by the criteria employed. The interest of this experiment lies in the fact that a real improvement in working efficiency was shown in a properly designed and controlled field experiment. The participants, moreover, were not convinced that such a result would emerge, so that the factor of morale did not appear to have contributed to the result. This last factor, however, is extremely important and various instances of its interfering effect can be cited. In one case, the speed of typing was increased when the noise level in the room was reduced, but remained at the higher speed when the original noise level was restored. In another case, Broadbent and Little (1960) studied the effect of noise level on the operators of film perforating machines; a reduction in level of about 10 dB produced an increased rate of working. The same workers were studied in rooms which had not yet received the acoustic treatment, but the accelerated rate of work persisted. The people presumably felt that something was being done for their welfare and their appreciation and improved morale was reflected in their rate of work. But a highly significant fact emerged. Although the rate of working in the acoustically treated and untreated rooms did not differ, the rate of machine stoppage, attributable to faults of the operators, was five times as great in the untreated as in the treated rooms. Such clear-cut and illuminating results as these are not often obtained, not least because the total amount of such investigations on which to draw is disappointingly small. This can be attributed to the difficulty of finding the right combination of circumstances in which to conduct field studies which are likely to yield unequivocal results.

Psychologists have thus sought to devise laboratory experiments in which the variables which beset field experiments can be eliminated, so making interpretation of a precise nature easier. There is, of course, sacrifice of realism, but this is usually the price which has to be paid for a well-controlled situation. A type of experiment that has yielded information in recent years has been much used in Britain, particularly by Broadbent and previously by Mackworth (1950), at the Medical Research Council Applied Psychology Unit at Cambridge; it has used broad-band noise with a fairly flat spectrum, electronically generated, at levels up to about 100 dB SPL overall. This noise, while not greatly dissimilar to some types of industrial noise, has no meaning to the listener and fulfils the requirement of a noise background devoid of particular distracting properties. With and without this background of noise, subjects have been presented with a variety of tasks, in which their performance, over an adequate period, which might be hours, is studied in detail. Tasks of short duration which are not long enough for the disappearance of short-term effects have been found to be useless, and the longer durations have revealed significant changes in performance with time. On the basis of such experiments, the following conclusions have been drawn.

Effects on efficiency have only been demonstrated with broad band noise of more than 90 dB SPL overall (Broadbent, 1957b). Thus if a noise of the type used by Broadbent has a level of 90 dB SPL or less, work is unlikely to be adversely affected. This was equivalent to approximately 90 dB(A); on a subjective scale of noisiness applied to motor vehicle noise (Committee on the Problem of Noise, 1963, p. 200) this would be regarded as intermediate between 'noisy' and 'excessively noisy' and would make speech communication difficult. In the intensity range where effects are possible, it has been found that high frequencies are more effective in impairing performance than low. Sudden bursts of noise produce an impairment which disappears in about half a minute, and there is some suggestion that a changing noise is more liable to cause impairment of performance than a steady noise. In one investigation a changing noise with an average value less than the limiting level discussed previously did give an effect (Sanders, 1961). Sudden alterations, up or down, in noise level produce momentary disturbance of work, and the degree of disturbance is determined by the extent of the change in level (Teichner, Arees & Reilly, 1963).

A distinction has been made between the rate of working and the accuracy of work. In general, the effect of noise is not to reduce the speed at which work is carried out, but to diminish the accuracy. Different types of task are affected by noise in different ways. In simple tasks, such as pressing a particular button when a particular lamp lights, the wrong button may be pressed (Broadbent, 1957b). Where the signals in a watching task are at lengthy intervals, the deterioration takes the form of increased delay in making the appropriate response, or failure to see the signal at all (Broadbent, 1954; Broadbent & Gregory, 1963). Failures to notice unexpected events are another feature (Carpenter, 1962). When noise is associated with other factors, some interesting and perhaps confusing results have emerged.

When an unfamiliar noise is first presented, a definite disturbance of work occurs which rapidly diminishes and disappears. Thus an initial improvement can be anticipated immediately after the introduction of the unfamiliar noise. The adverse effects previously discussed are quite distinct from the initial transitory effect. The effect of suggestion has already been mentioned. Suggestion may be effective either way; that is, better work may be done in high noise levels or minimal noise levels according to what the subject has been led to expect. At the same time, the findings that noise is generally detrimental to efficiency of work were obtained in experimental conditions designed to eliminate the factor of suggestion, and, as Broadbent stresses, some adverse effects of noise were obtained when the subjects expected a beneficial effect. The influence of motivation of subjects has also been investigated. It appears that those who are trying hardest are likely to be most affected; the result of this could be that a special effort might evoke a reduction in efficiency compared to work in which the subject was less concerned to perform well. Finally, a curious effect has been shown by Corcoran (1962) and by Wilkinson (1963) when lack of sleep and exposure to noise occur together. In these circumstances a person who has not slept the previous night shows less ill effect from noise, and in fact performance may even be better than in quiet surroundings. Carpenter (1962) speculates on the apparently opposite effects of these two factors, lack of sleep and noise, and invokes the concept of arousal, which, if diminished by loss of sleep, might be elevated to near normal levels by noise, so restoring the performance.

General effects on health

In any circumstances involving noise exposure to a community the question of harm to health almost invariably arises. There are repeated allegations that noise exerts ill effects on health, but on examination, they are vague. Frequently they are propounded by people on the basis of personal belief rather than on any grounds capable of medical verification, but remain difficult either to confirm or disprove. The question is nevertheless important, because these fears and beliefs are so widely held, and no doubt the assertions about ill effects on health will continue to be made.

It is first appropriate to enquire what is meant by health in this context. The report of the Committee on the Problem of Noise (1963) quotes the definition of the World Health Organisation: 'health is a state of complete physical, mental and social well-being and not merely an absence of disease and infirmity.' This enviable condition is obviously vulnerable in a number of ways. Noise can disrupt sleep, cause annoyance, interrupt conversation and trains of thought, as well as inflicting physical damage on the ears. Without invoking the last effect at all, it is quite clear that according to the World Health Organisation definition, health is prejudiced by interference with peace of mind, privacy, the pursuit of work or pleasure, or with the basic requirement of a natural and undisturbed nightly sleep. However, the impression seems to persist in the minds of some people that more subtle effects of noise, in the field of physical or mental illness, may exist. Carpenter (1962) in considering these factors, has summarised the situation from a medical viewpoint.

Mental illness

The question of mental upset as a result of noise centres round the proposition that noise can increase the probability of the appearance of symptoms of any nervous condition to which the particular person may be predisposed. A proper study of this possibility has apparently only been done in a systematic manner by the United States Navy (Davis, 1958). The effect of the noise of aeroplane jet engines on crew members on board aircraft carriers was thoroughly assessed but no evidence to support the hypothesis was obtained. The Committee on the Problem of Noise (1963) also sought possible relations between noise and mental upset, but likewise found very little specific evidence to support the possibility. The mental illness of one patient was attributed by his doctor to noise exposure in the vicinity of a major airport; but this was the only instance of its kind in the experience of general practitioners whose patients lived near the airport. No evidence was obtained that the consumption of sedatives by people in the airport area exceeded that in other areas where the doctors had practised. Subsequently, however, another attempt to clarify this difficult field has been made by Abey-Wickrama, a'Brook, Gattoni and Herridge (1969). They compared the details of admissions to a psychiatric hospital, over a period of two years, of patients coming from an area of maximum noise arising from London Heathrow Airport, with admissions of patients from outside this area. While such comparisons are inherently fraught with difficulties of interpretation it was shown that (particularly in certain categories of patient such as those with neurotic or organic mental illness) there was a significantly higher rate of admission from the area of maximum noise than from the quieter area. The authors do not assert that aircraft noise can cause mental illness, only that this noise may be a factor in increasing rates of admission to the particular mental hospital.

Physical illness

In considering the possible bearing of noise exposure on physical illness, it is logical to attempt to find effects on body systems other than the auditory system, and if these effects do occur, to attempt to extrapolate to possible associations with specific disorders. In the USA and in Europe, but to a lesser extent in Britain, responses of the body systems (i.e. somatic responses) to noise have been studied. In this text it is not intended to cover these aspects in much detail, and the reader is referred to the admirable reference work of Kryter (1970) *The effects of noise on man*, for details of such studies. Examples from a large literature are the following: Davis, Buchwald and Frankman (1955); Jansen (1969); Lehmann (1965); Oppliger and Grandjean (1957). The connection between possible somatic effects and exposure to noise suggests the occurrence of stress. Noise exposure may be regarded as an additional stress which could have measurable effects. As it has been claimed that stress may be a factor in the production of physical diseases such as thyrotoxicosis or peptic ulcer, the possibility has been suggested that noise might have a similar effect. In an attempt to demonstrate physical changes due to noise, heart rate, blood pressure, muscular activity, metabolic rate and other responses have been studied. It has been found that increase of muscular activity, heart and respiration rates, as well as changes in skin blood vessels occur when a loud unfamiliar noise begins, but they disappear even if the noise is continued. These changes would be expected in any situation where apprehension is present, but when the situation is resolved, or the noise accepted, they disappear. One effect, the constriction of small vessels, the arterioles of the hands, and possibly feet in addition, appears to be of much longer duration, and may cause paleness and sensory disturbances in some people. The significance of these changes is not yet clear. Again, we are left with little in the nature of definite clues which can be followed. Thus in certain circumstances noise is able to produce some signs of fear initially, and so presumably constitutes an element of stress; and there is the rather small and variable finding on blood vessel constriction in the extremities. As yet, little evidence has come to light to connect noise with identifiable and attributable physical disease. In this statement we exclude injury to the hearing mechanism. The situation is thus not a very clear and satisfying one; but on the evidence which exists it is difficult to see how major effects on health, if they existed, could so far have escaped detection.

The reader who wishes to pursue these topics further than is intended in this text may refer to Kryter (1970), pp. 491–516, on the following topics: physiological responses to noise; stress on health; adaptation to stress; possible non-auditory effects of industrial noise; and miscellaneous health problems associated with noise. The evidence available unfortunately does not appear to justify any firm conclusions for or against the involvement of noise in the causation of physical illness.

Disturbance from low-frequency noise

This is a somewhat poorly-documented and not very common source of disturbance; there is little doubt however that it is real, and can be a serious problem. The picture which usually emerges is of a subjectively unpleasant phenomenon, vaguely and variously described by different individuals. Not unexpectedly, some persons appear to suffer much more disturbance than the majority, amounting to actual distress. The isolation of the source of such low-frequency sound is sometimes very difficult. The intensities involved may be quite small, sufficiently to make the results of measurements inconclusive, while the lack of clear directional cues make subjective identification equally unsatisfactory. Each instance is an individual problem and the possible inclination to dismiss the energetic complaints of a few persons or even an individual as neurotic manifestations should be resisted, and the problem properly examined. This subject is discussed also in Chapter 14.

9
Measures to reduce disturbance

In this chapter, the term disturbance has the same broad connotation as in Chapter 8. We are here faced with the problem of trying to lay down conditions to eliminate or control the various possible undesirable consequences of noise on sleep, rest or activity.

The solution of this problem is essentially one of attempting to evolve scales of subjective reaction for comparison with scales describing some physical attribute, or combination of attributes of the noise, as broadly noted in Chapter 8. Experience has shown that certain noise measures may be used to gain an approximate indication of subjective reaction to sound environments. These environments have included community or environmental noise, usually with reference to road traffic noise, or to aircraft noise; noise of offices; noise of living apartments. Where noise is not uniform in level, but fluctuates according to road or air traffic, the number of occurrences and the magnitude of the fluctuations have been recognised as relevant. To encompass these factors, various elaborations of the primary measures (e.g. of sound level A, or of perceived noisiness) have been evolved. The situation has been needlessly complicated by a proliferation of such indices, many of which are essentially similar, and in this chapter we shall attempt to examine the various forms of disturbance, and the numerical means available for their evaluation and control. In this, it must be accepted that we cannot hope to encompass the personal idiosyncrasies of everybody, and that the elimination of all subjective disturbance from noise cannot in practice be achieved. Nevertheless it is practicable to try to establish maximum permissible sound environments for various purposes, described by measures of lesser or greater complexity. The following examples illustrate this approach: Rosenblith and Stevens (1953), Parrack (1957), Beranek (1971), Kosten and van Os (1962), the Committee on the Problem of Noise (1963). Included in these recommendations are noise specifications either in sound level A or by arbitrary sound spectra. We shall see later how more complicated indices have been used for particular purposes. We begin with the simpler sound level A or spectral descriptions of noise environments.

SOUND LEVEL A   The derivation of this measure has already been described (Chapter 4). Its usefulness has been demonstrated by an increasing acceptance over recent years in a number of different contexts. It is a useful simple means of describing interior noise environments from the point of view of habitability, community disturbance, and latterly hearing damage. Its great attraction lies in its direct use in measures of total noise exposure. These usages will be encountered frequently in subsequent discussions.

SOUND SPECTRA FOR NOISE RATING   This system, in various forms, consists essentially of the construction of a set of arbitrary sound spectra to serve as a frame of reference for rating noise environments. They are, of course, primarily intended for describing steady noises, that is those which do not vary markedly with time. The technique, described by Rosenblith and K. N. Stevens (1953), was originally evolved for the rating of outdoor community noise. Beranek proposed a similar method for rating office noise, having in mind the need to prevent difficulties in speech communication. Subsequently, Beranek modified the contours, and offered two versions each constituting a family of curves of sound pressure level of octave bands of noise centred at 8 specific frequencies, primarily for noise in buildings; the more rigorous (i.e. lower level) version he designated *noise criteria* (NC) contours, and a less rigorous, for situations presenting practical difficulties in noise reduction, which he called *alternate noise criteria* (NCA) contours (Beranek, 1957, 1960). The basis of each family is that, for acceptable speech communication, the loudness level in phons (Stevens' Mark I (1956)) should not exceed the *speech interference level* (see below) in decibels, by more than a certain value, which is 22 units for the NC series and 30 units for the NCA series. In use, the spectrum of the noise being assessed is measured, using the same octave bands as for the NC curves, and superimposed on the latter. The noise is then rated by the number of the curve (originally the SPL in the octave 1200–2400 Hz), which equals or just exceeds the noise spectrum, i.e. the point at which the noise spectrum is tangent to a particular NC curve. Reference to a table of values will then establish the suitability of the noise for various indoor conditions (Beranek, 1960, p. 523). This system originally used octave bands which have now been superseded by the ISO system of preferred frequencies (ISO, 1962; BSI, 1963; ANSI, 1967), but Schultz (1968) has published conversions of the original

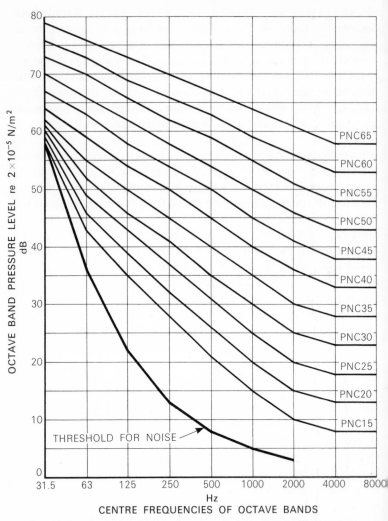

9.1   Preferred noise criteria (PNC) curves. For use see text and Table 9.1. Numerical values corresponding to these curves are given in Appendix L. (After Beranek, Blazier & Figwer, 1971; the noise threshold curve is after Robinson & Whittle, 1961.)

curves to the ISO preferred frequency system. No numerical changes in the NC or NCA ratings are required. These curves have been widely used for rating indoor noise and are employed by the American Society of Heating, Refrigeration and Air Conditioning Engineers (ASHRAE, 1967).

Recently, however, Beranek, Blazier and Figwer (1971) and Beranek (1971) have felt that the NC curves should be re-examined in the light of the preferred frequency octave bands now in use; the need for adjustment in the light of Robinson and Whittle's (1964) data on hearing thresholds for noise, and because of desirable changes in tonal quality. The outcome has been the development of *preferred noise criteria* (PNC) curves, which differ from their corresponding predecessors in three ways: (1) the speech interference level (based on 500, 1000 and 2000 Hz) is lowered by 1 dB; (2) the levels in the 63 Hz band are reduced by 2 to 7 dB; (3) the highest three frequency bands are reduced by 4 to 5 dB. These curves are shown in Fig. 9.1. The numerical values corresponding to the PNC curves are given in Appendix L.

In use, it is intended that a spectrum being assessed by this method should not exceed the values, in each of the octaves, specified in Fig. 9.1 and Appendix L. However, the noise may be allowed to exceed the PNC curve in one octave by up to 2 dB, provided the octaves above and below are not more than 1 dB below the PNC curve. The suitability for various indoor uses of different sound levels and PNC values are given, in NC, PNC and dB(A) from Beranek's new data, in Table 9.1.

In Europe, a similar system has evolved, particularly in the Netherlands (Kosten & van Os, 1962) which from the outset used the ISO octave bands and is designated the *noise rating* (NR) system. The tangent-to-curve method is again used to rate noises, and the system is widely used due to its association with ISO. By a curious anomaly, the curves (Fig. 9.2) have only recently passed the stage of a draft recommendation, in which status they have remained for some years. They are universally associated with ISO however, and will be referred to as 'the' ISO curves in this text. This situation has been principally due to the parallel claims of sound level A, and at present, the latter seems to be somewhat in the lead. The NR curves now appear, however, in an ISO recommendation (ISO, 1971) on environmental noise, as an additional method to the use of sound level A, of rating environmental noise. The corresponding numerical values are given in Appendix L.

The use of the noise rating curves, to recapitulate, can be seen from Fig. 9.2. Inspection of this figure will show that each curve in

TABLE 9.1

Recommended category classification and suggested noise criteria range for steady background noise as heard in various indoor functional activity areas

Type of space (and acoustical requirements)	NC or PNC curve	Approximate sound level dB(A)
Concert halls, opera houses and recital halls (for listening to faint musical sounds) . . .	Not to exceed 15	26
Broadcast and recording studios (distant microphone pickup used)	Not to exceed 20	30
Large auditoriums, large drama theatres and churches (for very good listening conditions) .	Not to exceed 20	30
Broadcast, television and recording studios (close microphone pickup used only) . . .	Not to exceed 25	34
Small auditoriums, small theatres, small churches, music rehearsal rooms, large meeting and conference rooms (for very good listening), or executive offices and conference rooms for 50 people (no amplification)	Not to exceed 35	42
Bedrooms, sleeping quarters, hospitals, residences, apartments, hotels, motels, etc. (for sleeping, resting, relaxing)	25 to 40	34 to 47
Private or semiprivate offices, small conference rooms, classrooms, libraries, etc. (for good listening conditions)	30 to 40	38 to 47
Living rooms and drawing rooms in dwellings (for conversing or listening to radio and TV) . .	30 to 40	38 to 47
Large offices, reception areas, retail shops and stores, cafeterias, restaurants, etc. (for moderately good listening conditions) . . .	35 to 45	42 to 52
Lobbies, laboratory work spaces, drafting and engineering rooms, general secretarial areas (for fair listening conditions)	40 to 50	47 to 56
Light maintenance shops, office and computer equipment rooms, kitchens and laundries (for moderately fair listening conditions) . .	45 to 55	52 to 61
Shops, garages, power-plant control rooms, etc. (for just acceptable speech and telephone communication). Levels above NC or PNC 60 are not recommended for any office or communication situation	50 to 60	56 to 66
For work spaces where speech or telephone communication is not required, but where there must be *no risk* of hearing damage . . .	60 to 75	66 to 80

the family of curves represents a spectrum composed of SPL values in the various octaves, from the octave centred at 31·5 Hz upwards. Each curve is designated by a number which is the SPL value in the octave whose mid-frequency is 1000 Hz. Any curve is then uniquely identified by its number, e.g. noise rating (NR) 80. The noise under investigation is measured in octave bands and the resultant values in dB SPL in each octave are superimposed on the family of curves, as described above for the NC curves. The spectrum being described

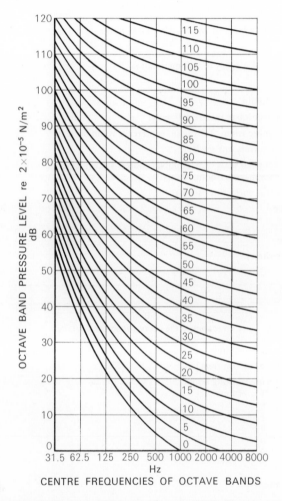

9.2 Noise rating (NR) curves of Kosten and van Os (1962). The curves are identified by the band pressure level for the octave centred at 1000 Hz. For use see text and Table 9.2. Numerical values corresponding to these curves are given in Appendix L. (After ISO, 1971.)

is then identified by the number of the NR curve which lies just above it. The properties of the noise may then be predicted from the known properties of the NR curves, in such terms as annoyance, interference with communication, or other parameters. In Table 9.2

TABLE 9.2

A

Recommended noise rating values for different types of accommodation
(Kosten and van Os, 1962)

Category of accommodation	Noise rating
Broadcasting studio	15
Concert hall, legitimate theatre 500 seats	20
Class room, music room, TV studio, conference room 50 seats	25
Bedroom (for corrections see B)	25
Conference room 20 seats or with public address system, cinema, hospital, church, courtroom, library	30
Living room (for corrections see B)	30
Private office	40
Restaurant	45
Gymnasium	50
Office (typewriters)	55
Workshop	65

B

Corrections to the above values for dwellings

Pure tone easily perceptible	−5
Impulsive and/or intermittent	−5
Noise only during working hours	+5
Noise during 25% of time	+5
6%	+10
1·5%	+15
0·5%	+20
0·1%	+25
0·02%	+30
Economic tie	+5
Very quiet suburban	−5
Suburban	0
Residential urban	+5
Urban near some industry	+10
Area of heavy industry	+15

are shown the recommendations of Kosten and van Os for various types of accommodation, in terms of NR. They resemble closely the numerical values of Beranek's NC curves.

We shall now examine how far specific aspects of disturbance can be related to noise measures. It will be found that for some of the former, elaborations of the latter will have to be used, so that the measures are of noise exposure rather than noise itself.

Disturbance of sleep

Accepting the idiosyncrasies of individuals, it is possible to conclude from general experience that a majority of people seem to find certain levels of noise compatible with undisturbed sleep. The maximum suitable level for bedrooms appears to be established with some measure of agreement between various different recommendations. In the noise rating system described by Kosten and van Os (1962), the noise rating number for bedrooms is basically 25 (Table 9.2A). This figure is raised or lowered by Kosten and van Os according to the particular circumstances, and the type of correction they employ is summarised in Table 9.2B. On this assessment, a suburban area is taken as the basic situation, with no intermittency, nor pure tone components of noise, and a bedroom should be not more than NR 25. Modifications of this figure are suggested by Kosten and van Os to allow for the other circumstances shown in Table 9.2B; for example, an urban residential bedroom would be permitted an allowance of +5, i.e. instead of NR 25, it would be NR 30. These authors also produce evidence that this system of rating for bedrooms is successful in their hands. Acceptability, judged subjectively, was broadly in agreement with NR 25, with corrections as in Table 9.2.

The Committee on the Problem of Noise (1963) also states that suburban sleeping accommodation, rated at NR 25 by Kosten and van Os (1962) is approximately equivalent to 30 dB(A). The use of the dB(A) scale is a convenient single figure expression, but it is not a complete substitute, in general, for a description in terms of a full frequency spectrum. The value in part lies in the ability it confers to check rapidly the approximate level of a sound; full frequency analyses must take more time in measurement and recording. The Committee on the Problem of Noise (1963) have restated the main points of the recommendations of Kosten and van Os (1962) for rooms in which people sleep, in terms of recommended maximum night-time levels of noise, in dB(A). These are

Country	25 dB(A)
Suburban	30 dB(A)
Urban	35 dB(A)
Busy urban	40 dB(A)

The gradual increase is based on the fact that the exterior, or ambient, levels are established and uncontrollable on any short-term basis, so that to some degree people may be expected to have become accustomed by reason of the area in which they live, to a generally higher level, for example, in busy urban than in quiet country districts. It was evidently felt, however, that a limit should be set to the level for sleeping. Beranek (1960) notes that room air conditioners manufactured prior to 1957 commonly produce levels of 40 to 55 dB(A) in bedrooms; these higher levels may, of course, be tolerated more willingly since they are the alternative to perhaps greater discomfort from excessive heat and humidity.

It would therefore appear that reasonable guidance on noise levels compatible with sleep is available, and an application of the corrections suggested in Table 9.2 will cover a considerable number of foreseeable circumstances including intermittency of the noise. Markedly variable noise may require the use of a special index as described later. Problems peculiar to individuals or localities will always remain.

The attainment of suitable conditions for sleep is thus in general an acoustic problem, but its dimensions are reasonably well known. Such aspects as the exclusion of external noise without sacrificing proper ventilation must be faced, and especially in warm climatic or seasonal conditions. Solutions are mentioned later on in the section on aircraft noise.

Annoyance

This rather wide term, defined for the purposes of the last chapter as the displeasure or resentment caused by sound, is a manifestation broadly deriving from urban living specifically due to the activities of neighbours, their children or their pets, to road traffic, industrial sources of noise in the vicinity, or to the proximity of air traffic; it must be simplified and subdivided to some extent to codify means of assessment and so of control. In this section we will emphasise urban community disturbance, and refer to specific sources; aircraft noise has certain unique features which are better discussed as an allied, but distinct problem.

A theme running through many of the noise situations which have been investigated is the need to recognise the factor of total exposure. Where the noise level varies little with time, simple indices such as weighted sound levels or spectrally-derived quantities as in the NR system, will suffice. If some fluctuation of level occurs, a statistical derivation, such as defining the proportion of total time during which a measured or defined sound level occurs, may be adequate. If large variations between background level and maximum level occur, as seen in extreme form in the case of aircraft overflying a given locality, or road traffic noise at close range, some more explicit recognition of the frequency of occurrence as well as its acoustic magnitude, becomes necessary. The acceptance of this requirement has caused an alarming multiplication of indices in different countries, and this number is now in double figures.

For the purpose of this chapter, we shall deal primarily with the main urban disturbances, industrial noise and road traffic noise. These of course need not essentially be urban; rural or at least isolated communities may be equally affected with industrial or with motorway noise, and aircraft noise will depend on airway patterns. The magnitude of the disturbance, however, will always tend, in terms of numbers of persons involved, to be greatest in urban areas.

Noise affecting indoor accommodation

The specification of acceptable indoor noise for various types of accommodation has been mentioned, in connection with disturbance of sleep, in the previous section. Reference back to Fig. 9.2 and Table 9.2 will show the Kosten and van Os (1962) proposals for a variety of indoor conditions, in different types of locality, duration and type of noise. Table 9.2 contains the essence of many of the working concepts underlying such specifications. Ranging from NR 15 for a broadcasting studio to NR 65 for workshop accommodation, the list includes living rooms, with the basic values of NR 30, and for bedrooms, NR 25. These basic values are then modified by the factors shown in Section B of Table 9.2. This allows progressively higher values (i.e. more noise) the shorter the duration of the noise, and likewise a progression of higher allowable values according to the noise environment of the area as a whole. Thus a bedroom in an area designated 'urban near some industry' is permitted $+10$, i.e. basic value 25, $+10$ equal to NR 35.

Other allowances suggest themselves, such as the effect of climatic conditions. Thus in warm conditions, more time would be spent outdoors than indoors, but outdoors more tolerance to noise might be expected. Another variable is the effect of windows, open or closed, on the reduction of outside noise to be expected in rooms. We shall return to this point later.

In addition to these Kosten and van Os recommendations, the Committee on the Problem of Noise (1963) on the basis of the survey of noise in London in 1961–2, together with the subjective reactions recorded, came to certain conclusions. They recommended that the noise levels shown in Table 9.3 should not be exceeded for more than 10% of the time, and specified separate values for day and night.

TABLE 9.3

Noise levels inside dwellings

Situation	Levels dB(A) which should not be exceeded for more than 10% of the time	
	Day	Night
Country areas . .	40	30
Suburban areas .	45	35
Busy urban areas .	50	35

These estimates are reasonably in accord with similar estimates made in other countries, such as those already quoted, although the method of defining the noise is not the same as in, for example, Table 9.2. The recommendations are also consistent with expectations for mutual disturbance by neighbours in flats and apartment houses, based on the insulation between dwellings attained with the British Building Research Station (BRS) Grade I and Grade II standards of insulation (Building Research Station, 1956). For example, it is assumed that noise originating in one flat will not often exceed 80 dB(A). With the BRS Grade II insulation (45 dB over the range 100–3200 Hz), neighbours would presumably receive levels of noise of some 35 dB(A). Social surveys have indicated that when the insulation is worse than BRS Grade II, neighbours' noise is the major source of complaint. When BRS Grade I insulation is used, giving some 50 dB of attenuation, neighbours' noise ceases to be a major cause of disturbance. This would correspond to about 30 dB(A) in an adjacent flat so that most people would appear to be reasonably satisfied with 30 dB(A), as the normal maximum value for sleeping, (Table 9.3), with 35 dB(A) as the value not to be exceeded.

Specific sources of annoyance, possessing disturbing effects unrelated to the actual noise intensity, which may be trivial, are nevertheless capable of causing much irritation and even distress. The control of such noise is frequently simple, and the aim to secure enough goodwill, sympathy and co-operation to achieve the desired end.

It will be noted that the Committee on the Problem of Noise recommendations used sound level A, and not the NR system, and this usage was continued in a British Standard which is, in a sense, a derivative of the Committee proposals. This is BS 4142, *Method of rating industrial noise affecting mixed residential and industrial areas* (BSI, 1967). It is intended to apply to noise from permanent industrial installations, and its primary function is to provide a rating for such noises in terms of their effects on persons whose homes are in the vicinity. It deals with noise measured out-of-doors in the neighbourhood of residential housing. It does not rate directly the noise measured inside the building or arising within the building.

This standard contains some new features. Instead of the previous simple statements of source levels, perhaps with a statistical qualification of the proportion of total time occupied by noise, this standard introduces a more comprehensive system of measuring outdoor noise. For this purpose three measured noise levels are recognised: (1) background or ambient noise in the absence of the potentially offending noise; (2) if the potentially offending noise is steady, its general level (L_1); if the potentially offending noise occurs in intermittent bursts, this elevated level (L_2). The levels L_1 and L_2 are then individually subjected to corrections according to their character and duration to yield the *corrected noise levels*. These corrections consist of additions or subtractions of stated values in dB(A) according to the tonal character, impulsive character, intermittency and duration. Such corrected noise levels are designated L_1' and L_2'.

Since the liability of a noise to provoke complaints depends primarily on the pre-existing background level, the noise is rated in principle by comparing it with the measured background. In BS 4142, an elaboration is introduced, whereby this comparison is made between the noise being rated (in the form L'_1, L'_2), and the background level, in conjunction with a derived level, the *corrected criterion*. This is derived by taking a *basic criterion*[1] of 50 dB(A) and making corrections according to a number of factors, including previous history of the locality with respect to industrial development; type of district (e.g. rural, suburban, industrial, etc.); time of day; season of the year.

Where, as frequently may be found, the background noise level is either difficult or impossible to measure, the comparison is between the noise being rated and the corrected criterion only. The comparison procedure thus resolves into three conditions, depending on the relative levels of the background and corrected criterion, and on whether a background level is available or not. The expectations in these three conditions are (1) where the background noise level is available and exceeds the corrected criterion by less than 10 dB(A), complaints from residents may be expected if either the corrected noise levels L'_1, L'_2 or both exceed the corrected criterion by 10 dB(A) or more. If the background noise level greatly exceeds the criterion and L'_1 or L'_2 are intermediate, and if they are 10 dB(A) or more below the background noise level there should not be complaints; (2) where the background noise level is available and is less than, or equal to, the corrected criterion, complaints may be expected if either L'_1 or L'_2 or both exceed the background by 10 dB(A) or more; (3) where the background level is not available, complaints may be expected if either L'_1 or L'_2 or both exceed the corrected criterion by 10 dB(A) or more. The converse, when both L'_1 or L'_2 are below the corrected criterion by more than 10 dB(A) complaints should definitely not arise. This standard has been in use for some time, and in order to provide a check on its effectiveness in practice, BSI has recently sampled user opinion. It would appear that very little difficulty has arisen, or that where it has, individual cases would require amendments that would cancel each other out, and that in general it is serving its intended purpose well. It must be emphasised again, that it specifically deals with measurements of

---

[1] The terms 'basic criterion' and 'corrected criterion' may be amended in a subsequent version of BS 4142 to more explicit forms, such as 'reference background level' and 'corrected reference background level'.

outside noise in order to predict indirectly the response of people indoors in houses in a specific noise environment.

Another document for the same purpose, sponsored in this case by ISO (1971) is entitled 'Noise assessment with respect to community response' (R1996). This recommendation is also for the purpose of rating (outdoor) environmental noise with respect to its compatibility or otherwise, with residential housing. It has a fairly strong resemblance to BS 4142, but with some significant divergencies. It is, like the British standard, based on sound level A, but in addition is specifies the noise rating (NR) system (Kosten & van Os, 1962) by which the actual spectra of noises under investigation may also be assessed by octave frequency bands. The general approach is quite similar to BS 4142, inasmuch as outdoor noise is measured, modified by the application of various numerical corrections to accommodate different conditions to yield the *rating sound level* L_r, and then compared with a criterion value in dB(A) again derived broadly in the same way as in the British standard. If L_r exceeds the corrected criterion value by 10 dB(A) or more, complaints may be expected.

Variations from BS 4142 occur in the corrections for variability of the noise. If the variations of sound level A with time occur in a more complicated manner than can be accommodated by the described corrections to the numerical value, a statistical analysis of the time-history of the A-weighted sound level of the noise is performed, and the *equivalent sound level* (L_{eq}) is derived thus:

$$L_{eq} = 10 \log_{10} \left[\tfrac{1}{100} \sum f_i 10^{Li/10} \right] \qquad (9.1)$$

where L_{eq} = the equivalent sound level in dB(A)

L_i = the sound level in dB(A) corresponding to the class-midpoint of the class i (for class intervals not greater than 5 dB(A) the arithmetic means can be used; for larger intervals logarithmic averaging should be used)

f_i = that time-interval (expressed as a percentage of the relevant time period) for which the sound level is within the limits of class i.

This indicates total energy and in fact represents the equivalent steady level, i.e. the energy mean noise level, of the period concerned. This value may also be corrected if the sound contains pure tones or is impulsive. The corrected criterion value with which the rating sound level L_r is compared is not specified exactly, but should be chosen appropriate to the community and living habits of the people. The suggested limits are 35 to 45 dB(A). The correction system for the basic criterion is again similar in principle to that of BS 4142. The foregoing concerns exclusively outdoor noise. An important extension embodied in ISO R1996 is the coverage of indoor noise also. This is accomplished by the simple procedure of deriving an indoor criterion for residential premises from that for outdoor noise by reducing the value of the latter by amounts corresponding to the approximate attenuation of the building with open or closed windows.

For the range of size of rooms in dwelling houses of normal construction, it is suggested that the attenuation outside-inside may be taken to be the following:

Windows open	10 dB(A)
Single windows shut	15 dB(A)
Double windows shut	20 dB(A)
or non-openable windows	20 dB(A)

Criteria in dB(A) for rooms other than residential are also given. Since NR values can usually be taken, for broadband noise, to be approximately equal for rating purposes to the sound level A (L_A) value less 5 dB (i.e. NR = $L_A - 5$) rough equivalences can easily be seen at a glance. Some of the indoor criteria, i.e. suggested maximum noise levels for indoor conditions, from ISO R1996 are given in Table 9.4. They apply only to noise originating outside the rooms, the noise being measured inside. It can be seen that the values in dB(A) are broadly similar to the values of Kosten and van Os (1962)

TABLE 9.4
Suggested maximum noise levels for indoor spaces from ISO R1996

Examples of type of room	Suggested maximum level dB(A)
Larger office, business store, department store, meeting room, quiet restaurant	35
Larger restaurant, secretarial office . . .	45
Larger typing halls	55
Workshops, according to function	45–75

quoted in Table 9.2. These values are to some extent judgements, and cannot be expected to be entirely consistent or immutable.

Indices of indirect effects

It will have been observed that, so far, limitation of noise for the purpose of specifying acceptability or otherwise of living or working accommodation has been recommended in the form of relatively simple noise specifications. These are sound level A, NR or other rating systems, and simple elaborations such as the equivalent sound level L_{eq}. The applicability of these measures has been extended by arbitrary corrections to meet different conditions or characteristics of the noise itself. This relatively unsophisticated approach, while highly desirable on grounds of simplicity, has been shown to be sometimes insufficient to meet real conditions. In order to evolve scales more meaningful in terms of subjective judgement of degree of annoyance or of some other designated aspect, it has been found that a more systematic attempt must be made to embrace three main aspects: the intensity of the noise; its duration; its variation with time. The latter characteristic varies greatly between different types of noise. It is seen in extreme form when a series of aircraft overfly an area of otherwise low noise level; road traffic noise will vary with time depending on the density, and office noise may be relatively steady, in large areas with many machines such as typewriters in operation. To meet these various demands, many investigations in different countries have resulted in the publication and, in varying degree, the use of at least a dozen different indices. Robinson's (1970) analysis of these indices should be consulted for more detailed information; here we may note these, and discuss certain aspects in more detail, in this chapter and elsewhere. The list includes the following:

> *aircraft exposure level* L_E; *aircraft noise exposure level* L_{exp}; *annoyance index*, AI; *composite noise rating*, CNR; *equivalent disturbance level*, L_q; *indice de classification*, R; *noise and number index*, NNI; *noisiness index*, NI; *noise immission level*, NIL; *office noise acceptability* L_A/TPI; *traffic noise index*, TNI; *Störindex*, $\bar{Q}$.

To this list must be added *noise pollution level*, L_{NP} (Robinson, 1970) which appears to combine the features of many of the above, and which holds promise of some rationalisation and unification of the present profusion of indices. It will be discussed later. In addition, the concept of *noise exposure forecasts* (NEF) (Galloway & Bishop, 1970) has been introduced for rating noise intrusion in communities by aircraft.

Of the indices above, the majority have been developed to meet the need for a measure of disturbance from aircraft noise (see Chapter 13). The indices used for this purpose, in abbreviated form, are AI, CNR, L_E, L_{exp}, NI, NNI, $\bar{Q}$, R, and NEF. Of the other indices, measurement of traffic noise disturbance is the object of TNI (Griffiths & Langdon, 1968); office noise in the case of L_A/TPI, a two-dimensional index (Keighley, 1966); speech disturbance L_q will not be assessed as such, in view of the coverage later of speech intelligibility. Hearing loss from noise can, within defined limits, be predicted from noise immission level (NIL), which will appear in Chapters 11 and 12 (Burns & Robinson, 1970).

In the context of this chapter, having arbitrarily removed to Chapter 13, for descriptive convenience, the question of aircraft noise, the most important source of disturbance is traffic noise. After a short general introduction of these indices, we may then discuss the restriction of disturbance by road traffic noise, including the use of the noise pollution level L_{NP} for this purpose.

Of the aircraft noise indices, Robinson (1970) notes their essential similarity, inasmuch as they contain two main terms, the noise intensity, and the number of occurrences of the noise. These two terms are combined, together with a multiplicative constant which will yield a greater or lesser product. Since the noise is in the form of a level, the number of occurrences is also expressed logarithmically, with a coefficient of varying value according to the taste of the originators. In symbols

$$\text{Index} = \bar{L} + A \log N$$

where $\bar{L}$ = mean of peak noise levels of each overflight, expressed, for example, in PNdB or EPNdB[1]

A = a coefficient, numerically in the range 6 to 24

N = number of occurrences.

If $A = 10$, the expression can be seen to yield a straightforward energy measure; if greater than 10, the index of disturbance due to

---

[1] For information on the various measures of perceived noise, see Appendix K.

repeated overflights increases, with increase of numbers of occurrences, at a greater rate than that determined by total energy. The duration of each noise episode due to a single overflight is sometimes included, sometimes not; the ISO index, *aircraft exposure level* (L_E), includes the duration by expressing L as *equivalent perceived noise level* (in EPNdB) thereby specifically including an allowance for duration and spectral irregularities (see Appendix K). The British noise and number index, NNI, on the other hand does not take account of the duration of the noise of the overflights in indicating exposure.

Traffic noise

The widespread nature of the disturbance due to urban traffic noise has been documented in the last chapter (e.g. Table 8.5). The traffic noise problem has been conspicuously displayed in the case of motorways, and particularly urban motorways, and these developments have illustrated again the incompatibility of motor roads and residential areas. Understanding of the conditions necessary for adequate restriction of road traffic noise is increasing (Road Research Laboratory, 1970) and this should prevent mistakes in urban road planning which could result in serious community disturbance. Such disturbance has regrettably occurred on occasion in the past, and the situation still tends to remain an area of some controversy between planning authorities and conservationists.

The approach to the restriction of road traffic noise disturbance must be attempted in a wide variety of circumstances. First, areas of residential housing should be prevented as far as possible from sustaining more than the stated levels of noise from any source, which has already been discussed. Thus if the conditions suggested for avoidance of community disturbance in the recommendations of Kosten and van Os (1962), Committee on the Problem of Noise (1963), BSI (1967), ISO (1971) can be implemented, disturbance should be avoided. However, conditions may develop gradually until disturbance clearly exists; a new road system may be projected, or housing in the vicinity of a road already in existence may be proposed. New types of vehicle may be introduced, or different types of road usage develop, suddenly or gradually. These need specific approaches, in which the general recommendations above are only generalisation, and not of specific assistance. Closer study of (1) road traffic noise as a source of disturbance; (2) road planning aspects; (3) vehicles as noise sources, is essential. These will be discussed in that order.

ASSESSMENT OF DISTURBANCE: TRAFFIC NOISE INDEX   The Committee on the Problem of Noise, in specifying acceptable dwelling house levels for noise, incorporated a kind of time-variable allowance inasmuch as they specified (Table 9.4) the A-weighted levels which should not be exceeded for more than 10% of the time. This level, usually designated L_{10}, gained considerable acceptance, and in the absence of full validation of more elaborate indices is used widely at the present time (Greater London Council, 1970; Scholes & Sargent, 1970). A systematic attempt in Sweden (Fog, Jonsson, Kajland, Nilsson & Sörensen, 1966) to correlate subjective judgements of noise disturbance with various noise measures did not show L_{A10} to be as good an index as the simple mean energy level in dB(A), over a 24 hour period. This index can be seen to be similar to the ISO aircraft exposure level L_E (see Chapter 13), except for the latter's use of EPNdB instead of dB(A). Pursuing the question of the nature of road traffic noise disturbance generated particularly from motorways, Griffiths and Langdon (1968) carried out an extensive investigation in the Greater London Area, in which the traffic noise measured outside dwelling houses was correlated with the subjective reactions of the inhabitants. It was found that disturbance was related to both the background level and the magnitude of the fluctuations of sound level. The simplest form of their traffic noise index (TNI) is:

$$\text{TNI} = L_{A90} - 4(L_{A10} - L_{A90})$$

where L_{A90} = sound level A exceeded for 90% of the time
L_{A10} = sound level A exceeded for 10% of the time.

Langdon and Scholes (1968) have described application of the TNI as a method of assessing noise nuisance. Their findings will be discussed later after other aspects have been covered.

OTHER INDICES: NOISE POLLUTION LEVEL   Robinson (1970) has reviewed extensively various indices of disturbance and has derived an index, noise pollution level (L_{NP}), which he discusses at length, comparing its performance in describing annoyance, against other recognised indices. On his very complete assessment, this index seems to have a wide applicability as a general index of disturbance. It reproduces the properties of a number of indices, and also can be shown to reconcile apparently somewhat contradictory findings such as those of the Swedish Road Traffic noise enquiry mentioned above (Fog *et al.*, 1968) and the Griffiths and Langdon (1968) data. Robinson's original derivation and analysis should be consulted for a full assessment of the index L_{NP}, but its basis will be summarised here.

The essential feature of Robinson's synthesis of the various experimental approaches to the measurement of annoyance is the recognition of three main findings. These are:

(1) that the amount of sound energy in the stimulus in a given period of time is a primary component;
(2) that annoyance due to increased numbers of noise episodes (e.g. aircraft overflights) increases at a rate greater than the increase of total energy of the series;
(3) that the range of variation in level of noise fluctuating about a mean value influences annoyance.

Considering first the aspect of total energy, we may take, for example, the various indices of aircraft noise disturbance, which apply to a situation in which packets of energy are delivered to the ear each time an aircraft passes. On a total energy basis these packets would be numerically additive and the total energy level would be given by the sound level of each overflight plus 10 times the logarithm of the number of overflights. In the British noise and number index (NNI) (Chapter 13), the coefficient selected was in fact 15, but in other work, coefficients of 6 to 15 for different kinds of annoyance have been used (Meister, 1968) and in fact, as we noted above, all the main indices of annoyance by aircraft noise use the same basic form: sound level plus 10 to 15 times the logarithm of the number of occurrences. The use of total energy does not violate known characteristics of annoyance or disturbance possessed by different frequencies of sound. This factor is accommodated by using weighted sound energy, the weighting being appropriate to the sound and purpose of the measurement. Noise immission level is measured in sound level A or a statistical derivative of it, while the aircraft noise measures use PNdB or EPNdB, the latter also making allowance for the duration of individual overflights and spectral irregularities. The other component, that of fluctuation of sound level, Robinson derives from the data particularly of Griffiths and Langdon (1968) and of Pearsons (1966).

The two components in the concept of noise pollution level appear as follows

$$\text{Noise pollution level } (L_{\text{NP}}) = L_{\text{eq}} + 2\cdot56\sigma$$

where L_{eq} = energy mean level over period
     σ = standard deviation of sound level fluctuations during the same period.

In practice $2\cdot56\sigma$, for Gaussian distribution of sound level as a function of time, happens in fact to equal $L_{10} - L_{90}$ but was set at this value after consideration of the data of Griffiths and Langdon (1968) on time fluctuations of noise; and of Pearsons (1966) on the relation of intensity to duration, in the causation of annoyance.

The particular weighting of the sound levels can be chosen appropriately, as dB(A), PNdB or possibly other variants, as the conditions suggest. The relation of dB(A) to PNdB is such that for most noises, $L_{\text{NP(A)}} = L_{\text{NP(PN)}} - 13$. For detailed numerical comparisons of noise pollution level with other scales of disturbance, Robinson's (1971) paper should be consulted, along with that of Bottom and Waters (1971).

We shall return again to the consideration of actual levels for acceptability for traffic noise for urban planning at the end of this chapter.

Road planning and traffic aspects

The design of major urban roads must include consideration of the noise they generate as a factor of prime importance. Failure to do this may necessitate the evacuation of residential housing in the immediate proximity of these roads. The design criteria are well enough understood to avoid such occurrences; all that is necessary is that advice on acoustic matters should be taken and heeded. The experience in Britain of urban motorways has come rather later than in some other countries, but has given rise to a useful body of scientific fact on which to draw. A recent symposium on 'Road and environmental planning and the reduction of noise' is available in volume 15 (1971) of the *Journal of Sound and Vibration*. In this collection of papers, Burt (1971) discusses methods of reducing traffic noise from various aspects of design and utilisation. Dennington (1971) deals with urban road construction as it affects noise, and Bor (1971) considers urban transport in general as a broad environmental problem. Sound insulation standards for buildings adjacent to urban motorways are discussed by Hardy and Lewis (1971). A highly practical guide to the design aspect of urban motor roads is given in *Urban design bulletin* No. 1 (Greater London Council, 1970). This publication uses the noise measure L_{10} (noise levels in dB(A) exceeded for 10% of the time) but it may be that an index of wider usefulness, such as TNI or especially L_{NP}, would be preferable in the future. This view is shared by the authors of an admirable review of road traffic noise (Road Research Laboratory, 1970) which merits close study. The coverage of the problem is wide but compact and an excellent bibliography is provided. In the Greater London Council (1970) publication is to be found a summary of information of immediate usefulness for assessing noise conditions in major urban motor roads, and guidance for their design. Local small-scale problems are not neglected; for example, the possible situation of a projected house near an urban road. Subjects include noise measurement and standards, the control of traffic noise from major roads by planning; the control of traffic noise in new building developments, estimation of noise reduction. For this purpose transparencies are included to assist in the graphic solution of specific questions on noise propagation from roads.

Reduction of road vehicle noise

Road vehicle noise can be subdivided conveniently into (1) roadside noise and (2) noise inside vehicles experienced by the occupants.

Roadside noise

RESTRICTION BY LEGISLATION   The subjective rating of motor vehicle noise has been studied by Mills and Robinson (1963) and standards for measurement of this noise have been issued (BSI, 1961, amended 1963 and 1966; ISO, 1964) for the measurement of motor vehicle noise. The Committee on the Problem of Noise (1963) on the basis of Mills and Robinson's findings recommended that, using the measurement procedure of the British Standards Institution (1961), BS 3425, which consists of an acceleration test at full throttle from a stated running condition, vehicles should be constructed so as not to exceed 85 dB(A), except motor-cycles and other two-wheeled mechanically propelled vehicles, which should not exceed 90 dB(A). For roadside checks maximum levels 3 dB above these levels were suggested. As a simplification of road vehicle noise measurement, Denby (1967) for control purposes, has suggested a less demanding technique. The proposals of the Committee were in fact relaxed for legislative purposes, and at the present time the legal position is as follows.

After preliminary circulation the following regulations became law. New vehicles of current manufacture, since April 1970, are required to be constructed so as not to exceed the following levels, measured as described in British Standards BS 3539: 1962 and BS 3425: 1961 (amended 1963 and 1966). The levels are as follows:

(i) Motor-cycles, cylinder capacity of the engine not exceeding 50 cubic centimetres	77 dB(A)
(ii) Motor-cycles, cylinder capacity of the engine exceeding 125 cubic centimetres	86 dB(A)
(iii) Any other motor-cycle	82 dB(A)
(iv) Goods vehicles of gross weights more than $3\frac{1}{2}$ tons, land tractor, works truck or a passenger vehicle for carrying not less than 14 occupants	89 dB(A)
(v) Passenger cars or other light motor vehicles	84 or 85 dB(A) according to definition.

Where levels are measured in roadside checks, a tolerance of $+3$ dB is allowed on the above levels.

At the present time, proposed new regulations are being circulated by the UK Secretary of State for the Environment for comment, as a preliminary to the formation of new requirements in respect of noise, smoke emission, plating specifications and power/weight ratio. The noise regulations now proposed contain some divergencies from the above (1970) regulations. The levels which vehicles must be manufactured not to exceed are:

Class or description of vehicle	Maximum level dB(A)
1. Two-wheeled motor-cycle (with or without sidecar attached) of which the cylinder capacity of the engine:	
(a) does not exceed 50 cubic centimetres	77
(b) exceeds 50 cubic centimetres but does not exceed 125 cubic centimetres	82
(c) exceeds 125 cubic centimetres but does not exceed 500 cubic centimetres	84

(d) exceeds 500 cubic centimetres	.		.	.	.	.		86
2. Passenger car, as defined		.	.	.	.	.	.	80
3. Light goods vehicles or light passenger vehicle .				.		.		82

4. Motor tractor, the weight of which unladen does not exceed
$1\frac{1}{2}$ tons 82
5. Any other vehicle not elsewhere classified or described in this
column of this Table of which the engine is:
(a) of not more than 200 brake horse-power . . . 86
(b) of more than 200 brake horse-power 89*

---

* No legislative action has occurred on these proposals at the time of writing.

RESTRICTION BY DESIGN OF VEHICLES The Road Research Laboratory (1970) has measured the various sources of vehicle noise, as has Priede (1967, 1971), and Waters, Lalor and Priede (1970). It should be realised that, at the present time, the major source of annoyance, that is the one affecting the greatest number of people in developed countries, is the internal-combustion reciprocating-engined road vehicle, primarily the heavy diesel-engined truck. Legislative pressures on motor vehicle manufacturers to increase safety and reduce exhaust emission of combustion products have had beneficial effects, and increasing pressure to reduce noise may likewise in the not far distant future markedly influence design particularly of engines. In the meantime, however, the tendency is still for diesel-engined truck noise to increase.

Cause of vehicle noise

Priede (1971) classifies the sources as follows: (a) engine, transmission and accessories; (b) road excitation; (c) air buffeting. At normal speeds (c) is relatively unimportant, leaving (a) and (b).

These noise sources are first, the outer surfaces of the vehicle structure, vibrating due to forces directly transmitted through the structure from engine or road sources, or in response to airborne noise from the engine; and second, direct radiation from engine exhaust, inlet, engine itself, transmission, cooling fan, tyre noise. These sources are usually able to radiate freely through the various paths: downwards, forwards and sideways through the gaps round the engine. Priede (1971) summarises these sources conveniently in Table 9.5. This clearly demonstrates the distinction which must be drawn between noise inside the vehicle cab and that heard at the kerbside.

TABLE 9.5

Origin of noise	Noise inside the vehicle	Noise outside the vehicle
Engine vibration	Major source of low frequency noise	Not important
Engine airborne noise and its transmission	Major source of high frequency noise	Major source of high frequency noise
Engine exhaust	Not important	Major source of low frequency noise
Engine inlet	Not important	Major source of low frequency noise following exhaust
Fan noise	May be noticeable	Can be significant in low and middle frequency ranges
Road-excited vibration	Major source of low frequency noise	Not significant
Road-excited tyre noise	Not significant	Significant

It is not our immediate purpose to study the engineering acoustics of road vehicles except in so far as it may shed light on possibilities of noise reduction. As a supplement to Table 9.5, however, it is instructive to observe the quantitative contribution of different sources, in an experiment involving selective reduction of individual sources, in the kerbside noise of a 10-ton truck (Fig. 9.3). The noise is high at the low frequencies, diminishing with increase of frequency at a rate of about 6 dB/octave. In the selective reduction or removal of noise in this particular vehicle, additional exhaust silencing gave a reduction of 10–15 dB in the frequency range up to 250 Hz. Engine inlet silencing gave a lesser reduction in the frequency range 50–500 Hz. Engine combustion noise was reduced by measuring the total noise with engine motored (i.e. no power being delivered). The reduction of noise was marked in a broad band in the audible higher frequencies around 1000 Hz. The importance of combustion noise radiated from the engine is seen in Fig. 9.3, and this is the component which is least amenable to improvement within the current technical restraints imposed by the economics of vehicle design. This factor of direct engine-radiated noise merits further attention.

ENGINE NOISE   The extensive and illuminating researches of Priede over a period of years, on the noise of automotive diesel engines is conveniently collected by Waters, Lalor and Priede (1970). Summarised, this paper confirms that the external noise of diesel-engined commercial vehicles is determined by engine speed and not road speed. The engine noise radiated by the structure of the engine, measured by the ISO test or the essentially similar BS 3425, has a level of 85 to 91 dB(A) in typical vehicles. All the other sources of noise together, i.e. engine intake, exhaust, cooling fan, gearbox and other noise sources, contribute a similar amount. The engine

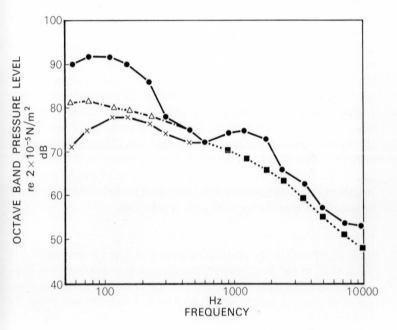

9.3 Kerbside noise of a 10-ton truck. The different spectra represent states of modifications of different components. Symbols: ●—●, unmodified; △–·–△, exhaust silenced; ×—×, inlet silenced; ■----■, engine airborne noise reduced. (After Priede, 1971.)

noise is principally determined by the cylinder bore, engine speed, and combustion system. Using these parameters Priede has derived an empirical formula which predicts engine noise levels to within ±2 dB(A). The sound level (A-scale) of an engine is predicted as follows:

$$dB(A) = D \log_{10} N + 50 \log_{10} B - (3 \cdot 5 D - 73 \cdot 5)$$

where N = rev/min

B = bore in inches

D = predominant slope of cylinder pressure spectrum in dB per ten-fold increase in frequency. For a normally aspirated engine $D = 30$, for a turbo-charged engine, $D = 50$.

Study of a variety of diesel engines, with in-line or V-formation of cylinders, 4, 6 or 8 cylinders, dry or wet liners, crankcases and sumps of different designs, and with different numbers of main bearings, and design of cylinder heads and rocker covers, are shown to have only secondary effects on the sound level A of the noise. Although the quality of the noise may vary, these differences can be accommodated in Priede's equation. The problem of quieting large diesel-engined commercial vehicles is primarily one of engine noise, i.e. noise radiated from the engine structure. Waters, Lalor and Priede (1970) note that in vehicles with engines of 6 to 12 litres displacement in both in-line 6 and V-8 forms, the engine noise itself, in the ISO test is about 88–91 dB(A). Thus in view of the 89 dB(A) actual or proposed legislative noise levels, the engine noise is a limiting factor in attempted compliance with such regulations.

REDUCTION OF DIESEL VEHICLE NOISE   Priede (1971) sums up the lines of promise in the possible reduction of diesel engine noise. He notes that, in the diesel engine, the difference in noise output from no-load to full-load is very small. Fig. 9.4 shows the difference compared to a petrol engine, which is quiet at low loads and relatively noisy at high loads. Thus high loading is not a source of extra noise. Inlet and exhaust, and fan noise are amenable to reduction, so that the possibilities for a quieter type of vehicle of conventional engine design are suggested by Priede as follows.

(1) Design for adequate attenuation of engine noise
(2) Suitable engine design, in terms of
    (i) limited maximum speed
    (ii) limitation of energy capacity
    (iii) increase of engine load
(3) Design for an engine structure which will have characteristics giving less radiated noise.

Priede believes that these factors should not be prejudicial to cost or engine performance or operation. Exhaust emission may, he sug-

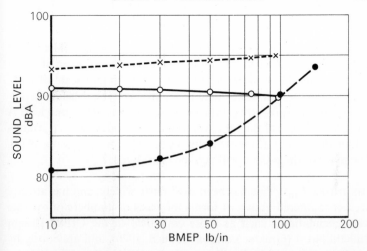

9.4   The effect of load, as indicated by the brake mean effective pressure (BMEP), on sound output for a diesel and a petrol engine. Symbols: ×——×, 9·6 litre diesel engine at 1000 rev/min; ○—○, 2 litre diesel engine at 2000 rev/min; ●——●, 3 litre petrol engine at 2000 rev/min. Note the lack of effect on the diesel engine noise of load variations, unlike the petrol engine, which becomes much noisier with increased load. (After Priede, 1971.)

gests, be reduced by high pressure-charging. In all, he feels that a reduction of 10–15 dB(A) in the noise of heavy vehicles is a practical possibility in the near future.

OTHER TYPES OF POWER PLANT   The present recognition of the unpleasant facts of atmospheric pollution and of noise, may be expected to stimulate design of engines with reduced exhaust emission and noise. The means for the latter are stated above, for conventional engines. Other prime movers exist. The possible revival of steam propulsion for exhaust emission reduction have been receiving some attention. This might be expected, even in reciprocating form, to be quieter than the internal combustion powerplant. The diesel version of the Wankel ('cottage-loaf') engine may have different noise characteristics in view of the different shape of the combustion spaces. The advent of the gas-turbine heavy vehicle completely alters the nature of the acoustic source, and although its characteristics are not known to the writer at this time, it is believed that the large gas-turbine driven truck will yield noise levels of not more than 85 dB(A) under ISO test conditions.

Restriction by monitoring of traffic noise

This is another aspect of legislation. Having limited vehicle noise and used proper road planning methods, the success of these measures should be monitored by some meaningful scale of noise measurement for the assessment of annoyance. The measures available for this purpose are (1) the index L_{10}, that is, the sound level A exceeded for 10% of the time; (2) traffic noise index (TNI) (Griffiths & Langdon, 1968); (3) noise pollution level (L_{NP}) (Robinson, 1970). The place of these three will now be considered.

THE INDEX L_{10} This has been used fairly widely, and has already been encountered above. It forms the basis of numbers of existing recommendations, such as those embodied in the Urban Design Bulletin No. 1 (Greater London Council, 1970), and previously by the Committee on the Problem of Noise (1963). Recently, Scholes and Sargent (1971) have discussed at some length the status of indices of annoyance originating in traffic noise. They believe that complex units, such as TNI or L_{NP}, are necessary to take account of both the continuous level and the variability of a noise, in quantifying annoyance, or dissatisfaction. They feel, however, that before such units can be fully utilised for setting standards, more information on the characteristics of traffic noise patterns is still needed. Until this is available, they favour the use of a simpler scale, and the one they suggest is the L_{10} value at the façade of buildings, between the hours of 06.00 and 24.00. They do emphasise, however, that their recommendations refer to motorway or other free-flowing traffic, and that for congested traffic or urban conditions, the use of an index of noise exposure which takes into account the factor of variability, is desirable. The authors review the attributes of the measures at present immediately available for traffic noise, viz. L_{eq}, L_{10}, TNI and L_{NP}. They point out the importance of variability of the noise, with time, in determining the method of measurement. High correlations of L_{eq} with subjective reaction are possible with steady noise climate but are not found with variable noise; TNI and L_{NP}, on the other hand correlate very well with subjective reaction for noise of variable nature. Design rules for these two indices have not yet been worked out, and accordingly for the present Scholes and Sargent suggest that for traffic noise, the noise exposure of dwelling houses should be the average value of L_{10} measured 1 m from the façade of the building in a weekday period 06.00 to 24.00 hours. For this purpose, the average value of L_{10} is defined as the mean of hourly samples of the noise over a short duration covering the passage of 50, and preferably of 100 vehicles. This sample might occupy from 1 to 12 min depending on traffic density. Estimates have been made of the noise levels from traffic to which houses in which people live are subjected in the UK (Road Research Laboratory, 1970) and it appears that 19% of the urban population occupy houses subject to L_{10} values, as defined for daytime above, of 70 dB(A) or more, equivalent to about 74 TNI or 72 L_{NP}. Scholes and Sargent suggest this L_{10} value as a basis for discussion on the grounds that 74 TNI is compatible with a reasonable measure of acceptability or at least toleration. They also include a discussion of the various factors in traffic noise and possibilities of their attenuation.

INDICES TNI AND L_{NP} The limitations of a simple statement of noise levels have been shown to be inadequate in the case of fluctuating noise, as we have already noted. Two indices already mentioned are candidates for use in the measurement of road traffic noise. They are the traffic noise index TNI and noise pollution level L_{NP}.

Griffiths and Langdon (1968) and Langdon and Scholes (1968) describe quantification of the subjective reaction to road traffic noise by means of TNI. This index takes into account the maximum levels and the background against which they are heard. A 6-point scale of dissatisfaction, derived from the opinions of 1400 residents exposed to road traffic noise obtained in the course of the London Noise Survey (Building Research Station, 1968), was correlated with derivatives of the noise measurements, using various combinations of L_{10} and L_{90}, in dB(A). Of these, the index giving the highest correlation was $L_{10} - 0.75 L_{90}$, and this was designated the traffic noise index. The most convenient expression for this is:

$$TNI = 4(L_{10} - L_{90}) + L_{90} - 30$$

The subtraction of 30 dB is only to give convenient values. Langdon and Scholes discuss the factors governing the choice of a quantitative noise limit for controlling traffic noise. The primary aim is to specify an acoustical limit consistent with absence of disturbance, and general acceptability. The attainment of an ideal environment would impose such restrictions that it would be impractical, and so a compromise, as is common in such decisions, is necessary. They have selected the level at which less than half of the residents profess to be suffering from any of a number of annoyances consequent upon the noise. It would not be opportune to discuss here the details of the construction of subjective scales and the information to be derived from them, and for this purpose, the original papers should be consulted. Suffice to say that the originators have chosen 47 TNI as indicating a reasonable standard of acceptability of outdoor noise in urban areas. At this level they state that there is only one chance in 40 that the average person living in the vicinity of main road traffic is likely to be dissatisfied.

Practical aspects of the use of TNI are discussed under the headings of distance, sound insulation of buildings, effects of barriers, effect of traffic volume, and again the original papers should be consulted for detail. It should be noted that TNI refers to outdoor noise, and in fact, the London Noise Survey data on which it was based were slightly increased (by about 3 dB) compared to free-field measures, due to the fact that the microphone was about 1 m from the face of buildings. For the derivation of TNI, values derived from free-field noise measurements may be used, but the TNI values should be increased by 3 units. Again, where TNI values are being predicted at different distances from the noise at a given point, allowance must be made for the different attenuation rates of the L_{90} and L_{10} values of traffic noise. The result is that TNI decays at a rate of 15 TNI per doubling of distance, except for very short distances from the roadway; the minimum distance from which predictions should be made is 7 m. The TNI in summary, is an index, capable of giving high correlations with a scale of disturbance, and designed to accommodate fluctuating noise. The noise pollution level (L_{NP}) of Robinson (1969, 1971) has been evolved with the express purpose of attempting to produce an index which will accommodate as many as possible of the recognised sources of annoyance, such as road traffic noise, aircraft noise, and possibly other components, such as railway noise, either singly or in combination. Its basis has already been explained, and here only the application need be considered.

Robinson's formula, it will be recalled, recognises two components; first, the equivalent continuous frequency-weighted sound level on an energy basis; and second, the fluctuation of the noise with time. He has examined the effectiveness of L_{NP} for traffic and for aircraft noise, against other indices, in various studies (McKennell, 1963; Griffiths & Langdon, 1968; Kryter & Pearsons, 1963; Pearsons, 1966; Coblentz, Xydias & Alexandre, 1967).

The L_{NP} values are shown by Robinson to express the disturbance of traffic noise as well as TNI. For aircraft noise, L_{NP} is concordant with NNI, and agrees qualitatively with the data of Coblentz *et al.* In addition, it does not possess the limitation of the assumed ambient level of 80 dB(PN) of NNI. It also predicts satisfactorily the relation between duration and level found in laboratory tests.

Robinson expresses the hope that, with further investigation and application, the index may be developed into a noise rating system which would have the highly desirable, and now actually necessary attributes of quantifying disturbance from numbers of different sources, simultaneously if necessary. A very satisfactory start to the practical investigation of the utility of noise pollution level (L_{NP}) has been made by Bottom and Waters (1971), who carried out a social survey in which dissatisfaction was correlated with aircraft and road traffic in various relative concentrations. They have shown that dissatisfaction (numerically scored) is dependent on both mean level and variability of noise, the concept on which Robinson's L_{NP} measure is based. Further, disturbance by aircraft noise is inversely related to the level of traffic noise. The mean dissatisfaction score due to simultaneous aircraft and road traffic noise indicates that increasing road traffic increases dissatisfaction at low aircraft noise (rated by NNI) but at high NNI values increasing road traffic noise reduces dissatisfaction. The mean dissatisfaction score plotted against L_{NP} in 9 cells showed a correlation coefficient of $+0.96$, a highly satisfactory result. Bottom and Waters felt that further studies should be made to continue the validation of the noise pollution level concept.

In view of the exploratory stage at present reached in the case of L_{NP} Robinson (1971) only advances tentative suggestions for the use of this index in the control of noise disturbance. He offers the value of $L_{NP(A)}$ (i.e. calculated from sound level A) = 72 or $L_{NP(PN)}$ (calculated from perceived noise) = 85. These are approximately equivalent to NNI = 38, in all cases for outdoor noise.

At this stage in this complex field, it is difficult to offer a short critical judgement. However the conclusion is inescapable that a great advance would be achieved if the uniformity and wide application conferred by the concept of noise pollution level could be realised.

Noise inside vehicles

PRIVATE CARS   Private passenger vehicles have in general been developed to provide a quiet interior, the degree of noise limitation depending on the particular section of the motoring public for which the vehicle is designed. High-performance cars may tend to provide higher interior noise levels than luxury vehicles in which quietness is a primary aim. In general, however, passenger cars have, to varying degree, been refined to give acceptable noise levels except at the lowest frequencies in some cases (Evans & Tempest, 1972). However, insulation of the passenger compartment is not related to the radiation of noise to the exterior, which may be considerable.

COMMERCIAL VEHICLES   Experience would suggest that acceptable noise conditions are attained in public service passenger vehicles. The driving compartments of diesel-engined trucks and commercial vehicles generally have been investigated as noted above by Priede (1967, 1971). He measured the driving compartment or driving cab noise in 15 types of commercial (goods) vehicles. The vehicles were driven at maximum rated engine speeds and the noise varied over a range of about 15 dB from the quietest to the noisiest vehicle. No relation was obtained in this series between the interior noise and vehicle size, engine capacity, engine speed or engine noise. The quality of the cab design was the most important factor, and some manufacturers were found to have been notably successful in the application of sound insulation principles while others did not provide acceptable interior noise levels. Increase of interior noise, mainly as a result of misuse, could however occur. The causes were disturbance of sound insulation, imperfections of acoustic sealing of engine cowls, or loss of sealing devices such as grommets between the vehicle interior and engine compartments.

The spectra of the noises in the various driving compartments were similar. The highest components were in the low frequencies and the spectrum sloped downwards at a gradient of 18 to 25 dB per 10-fold increase in frequency. Priede discusses the origin of this type of spectrum. In the low-frequency range, engine vibration is the source of low-frequency noise, modified by the structural features and vibration paths peculiar to the vehicle. Priede found the high-frequency noise was determined by the characteristics of the engine noise and the degree of sound attenuation between the engine compartment and the interior of the driving compartment.

The low-frequency noise was found to vary in an irregular way with vehicle speed, due to the resonance characteristics of the vehicle structure. The highest frequency noise was determined by the engine speed very predictably; above about 500 Hz, Priede found increments of about 11 dB per doubling of engine rev/min. This relation was true for either top or third gear, demonstrating that engine revolution rate and not road speed was the determining factor; and in fact the same relation between revolution rate and noise level could be demonstrated with the engine mounted on a test bed.

RISK TO HEARING IN COMMERCIAL VEHICLE DRIVING COMPARTMENTS Priede considers the possible effect on hearing of driving commercial vehicles. He found that about one third of his sample of 15 vehicle types, at the maximum rated engine rev/min, did not exceed the NR 85 contour. This contour is about the level at which small losses of hearing will occur after some years of exposure for about 8 hours per day in persons of average susceptibility to noise-induced hearing loss. This is discussed in detail in Chapter 11. The remaining two thirds of the vehicles examined exceeded this value, driven at maximum rated engine speeds, which provide the noisiest condition. If these levels were maintained for 8 hours per day, varying degrees of hearing loss, some significant, would be sustained. However, road vehicles do not normally operate at maximum rated engine speed, and these speed reductions would tend to reduce the noise exposure; but it must be remembered that speed reduction diminishes noise less in diesel than in petrol engines. In any case, both motorway driving and hill-climbing would be expected to maintain high levels of noise and to reduce quieter intervals. The driving compartments with the highest levels were excessively noisy, both on grounds of potential damage to hearing and possible reduction of efficiency in the performance of the driving task. The highest levels reached about 100 dB(A) or NR 95. This is unduly high even accepting the variable level of the noise, and compares most unfavourably, for example with the maximum levels of noise regarded as acceptable by British Rail for the cabs of locomotives, which will be mentioned below.

Noise in other surface transport

In other types of transport, the interior noise should be compatible with comfort, the retention of performance and efficiency by the driver, and freedom from risk of deafness. In rail transport, the authorities have shown themselves to be conscious of the need to consider the interests of locomotive crews and the appeal of quietness, freedom from vibration and good environmental conditions generally for the passenger. Design limits for noise have been employed both for passenger stock and for locomotives. British Rail has for the present stipulated the following maxima for various interiors of rolling stock (Koffman, 1967).

Luxury and sleeper coaches	NR 60 = 65 dB(A)
Dining and long distance coaches	NR 65 = 70 dB(A)
Suburban stock	NR 70 = 75 dB(A)

Similarly British Rail has specified a maximum permissible locomotive cab noise for design purposes of NR 75 = 80 dB(A) (Koffman & Jeffs, 1967). This is about 20 dB less than the levels in the noisiest road vehicles measured by Priede (1967). Such a level is compatible with freedom from risk to hearing, from loss of efficiency due to noise, and still permits the engine crew to be aware of any malfunctioning of their locomotive which might be indicated by a change in its characteristic sounds.

Interference with performance and efficiency

Enough has been said in the previous discussion of this topic to indicate the lines which must be pursued. These adverse effects have been noted at levels above about 90 dB(A), and the higher frequencies are the more liable to cause loss of efficiency. The level at which appreciable risk to hearing exists is thus similar to, or even below that capable of causing loss of efficiency. Thus any sound environment which is controlled, as it should be, to protect those working in it from hearing damage, may be expected at the same time not to impair efficiency. It is also fortunate that sound reduction treatments will, in general, be more effective at high than at low frequencies, and the experience of Broadbent and Little (1960) suggests that quietening procedures on rooms or equipment may show real savings in the shape of diminished error and wastage in processes.

Interference with communications

The quantitative problems in this field occur in a variety of ways. An existing situation may have proved unsatisfactory for voice or telephone conversation, or a completely new situation must be designed from the outset to provide a suitable environment for direct or telephone speech to be understood at reasonably low levels. There are a number of ways in which the intelligibility of speech of a given acoustical quality can be assessed. Beranek (1960) evolved a system in which a simple arithmetical average of the SPL in dB in the three octaves, 600–1200, 1200–2400 and 2400–4800 Hz, is obtained and the resultant value, designated *speech interference level* (SIL), is compared to a table of values which indicates the maximum distance at which speech at different levels is of specified intelligibility. This method has been used extensively but many analysers will not now provide the octave bands from which the speech interference level is derived and a revision is now available (Table 9.7). In addition, the use of the proposed ISO NR curves, as noted in the last section, was also envisaged for the specification of speech intelligibility, and this application has also been used.

Other methods of assessment of speech communication exist and a measure widely used is the *articulation index* (Kryter, 1962; Kryter, Licklider, Webster & Hawley, 1963). This is derived from the SPL of the speech and of the noise, and the acoustic properties of the former must be appreciated. Speech is of course a discontinuous sound with marked variation in level. For the purpose of deriving the articulation index the maximum values of the speech level are used. Various methods, of different degrees of complexity, are available, and perhaps the simplest is the so-called weighted octave method (Kryter, Licklider, Webster & Hawley, 1963) using five octaves. Broadbent (1966) has suggested a modification of this method which uses ISO octave frequency bands and is a simple and quick way of making the calculation. His method is as follows.

The frequency spectrum of the speech and of the noise is derived in each case from the SPL in the octaves centred at 250, 500, 1000, 2000 and 4000 Hz; the highest values of SPL of the speech sounds are to be used. The difference between the level of the speech sound and the noise in dB, is recorded for each octave. If the speech level is greater than that of the noise by more than 30 dB, it is recorded as 30 dB. If the noise level is greater than the speech level, the difference is recorded as zero. The decibel difference values for each octave are multiplied by a factor which is different for each octave, and the resulting values added together. This summation will give a figure between zero and one. The factors for each octave are given in Table 9.6.

TABLE 9.6

Calculation of articulation index

Octave centre frequency Hz	Contribution to articulation index
250	$(S - N) \times \dfrac{3}{1200}$
500	$(S - N) \times \dfrac{6}{1200}$
1000	$(S - N) \times \dfrac{9}{1200}$
2000	$(S - N) \times \dfrac{12}{1200}$
4000	$(S - N) \times \dfrac{10}{1200}$

S = speech peak level (dB) in a particular octave
N = noise level (dB) in the same octave

Both values should be expressed as octave sound pressure levels.

The relative importance of the octaves is indicated by the size of the factor used. The articulation index, derived by the method of Table 9.6, should be used on the assumption that the following conditions are observed. The first is that the listener does not see the talker's mouth; if the mouth is visible, enough lip-reading occurs to raise the articulation index by 0·1. The second condition would not normally apply to direct speech, but to telephone or other amplified speech; it is that the speech shall not consist of narrow bands of frequencies. If it does, other methods of calculation

(Kryter, Licklider, Webster & Hawley, 1963) must be used. The third condition is that the noise is continuous. Interruption of the sound will normally increase the intelligibility and with it the articulation index.

Having obtained the value for the articulation index, it may then be used to indicate the probable level of intelligibility. If all octaves have a difference between the speech peak level and the noise level of 30 dB, the articulation index would be 1·0, and all words would be heard correctly even from an unrestricted vocabulary, assuming that the listener has unimpaired hearing. An index of 0·5 will mean that about 75% of isolated, phonetically balanced (PB) words will be correctly understood, which may be taken as meaning that ordinary sentence conversation can usually be understood, but understanding will depend on the listener and on the talker. If it is necessary for 90% of isolated words to be heard, the index will have to go up to 0·75. The role of vocabulary size must however be remembered. If the vocabulary is restricted to 32 definite words, 90% correct response can be obtained with an articulation index of only 0·2. For further details of interpretation, Kryter, Licklider, Webster and Hawley (1963) should be consulted.

Webster (1965) has summed up a long series of his researches on speech in noise. He concludes that a version of the speech interference level is a very good index of noise from the viewpoint of prediction of speech interference in communication. The particular SIL he employs is that derived from the arithmetical mean of the overall SPL value of the octaves centred at 500, 1000 and 2000 Hz; this version, since it employs the preferred frequencies, is usually termed *preferred speech interference level* (PSIL). Three communication situations are examined: face-to-face speech, sound-powered telephone, and amplified speech. Only the first will be mentioned further here. Face-to-face conversation in noise is limited by the distance between the listener and the talker. To maintain the same level of intelligibility, each doubling of distance requires that the level of noise be reduced 6 dB. The noise on which Webster's findings are based is a thermally generated white noise falling in intensity with ascending frequency at a rate of 6 dB per octave. It is comparable to noises used in other investigations of speech intelligibility, including office noise (Beranek, 1956, 1957), ship noises (Klumpp & Webster, 1963) and in laboratory studies (Kryter, 1946). Such a noise is therefore a good practical compromise for

indicating the general effects of realistic situations. The values for
PSIL in Table 9.7 (Webster, 1969, personal communication) based
on the octaves centred at 500, 1000 and 2000 Hz indicate an articulation index of about 0·4 (Beranek, 1971, p. 559). In Webster's data,
lip-reading can be considered not to have influenced intelligibility.

TABLE 9.7

*Preferred speech interference levels of noise that just permit conversation with marginal
reliability at the distances and voice levels indicated*

Distance from listener		PSIL dB*			
m	ft	Normal voice	Raised voice	Very loud voice	Shouting
0·15	0·5	74	80	86	92
0·3	1	68	74	80	86
0·6	2	62	68	74	80
0·9	3	58	64	70	76
1·2	4	56	62	68	74
1·5	5	54	60	66	72
1·8	6	52	58	64	70
3·6	12	46	52	58	64

* Average SPL in dB of noise in the octave bands centred at 500, 1000 and 2000 Hz.

This SIL is thus a very simple and effective method of assessing a
noise situation. Only the SPL values in the three octaves need be
measured, averaged and compared with the values in Table 9.7.

Having assessed a situation by the methods suggested, various
devices are available which can improve intelligibility. Obviously,
acoustic means of increasing the difference between the level of the
speech and of the noise can be exploited by reducing the noise level.
For direct speech this must be done by reduction at source. For
telephone purposes exclusion of noise by telephone booths, or
partial exclusion by hoods over telephone positions is useful. The
amount of improvement possible by increasing the speech level is
limited by an overloading effect on the ear which merely makes the
speech sound louder without increasing intelligibility. The effect on
speech intelligibility of wearing ear protection will be discussed in
Chapter 12. Other means of improving speech communication are
possible, without resorting to acoustic means. The most obvious
is to restrict the vocabulary. Here, if words from a set of 32 are used

only, 90% of words will be heard correctly when only 20% would be achieved with an unrestricted vocabulary. The choice of the best words is an important factor in devising communication procedures, and in general long words with many syllables, especially if they are in common use, are correctly heard more than short words. This principle is used in the identification of the letters of the alphabet by standard words: for example, in speech procedures employed by the International Civil Aviation Organisation (ICAO) and many other bodies, the word-spelling alphabet (Moser & Bell, 1955) shown in Table 9.8. is in standard use. It yields high intelligibility, even when used by talkers of different language backgrounds, and may be used in any circumstances for the identification of letters (e.g. call signs or aircraft identification) or to spell out words; the latter process is of necessity slow.

TABLE 9.8
International word-spelling alphabet

A	Alfa	N	November
B	Bravo	O	Oscar
C	Charlie	P	Papa
D	Delta	Q	Quebec
E	Echo	R	Romeo
F	Foxtrot	S	Sierra
G	Golf	T	Tango
H	Hotel	U	Uniform
I	India	V	Victor`
J	Juliet	W	Whisky
K	Kilo	X	X-ray
L	Lima	Y	`Yankee
M	Mike	Z	Zulu

Technical postscript

This book is not intended to provide guidance on the practical aspects of noise reduction in structures and machines. Some of the sources of such information have been cited and are obviously in specialised mechanical fields, and thus the concern of acoustical engineers. For building purposes, however, it is useful to note that valuable publications exist, including Chapter III of the British Standard Code of Practice, entitled 'Sound insulation and noise reduction.' It may be obtained separately from the British Standards Institution (BSI, 1960). In addition, Parkin and Humphreys (1969) and Bazley (1966) are standard sources of information in this field.

Future possibilities

From a long-term viewpoint, other types of prime mover may alleviate or at least alter the noise of road vehicles. A new factor already present is the Wankel engine, which employs a lobed rotor in a casing instead of cylinders and pistons, to produce external work from the combustion of the fuel. Improved storage batteries and electric drive could eliminate the type of noise now produced by petrol and diesel engines. Gas turbines produce different types of noise from that of piston engines. Thus road vehicle designers are likely to continue to face noise problems in the future, and the most attractive possibility is some form of electric drive.

10
Temporary effects of noise on hearing

We are now in a position to survey the more important known facts on changes in the hearing mechanism during, and following, stimulation by sound. We have already noted the various meanings of the word noise, and in the present connection we shall use it as signifying any sound, wanted or not, which may affect the performance of the hearing mechanism, as well as in the physical sense of unpitched sound composed of many frequencies (Chapter 3). Noise may not be devoid of use, since it may give essential information to the operator of a machine, for example, and although his hearing may be at risk as a result, the sound could not be classed as unwanted, since the safety of the machine, and even of the operator and of others, may depend upon the continual flow of information produced by the noise.

At this stage discussion will be helped by establishing the meaning of certain terms used in the description of the effect of noise on people. *Noise exposure* means the total sound stimulus received by the ear, or by the body as a whole. We recognise the implication of total stimulus by including two aspects, or dimensions, of the noise exposure; one is the physical characteristics of the sound and the other its duration, or time pattern. In many situations the precise specification of noise exposure, due to the complexities of the acoustic and time elements, is sufficiently difficult to make any simple and significant specification impossible.

The terms hearing level and hearing loss are used in this text in the sense of Davis, Hoople and Parrack (1958a, 1958b) whose definitions have greatly clarified this terminology; these terms have already been introduced in Chapters 6 and 7, respectively. Another term, *threshold shift*, is also defined by the same authors. It signifies a change in hearing level with respect to a previously measured level, in a particular person. It is assumed, unless otherwise stated, that the change is one of deterioration of hearing. Temporary (TTS) and permanent (PTS) threshold shifts occur. The meanings of these terms will be discussed later. Sometimes, when a truly permanent shift is not necessarily established, the term persistent may be used.

Categories of effects of noise on hearing

Noise may affect hearing in ways that are broadly divisible into three categories: temporary threshold shift; permanent threshold shift; and acoustic trauma. In that order they indicate in a general way the degree of severity of the noise exposure which caused them. Temporary threshold shift is a short term effect which may follow an exposure to noise, and as its name indicates, the elevation of the hearing level is reversible. The effects of a particular noise exposure in terms of temporary threshold shift are dependent on individual susceptibility. If, for any particular person and noise exposure, the limit of exposure compatible with recovery is exceeded, recovery from temporary threshold shift will not proceed to completion, but will effectively cease at some particular duration after the end of exposure. This residue is known as permanent threshold shift.

The term persistent threshold shift is used to denote the threshold shift remaining after at least 40 hours. The word permanent is reserved for conditions which may reasonably be supposed to have no possibility of further recovery, since some recovery of hearing may be found after a week-end away from noise (Atherley, 1964), and probably after. The measurement of hearing level after periods of months, as well as repeated audiometric examinations is desirable where a disability or alleged disability is being investigated. Finally, acoustic trauma is a condition of sudden aural damage resulting from short term intense exposure or even from one single exposure; explosive pressure rises are often responsible, and these may arise from fireworks (Ward & Glorig, 1961), small arms fire, gunfire or major explosions. In Denmark, the sale of fireworks is restricted because of the danger to hearing.

These effects, each detectable in man by an elevation, temporary or otherwise, of the hearing level, are presumably attributable to movements of the various vibrating parts of the ear. These moving parts are seldom, if ever, at rest. Mechanical activity is their normal condition. Sound apparently leaves an aftermath of diminished response, in the case of temporary threshold shift, which soon disappears. We can demonstrate this effect by the audiometer, or by electrophysiological means (Chapter 5), as well as by the subjective evidence of our own ears that the hearing has been dulled by noise. There is another effect, the singing or ringing in the ears, known as tinnitus, which is also an aftermath of noise exposure. This has been thought to be due to the discharge of nerve impulses in the fibres of the auditory nerve, a sort of irritative after-effect of the intense stimulation. But it is not properly understood and Kiang, Moxon and Levine (1970) offer a different explanation. These effects occur apparently in the ear mechanisms, not in the higher analytical levels of the brain. In the case of noise-induced damage, however, it is possible to see microscopically some of the results.

The precise nature of the changes induced in the various components of the organ of Corti by damaging sound stimulation is not clear; in any case, the variety of structures in the functional pathway, and the great variety of possible structural and metabolic changes, make simple explanations unlikely. This damage can be of various degrees, recognisable by the histologist; from minor changes particularly in the hair cells to obvious damage and even complete breakdown and disruption of the organ of Corti. People who have been near explosions or even very intense noise may sustain ruptured ear drums and damaged ossicles; this middle ear damage may be associated with the damage, referred to above, to the cochlear structures. In the present context, we are most concerned with the effect of less dramatic noisy episodes, or of long-continued noise, and such noise mainly involves the cochlea. Particularly are we concerned with occupational hearing loss, a long-term cumulative effect leading to permanent changes for the worse in the hearing of persons subjected to noise in the course of their work over long periods, of months or years, perhaps throughout the entire working day. Before considering occupational hearing loss, the characteristics of temporary threshold shift can profitably be examined.

Temporary threshold shift

This component of the total effect of noise on the ear is virtually always present to some extent during and following exposure to noise, irrespective of whether permanent threshold shifts ensue. Almost any sound stimulus applied to the ear will result in an alteration in the hearing level and this may last, depending on the nature of the exposure in any given individual, for a period which may be measured in seconds, hours, days or apparently even months after the cessation of the sound. Throughout these widely separated values of duration, however, it seems unlikely that the same mechanism underlies the threshold changes and their disappearance. The

characteristics of temporary threshold shift are of great fundamental, as well as practical, interest, and have been extensively
studied in man. Some of the important relevant publications are
those of Ewing and Littler (1935), Hood (1950), Davis, Morgan,
Hawkins, Galambos and Smith (1950), Hirsh and Bilger (1955),
Spieth and Trittipoe (1958), Ward, Glorig and Sklar (1959a), and
Ward, Selters and Glorig (1961). The earlier work has been reviewed
by Hinchcliffe (1957), and in a later review by Ward (1963). There
are various other publications which will be cited in the text. Such
work has demonstrated that relations exist between the frequency,
intensity and duration of the sound to which the ear has been
previously exposed, and the degree of temporary threshold shift,
the audiometric frequencies at which shifts occur, and their course
of development and recovery. The relations between temporary
and permanent threshold shifts have commanded attention for
many years, because of the desirability, mentioned above, of
finding some predictive method for indicating the possible long-
term effects of noise on hearing. The subject is profoundly complicated by interaction of the various factors mentioned above, by the
action of the aural reflex (Chapter 5) and by other effects such as
individual susceptibility to temporary threshold shift.

These various aspects require individual discussion.

Relations of exposure frequency to frequency of temporary threshold shift

The usual procedure in laboratory experiments on temporary
threshold shift is to determine the resting threshold of the subject, to
expose the ear to a known sound (in terms of characteristics of the
sound and the duration of its application) which will be called the
exposure sound, and then to follow the threshold usually by means
of Békésy-type audiometry in the period immediately following the
exposure sound. The temporary threshold shift is the difference, in
dB, between the resting threshold and the post-exposure threshold.
No statement of temporary threshold shift is adequate without an
indication of the duration of the interval between the cessation of
the exposure sound and the measurement of the temporary threshold
shift. Earlier work (Caussé & Chavasse, 1947) had shown that for
low-level exposures the maximum threshold shift occurred at the
exposure frequency. Hood (1950) confirmed this finding, but like

Davis *et al.* (1950), found also that high level exposure yielded maximum threshold shifts at a frequency about one half, or more, of an octave above the frequency of the exposure tone or frequency band of noise. Hirsh and Bilger (1955) have since confirmed that for higher levels of stimulation the temporary threshold shifts resulting from tones of 1000 and 2000 Hz were greater at 1400 and 2800 Hz respectively (the half-octave values) than at 1000 and 2000 Hz. These are important findings. In practical situations where temporary threshold shift may be used as an indication of the effect on hearing of a particular sound, the levels will usually be such that the upward displacement in frequency will occur. This is not to suggest that the temporary threshold shift resulting from a tone is confined to a narrow band of frequency; it is not, and an appreciable spread in frequency occurs, as would be expected from the pattern of vibration of the basilar membrane. Thus, following exposure to a pure tone, a considerable range of frequencies suffer temporary threshold shift, but the maximum shifts, for the higher levels of stimulation, occur about half an octave higher than the frequency of the tone to which the ear had been exposed. In the case of noise composed of broad frequency bands the same sort of effect, a spread extending upwards in frequency relative to the frequency band of the exposure noise, is seen. An illustration of the effect of experimental production of temporary threshold shift by stimulation by a pure tone is provided by Ward (1962a), and is shown in Fig. 10.1. While the upward spread of the frequency of temporary threshold shift with respect to the frequency of the exposure sound is generally found, the audiometric frequency range 3000–6000 Hz is particularly susceptible, and in conditions of stimulation by broad-band noise it will be preferentially affected. As will be apparent later, the same effect is seen in the deterioration of hearing as a result of occupational exposure to noise; here, almost irrespective of the exact frequency distribution of the noise, the deterioration of hearing occurs first in the 3000–6000 Hz region and particularly commonly at the 4000 Hz audiometric frequency. The reason for this susceptibility is not clear, despite the interest it has aroused. Possible factors are the action of the aural reflex which protects the ear to some extent against high levels of sound at the lower frequencies, together with the fact that hearing is, in this frequency range, intrinsically more sensitive due to the characteristics of the cochlea and middle ear itself. The importance of the cochlea in determining the dependence of the auditory threshold on frequency has been shown by Békésy (1949a). Sensitivity and susceptibility to damage are not of necessity associated, but it is not impossible that the most sensitive elements might be most easily damaged, and that the most marked effects of noise in the range 1000–3000 Hz, where maximum sensitivity lies, might be displaced upwards in frequency

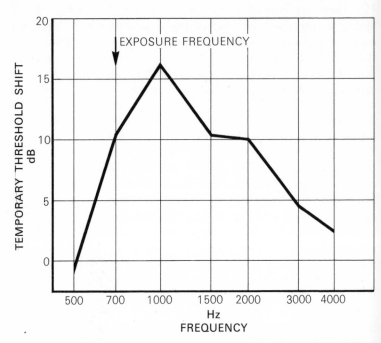

10.1   Temporary threshold shift at different frequencies, 5 min after the end of exposure to a tone of 700 Hz, for 5 min. Note the displacement upwards in frequency relative to the frequency of the exposure tone. (After Ward, 1962a.)

by the action of the aural reflex, which is protective against frequencies up to 2000 Hz (Reger, 1960). The actual disparity in the frequency of the exposure sound and the frequency at which the maximum audiometric temporary threshold shift can be found has not been adequately explained. It is presumably determined by the pattern of vibration on the basilar membrane, but a precise explanation is still lacking.

Effect of exposure intensity

The amount, in decibels, of the temporary threshold shift increases with the intensity of the exposure sound. The latter may be designated in SPL or as *sensation level*, which is the level in dB of the exposure sound above the subject's resting threshold. The effect of

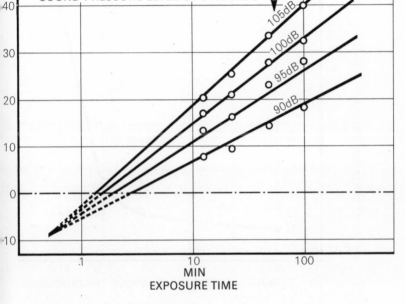

10.2   Temporary threshold shift at 4000 Hz, 2 min after the end of exposure to a band of noise of 1200–2400 Hz at the sound pressure levels and durations indicated. Points are observed data, lines from the equation developed by Ward. (After Ward, Glorig & Sklar, 1959a.)

exposure intensity is complicated by the operation of other effects, such as the duration of the exposure sound, and the interval between the end of exposure and the threshold measurement. If this interval is 2 minutes (for reasons which will emerge later) the relation of exposure intensity to temporary threshold shift is substantially linear up to a certain level of temporary threshold shift. Fig. 10.2 shows, in the form of average values from 26 ears of 13 men from Ward, Glorig and Sklar (1959a), the way in which higher exposure levels produce progressively greater temporary threshold shifts. Above a certain level, temporary threshold shift increases rapidly. There is reason to believe that repetition of exposures producing this rapid increase of temporary threshold shift would, if continued, result in permanent damage to hearing. Fig. 10.3 shows this effect from Hood's (1950) data; the discontinuity in the curve is obvious,

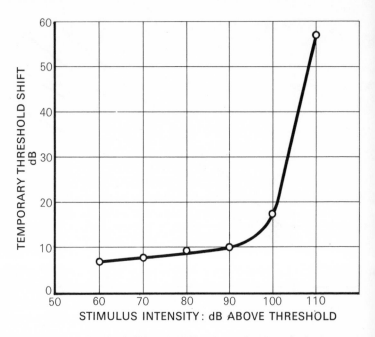

10.3   Temporary threshold shift at 2000 Hz, 10 sec after the end of exposure to a tone of 2000 Hz at the sensation levels (dB above threshold) indicated. Note the sharp increase of effect between 90 and 100 dB SPL of the stimulus tone. (After Hood, 1950.)

and in general it occurs at exposure values above 90 dB sensation level in the conditions of Hood's experiments. Above this level the shift was large and the recovery prolonged. The same effect appears in Ward's (1960) study of the recovery from high values of temporary threshold shift. In previous work Ward, Glorig and Sklar (1958, 1959a) had found that for values of temporary threshold shift not exceeding about 40 dB measured 2 min after the cessation of the exposure sound, the recovery of hearing conformed to a uniform pattern, which will be discussed later (p. 198). However, if the temporary threshold shift 2 min after the end of the exposure reached 50 dB on average, a markedly delayed form of recovery occurred. Although the mechanism of these different recovery patterns is not known, it is obvious that at levels of temporary threshold shift of over 40 dB measured at the 2-min interval, some significantly different response is being elicited, which must be interpreted as a greater potential hazard to the hearing mechanism, and which must be viewed seriously.

The general tendency, therefore, is that with certain reservations the amount of temporary threshold shift is determined by the level of the exposure tone or band of noise. One exception, however, is that at very high levels the opposite effect may be found; for example, temporary threshold shift may be diminished by increase of intensity, this change in response occurring at about 120 dB SPL. This appears to have been noted first by Davis and his collaborators (1950) in their studies of damaging exposure levels. The explanation would appear to be the alteration in the mode of vibration of the stapes, already mentioned in Chapter 5 and illustrated by Békésy (1949b), which reduces the pressure transmitted to the cochlear fluids at high vibrational amplitudes. In this mode, the stapes vibrates about the long axis of the footplate instead of the normal hinge-like movement on an approximately vertical axis, through the lower, posterior pole of the footplate. Due to the elongated shape of the footplate the former mode is much less effective in the transmission of pressure to the cochlear fluids. It has been suggested that the aural reflex is involved in this change in mode of vibration of the stapes.

Effect of exposure duration

On the average, temporary threshold shift increases with increased duration of exposure. Specifically the value of temporary threshold shift in dB is linearly related to the logarithm of the duration of exposure. In Fig. 10.2 is shown the temporary threshold shift measured at 4000 Hz 2 min after the end of the exposure, as a result of band noise, extending over the octave 1200–2400 Hz. The points for the durations of the exposure noise can be seen to be capable of being fitted to straight lines by the equation describing the growth of temporary threshold shift with time, developed by Ward. A feature of these data, as of other similar plots, is that the apparent origin of the lines at zero time does not give the average zero temporary threshold shift, i.e. the average resting hearing level of the group, but a negative value. The significance of this is not entirely clear. Ward and his collaborators (1959a) speculate on the possible reason, but the fact remains that the growth of average values of temporary threshold shift, is well described by a straight line relating dB of temporary threshold shift with the logarithm of exposure time, and the slope of the line is determined by a variety of factors, including the SPL value of the exposure sound. The eventual amount of temporary threshold shift will be determined by the SPL of the exposure sound, provided the other dimensions, e.g. duration of the exposure, remain constant.

Recovery

The recovery of the threshold after temporary threshold shift displays a number of different patterns according to the nature of the exposure and the way in which the recovery is measured. In addition, contrary to previous belief, the course of recovery is influenced by the way in which the TTS was produced (Ward, 1970). The least complicated recovery pattern following moderate exposures has been mentioned previously in the section describing the relation of exposure intensity to the temporary threshold shift produced; here we have a mode of recovery which is like the rate of growth. Recovery is more rapid at first and slower later. Thus, as the recovery proceeds, at any time the value of the remaining temporary threshold shift, in dB, is linearly related to the logarithm of time since the cessation of the exposure sound. This type of recovery is apparently only found if the temporary threshold shift is measured at an interval not less than 2 min after the end of the exposure sound, and is illustrated in Fig. 10.4. This can be viewed in conjunction with the growth of temporary threshold shifts of the same 13 subjects, using both ears (Fig. 10.2). The points are well fitted by the lines, which have been fitted by the statistical procedure of least squares and there is a tendency for the higher values of temporary shift to diminish somewhat more rapidly than the lower, but nevertheless from this and other data, it is clear that the greater the initial value of temporary threshold shift, the longer will be the recovery time. It can be seen, however, for the exposure to noise at 105 dB

SPL, that recovery is markedly delayed. Some of the values of temporary threshold shift from which the 105 dB recovery line was derived exceeded 50 dB, so that delayed recovery is to be expected. This specific effect which has been noted above in the section on the effect of intensity, is illustrated in Fig. 10.5, from Ward's (1960) data on recovery from high values of temporary threshold shift. He found that after production of 50 dB or more of temporary threshold

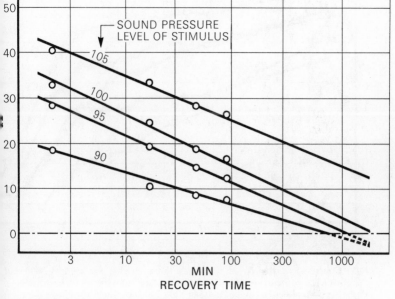

10.4   Recovery from temporary threshold shift at 4000 Hz, at various intervals after the end of exposure to a band of steady noise of 1200–2400 Hz, at the sound pressure levels indicated, and for the same exposure duration. The lines are fitted by the method of least squares. (After Ward, Glorig & Sklar, 1959a.)

shift at either 3000 or 4000 Hz, by noise in the octave 1200–2400 Hz, measured at 2 min post-exposure, the recovery was prolonged. Initially, the recovery appeared to be on the same basis as before, in which dB of temporary threshold shift are directly related to the logarithm of time. Later, however, the recovery process becomes more rapid and the relation could be described as a linear one of dB of temporary threshold shift against actual time during the recovery period. Fig. 10.5 shows the transition from a slow to a more rapid type of recovery, and also the variability between individuals which is a feature of temporary threshold shift effects. Ward (1963) emphasises that the recovery process following exposure to continuous noise, after 2 min, for amounts of temporary threshold shift in the order of 40 dB or less, is on average a straight-line relationship

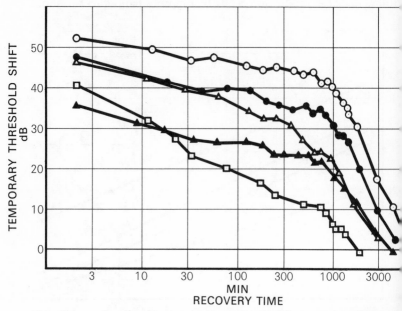

10.5   Course of recovery from temporary threshold shift averaged at 3000 and 4000 Hz, both ears combined, for 5 listeners (separate symbols for each listener). Exposure was to a band of noise of 1200–2400 Hz, SPL 105 dB, and duration sufficient to give 50 dB of temporary threshold shift measured 2 min post exposure, in one of 2 ears of a given observer, at either 3000 or 4000 Hz. (After Ward, 1960.)

on a graph of temporary threshold shift in dB against the logarithm of time. Only the value of temporary threshold shift at 2 min post-exposure is needed to predict the value at any subsequent time.

There are exceptions to this type of recovery. In the very early stage of recovery, instead of being smooth, the recovery is characterised by an undulating threshold during the first 2 min. First described by Bronstein (1936), an example of this type of behaviour from Hirsh and Ward (1952) is shown in Fig. 10.6. The rapid immediate recovery from the temporary threshold elevation followed by a deterioration maximal about 2 min, has been called a 'bounce' effect by these authors. The changes have never been explained adequately, but the practical implication is obvious; temporary threshold shift should be measured at an interval of not less than 2 min post-

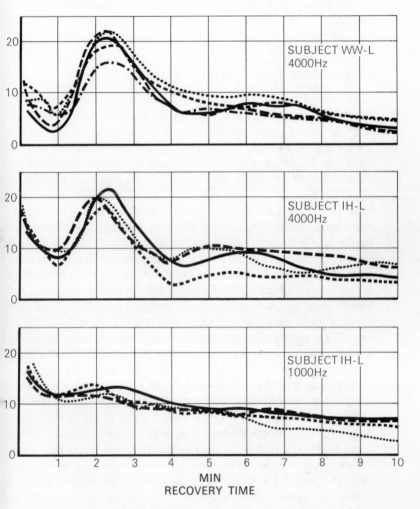

10.6 Recovery curves following temporary threshold shift, in left ears of two subjects at 1000 and 4000 Hz. Each set of curves represents the same experimental conditions repeated on four or five different occasions, represented by different types of line. Exposure was to a tone of 500 Hz at a sound pressure level of 120 dB for 3 min. (After Hirsh & Ward, 1952.)

exposure (TTS_2) to secure consistent results. There are now strong reasons why TTS_2 should not be used indiscriminately (Ward, 1970); this will be discussed later.

Ward has derived equations to predict the rate of growth and decay of temporary threshold shift. These are interesting and valuable extensions of the usefulness of temporary threshold shift as a means of assessing the effect of different forms of sound stimulation.

Effect of frequency of exposure sound

The effectiveness of a particular sound in producing temporary threshold shift is dependent, in addition to the factors already described, on its frequency content. Particularly noticeable is the lesser effect of the lower frequencies. A fair generalisation is that the higher the frequency of the exposure sound the greater the temporary threshold shift produced. This holds good at least up to 4000 or 6000 Hz (Ward, 1963). The relative effects of exposure to bands of noise or to pure tones have given rise to much discussion in the past, the original assumption being that the pattern of excitation of the cochlea for a pure tone would be more intense than if the same energy were spread over a range of frequencies, such as an octave or more. The concept of the critical band of frequency was invoked to support this contention by Kryter (1950), but clear evidence has been lacking, and on the other hand Davis and his colleagues in their classical work on temporary threshold shift (Davis and collaborators, 1950) specifically state that 'a band of frequencies an octave in width produces the same hearing loss as a pure tone of the same intensity, as measured by our sound-level meter, and of frequency corresponding to the middle of the octave'. Later work by Ward (1962a) has corroborated this statement for exposure frequencies above 2000 Hz. He finds that pure tones will give the same temporary threshold shifts as octave bands of noise of the same intensity, whose upper frequency limit is the same as the frequency of the pure tone. For exposure to pure tones or bands of noise one octave wide, below 2000 Hz, Ward shows that the pure tone does indeed produce greater temporary threshold shift than corresponding bands of noise, and this effect is associated with the action of the aural reflex (Chapter 5). This reflex, it will be recalled, is activated by moderately intense sounds over a wide range of frequencies and causes contraction of the stapedius and tensor tympani muscles, so that the vibration of the ossicles is reduced and the internal ear protected to some extent against intense sound, at frequencies below about 2000 Hz (Jepsen, 1963). Thus at these frequencies the reflex may intervene to limit the energy reaching the inner

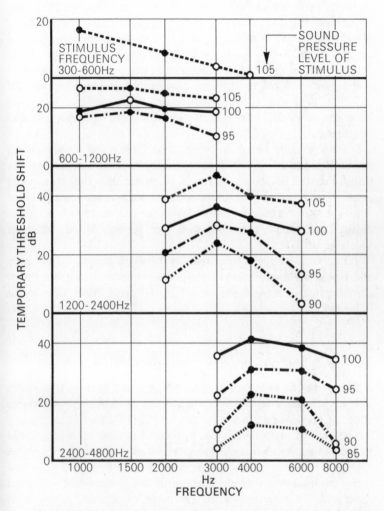

10.7 Temporary threshold shift measured at various frequencies, 2 min after the end of exposure, to the bands of noise indicated on the sections, for a duration of 100 min, at the sound pressure levels noted on the individual curves. Closed circles are experimental data; open circles are calculated by Equation 7a of Ward, Glorig and Sklar (1959a).

ear and so the production of temporary threshold shift. This action, however, is influenced by the nature of the stimulus; pure tones are effective at first, but the muscle contraction is not maintained and the protection rapidly falls away. On the other hand, bands of noise are capable of maintaining the reflex better; it may be supposed that the changing nature of the stimulus in the case of noise bands is more effective, and so maintains the muscles in a state of more sustained contraction. The result is that noise bands produce less temporary threshold shift in the lower frequencies than do pure tones. Ward, in addition, showed that the temporary threshold shift due to a pure tone in one ear can be reduced by noise in the other ear, since the reflex will be elicited in both ears by a stimulus in either ear.

It must also be noted, however, that Carter and Kryter (1962) found that even above 2000 Hz pure tones produced more temporary threshold shift than octave bands of noise. For this reason the permissible levels of pure tones were set 5 dB lower than those of octave bands in specifying noise levels for the restriction of noise-induced permanent threshold shift (see Kryter, Ward, Miller & Eldredge, 1966).

The general picture of the relation between exposure frequency and frequency distribution of temporary threshold shift, for exposure to bands of noise, is shown in Fig. 10.7 from Ward, Glorig and Sklar (1959a).

Susceptibility to temporary threshold shift

This topic has been recurring in the literature for many years. Its interest is in the quest for a predictive test to indicate the probable effect of occupational noise on the hearing of a particular person. So far, no clear solution to this problem has been found. It would be convenient if individuals were grouped into two categories, those with ears particularly susceptible to temporary threshold shift, and those minimally susceptible. Further, if this susceptibility reflected accurately the response of the ear to years of daily exposure to occupational noise, great benefits would accrue in the ability to identify those with ears liable to damage by noise. Unfortunately, the first of these two possibilities does not appear to be true, and the second, while apparently possessed of some validity in principle

(Burns, Stead & Penney, 1970; Burns, 1971) still presents difficulties in its practical application.

The grounds for a provisional acceptance of a relation between temporary and permanent threshold shifts depend on evidence such as that obtained by Jerger and Carhart (1956), Kylin (1960), Glorig, Ward, and Nixon (1961) and Burns, Stead and Penney (1970). The general postulate reached by Kryter, Ward, Miller and Eldredge (1966) after study of the available data is that the average permanent noise-induced threshold shift resulting from nearly daily exposure, 8 hours per day for about 10 years, to a particular noise, is approximately equal in dB to the average temporary threshold shift (measured 2 min post-exposure) produced in young normal ears by an 8-hour exposure to the same noise. This relation will be referred to again in more detail in discussing permanent noise-induced threshold shift and its prevention, but it must be repeated, this relation is not established for individuals, only on a statistical basis for groups.

The variability in the exact characteristics of temporary threshold shift between individuals has been noted by all who have worked in this field, but a given person will show less variation in successive tests than the variation between individuals. Repeated recovery curves, in the same person, can be seen to have a certain resemblance, as evident in Fig. 10.6. Ward (1965, 1968, 1970) examined the usefulness of temporary threshold shift, as an indication of possible permanent effects of exposure to occupational noise; his conclusions, together with other evidence (Burns & Robinson, 1970) will be discussed in that context (Chapter 12).

Intermittent exposure

If the exposure sound is interrupted at intervals, various new factors operate to determine the characteristics of the temporary threshold shift produced.

Intermittency in this context means exposure of the ear to a sound which is not continuous with time, but consists of periods of noise separated by periods of relative quiet. To describe such exposure the characteristics, in terms of frequency spectrum and SPL, of the noise can be stated in the usual way, but the durations of noise (*on-time*) and of relative quiet (*off-time*) must also be specified. These durations, in practical situations such as occupational noise, may vary infinitely. For example, the 8-hour work day might include one on-time period of 4 hours and 4 hours off-time; or at the other extreme, the day might consist of 5 min on-time, followed by 5 min off-time, recurring 48 times in the 480 min day, or any other of the infinite number of regular or irregular combinations. In this brief summary, only experimental laboratory exposures are considered, in which rigid control of the exposure and of the audiometric findings can be exercised.

Ward (1963) has reviewed this subject, and has subsequently published new data (1970) which modify the previous concepts somewhat. Thus the background and the recent findings must be taken in conjunction for an appreciation of the present position.

In determining the TTS due to discontinuous noise exposures, direct measurements of TTS in laboratory experiments are the primary sources of information. The total final TTS will be compounded of the growth and recovery, which will depend on the magnitude and the time pattern of the particular exposure, together with the interval between the end of the exposure period and the hearing measurement.

Direct data of Ward, Glorig and Sklar (1959a) for growth and recovery rates has already been given in Figs. 10.2 and 10.4 respectively for continuous noise. The recovery (Fig. 10.4) excluding the highest values of TTS such as that seen after the 105 dB SPL exposure, present an orderly converging pattern of TTS (actually TTS_2) with the logarithm of time. Thus the recovery rates can be calculated in suitable conditions with confidence for continuous noise (Ward, 1966). Essential to the calculation of the recovery after a single continuous exposure is the demonstrated fact that the rate of recovery is independent of how the TTS was produced (Ward, Glorig & Sklar, 1959b). Thus exposures consisting of single continuous noises of different duration and SPL, giving the same TTS, showed similar recovery rates. This finding applies irrespective of exposure frequency or test frequency and has been confirmed by Smith and Loeb (1969). This type of recovery does not occur with some types of intermittent noise (Ward, 1970), necessitating revision of previous concepts of the potential hazards of such noise. This important aspect, affecting the safety of noise environments is dealt with in Chapter 12. In the meantime, the known facts of generation of TTS from sounds are summarised.

After exposure to higher frequencies, the TTS_2 from intermittent noise is less than if the exposure had been to continuous noise. This reduction (Ward, Glorig & Sklar, 1959c) in terms of TTS_2, is proportional to the ratio of the time occupied by noise and relative quiet. For example, if the noise on-time is 2 min, off-time 2 min (i.e. on-fraction $= \frac{1}{2}$), the average TTS_2 would be half of the value, in dB, which would result from a continuous noise of the same SPL. Ward (1970) has called this the 'on-fraction rule'. It applies for octave band noise above approximately 1200 Hz; lower frequency bands elicit relatively less TTS, possibly due to greater activation of the middle ear muscles (Ward, 1962b).

Calculations of TTS_2 for various on-times and off-times have been made, using growth and decay curves, by Ward, Glorig and Sklar (1959c), in which it was assumed that recovery from a given value of TTS_2 was independent of how it was produced. Anomalous results were produced by this procedure for certain types of stimulation, and in a later investigation Ward (1970) has shown that indeed TTS_2 is influenced by the nature of the exposure, and in addition that certain exposures may give extremely prolonged recovery, particularly in the apparently more susceptible subjects. The latter finding casts great doubt on the validity of TTS_2 as an index of TTS, since it obviously cannot show the significant aspect of protracted recovery. The calculation procedure for growth and decay of TTS is successful (as well as for single exposures) for intermittent exposures in which the on-time and off-time are short; specifically, in the region of 3–5 min. However, if the on-time of repeated bursts of noise is in the region of 10 min or longer, the TTS is greater than would be predicted on the above assumptions. In addition, intermittent exposures, with short or long on-times, to noise at 105 dB in the 1400–2000 Hz frequency band, which produce not less than 15 dB of TTS_2, were found by Ward to have incompletely recovered 16 h or longer after the end of the exposure. This is illustrated in Fig. 10.8. This kind of pattern of delayed recovery has already been illustrated (Ward, 1960) in Fig. 10.5, following production of high values of TTS ($TTS_2 = 50$ dB at 3 or 4 kHz), by exposure to continuous noise. Apart from the delay in recovery in the most affected subjects in Fig. 10.8, the separation of the different degrees of susceptibility is very marked indeed. The explanation of these effects of repetitive noise bursts (in this case on-time $= 3$ seconds, off-time $= 7$ seconds) is not clear. It is possible that the sudden start of each noise burst, occurring in this case every 10 seconds, might be occurring too quickly for the middle ear reflex to respond, and so increasing the effect on the ear. Ward (1970) concludes that no simple explanation is available; he feels that TTS_2 is not always a useful index and that in general TTS measured 30 min (TTS_{30}) or even 1000 min (TTS_{1000}) after the end of exposure, will be a better indicator of the potential harmfulness of a noise.

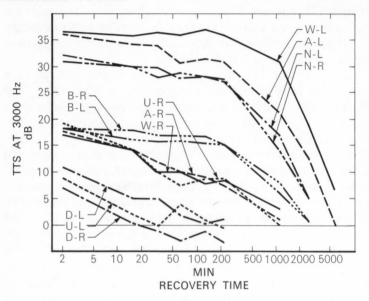

10.8   Recovery from TTS at 3 kHz produced by exposure for 6 h to intermittent noise (3 s on, 7 s off) in the frequency range 1400–2000 Hz, in the two ears of 6 subjects. Each curve is identified by the initial R or L ear. The most affected ears show delayed recovery, and the variation in susceptibility is wide. (After Ward, 1970.)

The implications of these findings will be discussed, in the context of noise-induced hearing loss and its prevention, in Chapter 12.

Relation of exposures in different frequency bands

The fact that exposure below certain base levels is incapable of affecting the characteristics of temporary threshold shift produced by more intense exposure naturally leads to the question of how much interaction occurs between exposures of different frequency. Ward (1961) finds that interaction does not occur provided the frequencies of the exposure sound are sufficiently separated. For example, exposure to a band of noise of 2400–4800 Hz was preceded or followed by exposure to a band of noise of 600–1200 Hz. Both noises were sufficiently intense to produce considerable threshold shifts, but neither affected the rate of growth, or the rate of recovery, of the other. Apparently, from this data, provided that frequency ranges of the exposure sound are sufficiently separated, effects on hearing manifested by temporary threshold shifts can be regarded as substantially independent. This is an important finding with implications for the assessment of the effect of occupational noise exposure. The practical applications of this will be seen in Chapter 12.

Temporary threshold shift due to impulsive noise

This type of noise occurs as a result of gunfire, of various hammering, stamping, pressing, chipping and riveting operations. It is characterised by sudden short duration noises, which may be repeated at various frequencies. The repetition frequency is important from the viewpoint of temporary threshold shift and presumably of damage to hearing. The aural reflex has a latent period (the interval between the onset of a sound stimulus and the beginning of the reflex response) of about 10 ms, measured as the interval between the onset of the sound and the start of the action potential (the electrical response) of the stapedius muscle in man (Perlman & Case, 1939). Activity in the tensor tympani appears to have a slightly longer latency. These time intervals are, however, insufficient to establish fully the protective function. By using measurements of the acoustic impedance of the tympanic membrane, which gives an indication of the degree of activity of the middle ear muscles, a realistic indication of the characteristics of the aural reflex may be obtained. By this means, latency was indicated by the interval to the time of beginning of the change in acoustic impedance of the tympanic membrane. The interval is shorter the more intense the sound stimulus, and Terkildsen (1960) found that for fairly intense sounds the mean value of the latency occurred at 44 ms, and the standard deviation in a series of persons was 11 msec. The significance of these findings is that one single intense, sudden sound can damage the ear before the protective reflex has time to act. Jepsen (1963) remarks on this imperfect protection against sudden sounds and notes that natural sounds are less inclined to be of this sudden type than the man-made sounds, which are known to be damaging to the ear. When a sudden short-duration noise like that from small arms is repeated at a sufficient frequency, the contraction of the middle ear muscles will be maintained. The interval for this must be less than about 1 s. Murray and Reid (1946) showed this and demonstrated how 28 single rounds several seconds apart produced substantial temporary threshold shift, whereas the same number of rounds fired in quick succession from an automatic (Bren) gun, gave insignificant temporary threshold shift. Protective effects of the aural reflex have been described by Fletcher and Riopelle (1960) and by Chisman and Simon (1961).

The understanding of the response of the ear to impulsive noise is beset with difficulties, due largely to the impossibility of making generalisations in a manner similar to those possible with temporary threshold shift for continuous noises. Loeb, Fletcher and Benson (1965) have drawn attention to the difficulties imposed by the almost infinite variety of acoustic characteristics of impulsive noise met with in practice. The variables include magnitude in terms of peak pressure; pulse duration; rise and decay time; interval between pulses; direction of pressure change with respect to ambient pressure; total number of impulses in a given exposure; and rate of repetition. They have used a spark-gap impulsive noise generator producing a short (length 40 μs) positive pressure pulse, followed by an almost equal negative pulse and a train of smaller oscillations. the whole disturbance lasted about 800 μs, and the peak SPL was 156 dB. The maximum energy per cycle occurred at about 100000 Hz. The temporary threshold shift from this noise, in terms of numbers of impulses and previous activation of the aural reflex showed some differences from previous work in which the pulses were longer. The growth of temporary threshold shift with number of pulses was not linear, as found by Ward (1963) and others, but less than logarithmic. Exposure to 400 pulses at a rate of one per second produced more temporary threshold shift than at a rate of 10 pulses per second. The temporary threshold shift was maximal at 6000 Hz and above. The conclusion was that apparently the aural reflex, if sufficiently activated by this stimulus, is effective at high frequencies. This does not correspond with more conventional concepts, but this type of stimulus has not previously been investigated.

The picture on the temporary threshold shift from impulse noise is thus far from clear. These problems appear again in the permanent effects arising from impulse noise and in the search for measures to counteract them.

As in other aspects of TTS, the practical applications lie in prevention of damage to the ear from impulsive sounds, and this is covered in Chapter 12.

Effect of prolonged exposure

Little information exists on the TTS resulting from prolonged exposure in man. Mills, Gengel, Watson and Miller (1970) have performed two exposures on one subject. The noise, an octave band centred at 500 Hz, lasted for 48 h at 81·5 dB SPL and for 29·5 h at 92·5 dB SPL, respectively. Five weeks separated the two exposures.

Various tests of auditory function were performed before, during (in quiet intervals) and after the exposures. The auditory threshold

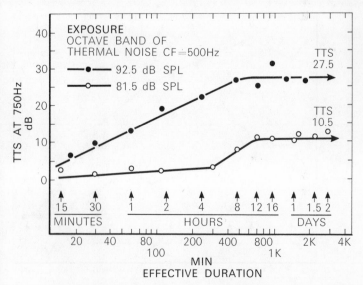

10.9 The course of development of temporary threshold shift. TTS was measured at 750 Hz, 4 min after the end of sound stimulation to an octave band of noise, centre frequency (CF) 500 Hz, at 92·5 dB SPL, or 81·5 dB SPL, for various durations of exposure from 15 min to 2 days. The effective duration of exposure is total time in noise plus 3 min for each 7 min out of noise (for audiometry). Note the attainment of a plateau in TTS values after about 8 h. (After Mills, Gengel, Watson & Miller, 1970.)

was most elevated at 750 Hz, and the TTS, measured after 4 min of quiet (TTS_4), reached and maintained a maximum during the stimulation. The increase of TTS_4 proceeded for 8–12 hours, and the plateau was at $TTS_4 = 10·5$ dB for the 81·5 dB exposure, and at $TTS_4 = 27·5$ dB for the 92·5 dB SPL exposure. Recovery from the lesser TTS took about 3 days; from the greater, about 6 days. The course of growth and recovery is shown in Figs. 10.9 and 10.10, and the growth data, in an audiogram-like presentation, in Fig. 10.11. The implications are discussed, but one observation on the audiometric, physiological and anatomical findings in the chinchilla ear following sound stimulation is of the utmost importance. This arises from the work of Miller, Eldredge and Bredberg (1970) who, in personal communications to Mills *et al.* and to the author, have shown that, after prolonged sound stimulation (1 to 9 days) permanent injury in the form of loss of hair cells can occur in the cochlea without permanent threshold shift. The implications are of

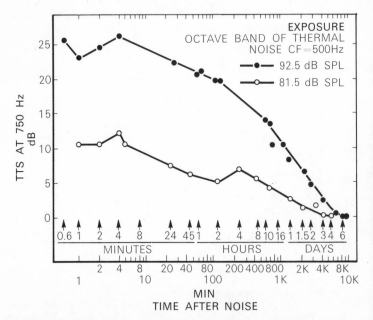

10.10   The course of recovery of temporary threshold shift. The recovery is following the growth of exposure shown in Fig. 10.9. Note that the recovery to 0 dB TTS took about 3 days for the 81·5 dB exposure and 6 days for the 92·5 dB exposure. (After Mills, Gengel, Watson & Miller, 1970.)

the utmost importance both scientifically and ethically. Mills *et al.* are consequently reluctant to expose repeatedly any listener to a noise producing TTS which persists for more than 16 h after the termination of the noise, in the belief that if recovery in excess of this period occurs, it is due to the healing of injury of some kind. They have consequently refrained from repeating the long exposures on a

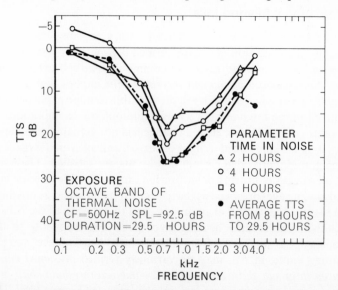

10.11 The growth of TTS at different frequencies. The TTS_9 following the 92·5 dB exposure of Figs. 10.9 and 10.10 is maximal about 750 Hz. Parameter: exposure durations. (After Mills, Gengel, Watson & Miller, 1970.)

human subject. While the findings on damage without permanent threshold shift admittedly apply to a lower mammal, it would appear that their intentions and reservations are proper and indeed, corroborate the general belief that TTS should not, in occupational exposures, persist until the beginning of the next day's work. Should this occur a cumulative situation would appear to exist.

11

Permanent effects of noise on hearing

Occupational hearing loss

Certain kinds of work are sufficiently noisy to cause damage to hearing. The condition is usually termed *occupational hearing loss*. *Noise-induced hearing loss* is also a commonly used self-explanatory term. *Noise-induced permanent threshold shift* implies that an irreversible increase in hearing level, attributable to noise exposure, has occurred in the interval between two audiometric examinations in the same person. Where such a change has not actually been measured, it may be presumed to have occurred, and an estimate made of its magnitude, provided certain conditions are satisfied. These are that all sources of hearing deterioration except that due to noise and to advancing age (presbycusis) have been excluded, as far as possible, by a history and clinical examination, and that the age of the person is known. The median hearing level in dB corresponding to the person's age is subtracted from the measured hearing level and the resulting value, in dB, may be variously termed *presumed noise-induced threshold shift*, *estimated noise-induced threshold shift*, *presumed noise-induced hearing loss*, or *age-corrected hearing level*. The first or last of the above terms will normally be used in the ensuing discussion, with the implication that noise-induced deterioration is for practical purposes permanent.[1] The diagnosis of noise-induced hearing loss is a medical responsibility, and should be made preferably by an otologist, since various specialised clinical criteria are involved and dangerous conditions could be missed if a full examination has not been made.

The existence of occupational hearing loss, as noted in Chapter 1, has long been appreciated. Quantitative studies, however, had to await the development of audiometry, and of accurate means of measuring the industrial noise which normally is the cause of this type of deterioration of hearing. Among early investigations were

---

[1] Absolute permanency must be established by periodical audiometric examinations after removal from noise, since absence from noise may result in minor improvements in hearing acuity over a period of weeks or months.

those of Crowden (1933) who used speech audiometry to investigate the hearing of those exposed to the noise of riveting, and Dickson, Ewing and Littler (1939) who studied by pure-tone audiometry the hearing of those exposed to aeroplane engine noise.

Since the end of World War II the relation of hearing loss to noise exposure has been quite widely studied, but the data directly of use for the prediction of the damaging effect of different noise levels, for example, are not very numerous, and the complexities of the situation are such that there are yet many uncertainties.

In attempting to formulate relations, in a variety of noise conditions, between exposure and hearing, the data which are of direct use are not as plentiful as one might hope. The difficulties of field work, its practical and ethical limitations, and the impossibility of setting up satisfactory controlled situations, have all contributed to the rather gradual development of knowledge in this field, and its persisting deficiencies. A great amount of work has to be done, in which it has been shown that occupational noise can damage hearing, but the difficulty arises in obtaining sufficiently precise data on noise exposure and auditory status. The very real difficulties are clearly shown in the review of hearing in the textile industry by Atherley and Noble (1968), which shows how much labour can be expended on studies yielding results more qualitative than precisely quantitative. In the category of data of immediate quantitative utility, one can include the following; the list is certainly not exhaustive, but indicative of the amount of reasonably precise data available. These sources include ANSI (1954); Rosenwinkel and Stewart (1957); Rudmose (1957); Kylin (1960); Nixon and Glorig (1961): Glorig, Ward and Nixon (1961); Burns, Hinchcliffe and Littler (1964); Gallo and Glorig (1964); Taylor, Pearson, Mair and Burns (1965); Hickish and Challen (1966); Baughn (1966); Burns and Robinson (1970a); Atherley and Martin (1971).

The characteristics of the condition with which we have to deal are these. As a result of work in a noisy situation, deterioration of hearing occurs at a rate mainly determined by the level of the exposure, but usually it is initially unnoticed. Temporary dullness of hearing after exposure to the noise at work—for example, at a lunch break or at the end of a day's work, with perhaps some noise in the ears (tinnitus) are the usual signs that damage is being done to hearing. Tinnitus takes various forms, and may be of a rushing or hissing nature or a more musical type of noise. Obvious tinnitus persists for periods of minutes or hours after a severe noise exposure, and is a sign of damage or potential damage. It is also found (Chapter 7) in ear disease. Since noticeable dullness of hearing and tinnitus tend to disappear after some hours away from noise, they usually make little impression and are often ignored. This attitude of indifference is increased by the fact that many noise environments

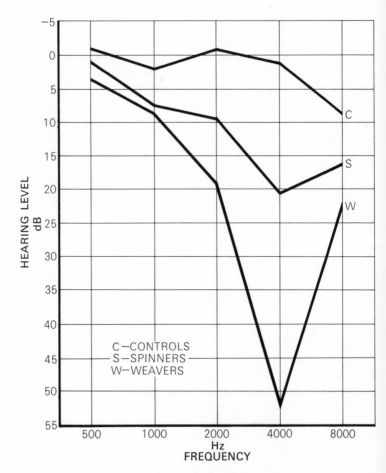

11.1   Median hearing levels of small groups of subjects employed at the same textile mill. Controls had not been exposed to noise; spinners and weavers had both spent 10 years, on average, in their respective occupations. (After Burns, Hinchcliffe & Littler, 1964.)

which produce occupational hearing loss are not regarded as very noisy, and the symptoms of dullness of hearing and tinnitus tend to diminish or disappear with continued exposure. A stage is reached, however, when the person begins to realise that his hearing is impaired, by which time little or nothing can be done to reverse the condition. The audiometric picture resulting from occupational noise exposure can be seen at a glance in Fig. 11.1. The difference in the median hearing levels of three groups of persons from the same textile plant engaged in quiet occupations, in spinning, and in weaving respectively are clearly shown. The effect of age has not contributed greatly to the differences. The higher hearing levels at 4000 Hz of the weavers compared to those of the spinners or the control population are self evident. What is needed, therefore, is sufficient knowledge of the relations between noise exposure and hearing loss to anticipate and to forestall this loss of hearing. Prevention of occupational hearing loss will be covered in Chapter 12.

Research in this field aims at making a comparison which is quite simple in principle. The basic data consist of information on the state of hearing and on the noise exposure sustained by a particular person. The results would be expected to yield information on the hearing deterioration likely to be sustained by the average person, without ear disease or other causes of deafness, as a result of exposure for a given number of years to a given noise. The difficulty in reaching this sort of general solution stems from a number of factors, which will emerge in the following discussion.

Investigation of occupational hearing loss

Laboratory studies which deliberately cause noise-induced permanent threshold shifts in man are obviously impossible. Laboratory studies which produce temporary threshold shift are feasible, but can only be used as a means of comparing the abilities of different exposures to produce temporary threshold shifts, and thereafter to attempt to assess the probable ability of such exposures over a long period, to produce permanent threshold shifts.

Specific information on permanent threshold shifts must therefore be obtained from field studies, in which the experimenter has little or no control, and must utilise a situation as he finds it. Two varieties of this field approach are possible: retrospective studies and prospective studies.

RETROSPECTIVE STUDIES   This method involves the measurement of the hearing of numbers of persons who have been exposed to a known noise for a known period of years. It is the only practicable method of estimating the degree of noise-induced hearing loss over a considerable number of years. If it were possible to perform audiometric examinations on numbers of persons, say one per year for perhaps 40 years in the same noise, it would still be unethical to continue with such a study if significant loss were occurring. If, however, the situation already exists and there are persons with permanent losses originating from many years of exposure in the course of their occupations, they may be examined audiometrically and an attempt made to estimate the degree of hearing deterioration if such exists, which is attributable to the exposure to noise.

We have noted above how the estimate of presumed noise-induced threshold shift is arrived at, through the subtraction of an assumed hearing level due to age, from the recorded level. This estimate, specifically of the median hearing level appropriate to the age of the particular person, implies some inaccuracy, since, apart from the basic uncertainty on what constitutes normal presbycusis, the particular individual's actual age deterioration may of course occur anywhere in the normal range and not precisely at the value for his age indicated on the median curves of Fig. 6.3. The other assumption, that deteriorations due to age and to noise, if each are of moderate extent, are additive, has good support (Glorig & Davis, 1961). However, before presumed noise-induced threshold shift can be derived, the conditions summarised above must be met; if they are not, the whole research may be invalidated. These conditions involve both the physical and biological sides of the problem. From the physical aspect, no useful information on relations between noise exposure and hearing loss is likely to emerge unless the nature of the exposure, in terms of the physical characteristics of the noise and the duration and pattern of the exposure is known with some precision. A noise which is continuous, does not vary in level or spectrum significantly with time or position in the work area, and which lasts for a known time each day, day after day for periods of years provides a suitable type of exposure. However, this is an almost ideal situation, and many occupational noises are variable in time and space, or are discontinuous, and these changes may occur in an irregular manner. The precise specification of such noise is exceedingly difficult, and will probably require special equipment.

From the biological point of view, similar difficulties are found. In the first place, uncertainty must always exist about the state of hearing before exposure to noise. The variability of hearing-level measurements by conventional audiometry with earphones introduces another source of uncertainty. It is thus impossible to say with accuracy how much the hearing of a particular person has deteriorated due to noise. Data must therefore mainly be derived from averages of groups of individuals.

In basic research on occupational hearing loss, the first objective is usually to ascertain the relations of hearing loss to noise exposure in persons without ear disease. Thus a preliminary otological examination is necessary to identify and exclude aural pathology. Exclusion might also be necessary on a medical history, e.g. of head injury or previous exposure to gunfire noise, which would be liable to invalidate the results. Finally the history must yield information which will ensure that only those with known durations of exposure to one noise are included.

These requirements are exceedingly difficult to meet. The tendency for many people to move from one job to another or one industry to another will disqualify many potential subjects. Further, the characteristics of many noise environments will make their use difficult or impossible, because of uncertainty of the actual exposure. It is therefore not surprising that the total numbers of such studies is not great.

PROSPECTIVE STUDIES In this type of investigation, individuals are examined periodically and changes in hearing are measured. The ideal would be to examine the hearing of people with 'normal' hearing before starting work and thereafter at intervals as long as they continue to work in noise. There are again major limitations of an ethical nature, for the same reasons as before, that it would be improper to allow significant changes in hearing to go unchecked. However, evidence of the first small changes is available from the audiogram before the slightest subjective impressions of hearing change are noticed. These early warning signs can be identified and the point noted in terms of exposure, so providing an indication of far greater reliability for individuals than is possible with the retrospective method. In practice, however, attainment of these apparently simple requirements is again beset with difficulties of a practical nature. The result is that the amount of information of scientific validity derived from this method is, at present, rather small. It is of course possible to combine the retrospective and prospective methods. Thus numbers of persons can be examined audiometrically by the retrospective method, and thereafter periodical audiograms can be obtained at intervals of perhaps one year. This is quite a satisfactory method, but it will not normally produce audiograms from individuals before any noise exposure has been incurred. The fundamental weakness of the retrospective approach, of uncertainty about the pre-exposure hearing level, thus remains.

Characteristics of occupational hearing loss

The characteristics of this condition will be considered in the light of data available from studies in industrial situations. In attempting to formulate relations between noise exposure and hearing loss a number of factors must be considered.

The factors influencing the response of the ear to prolonged exposure to noise include the following: (1) overall intensity of the noise; (2) total duration in months or years, of the exposure; (3) frequency characteristics of the noise; (4) susceptibility of the individual to noise-induced hearing loss; (5) differences of susceptibility at different audiometric frequencies possessed by the ear; (6) characteristics of the growth of presumed noise-induced threshold shift with time. These relations are best documented for continuous (i.e. non-intermittent) noise, of continuous spectrum type, without pure tones and lasting for the whole working day. For most steady industrial types of noise, items (1) and (2) above, have been found to be capable of combination to give an energy (Robinson, 1970) which is a primary determinant of the extent of noise-induced hearing loss. Complicated time patterns, for instance noise which lasts for some minutes, followed by a cessation for some minutes and so on throughout the working day, perhaps irregularly, are not easily amenable to field investigation because of the difficulties of securing satisfactory situations. Impulsive noise (Atherley & Martin, 1971) is now under investigation, with encouraging results, but the effects of pure tones are still elusive. Resort to various degrees of empiricism is still inevitable in some of the relations between noise exposure and hearing loss as in the case of pure tones. Taking the above factors individually, the present state of knowledge will be summarised.

Overall intensity of exposure

The overall intensity of the noise is an important factor in determining the extent of permanent threshold shift; the greater the intensity the greater the permanent threshold shift. This generalisation must be qualified, however. Noise exposure, it must be emphasised again, implies a complex made up of the physical characteristics of the sound and the duration and time distribution of its application to the ear. It would thus be more precise to say that provided all the other factors in the exposure are held constant, greater permanent threshold shift would be expected to result from greater intensity, and vice versa.

As an illustration of the effect of intensity, certain retrospective studies can be quoted. The American National Standards Institute (ANSI, 1954) obtained from industry some 7000 audiograms. Of these, 200 were found to be suitable for a comparison of hearing and noise exposure. It was found that the SPL in the octave 300–600 Hz correlated best with the estimated noise-induced permanent threshold shift at 1000 and 2000 Hz, while the SPL in the octave 1200–2400 Hz correlated best with the threshold shift at 4000 Hz. The authors, with great prudence, claimed no definite cause-and-effect relationship, and presented the results in the form of 'trend curves' from which presumed permanent noise-induced threshold shift at a specified frequency can be read, for various durations of exposure in years, provided the SPL in the appropriate octave is known and the spectrum of the noise was within certain limits. In these results, for example, the threshold shift at 2000 Hz for a range of durations of exposure increased with increase in SPL in the octave 300–600 Hz. Specifically, at 80 dB SPL, no apparent threshold shift occurred at 2000 Hz, while after 10 years of exposure, 88 dB SPL produced 9 dB of threshold shift, and 95 dB produced 15 dB of threshold shift. A similar relation was found between the SPL in the octave 1200–2400 Hz and threshold shifts at 4000 Hz.

The effect of different intensities of the exposure noise can also be seen in data from Nixon and Glorig (1961). In this investigation, using the same approach of correlating estimated noise-induced permanent threshold shift with noise exposure, the effects on hearing at 2000 and 4000 Hz of three different noises of different intensities and of rather similar spectra were examined. Again an orderly increase of threshold shift with increase of exposure noise level was found at all durations over about three years, at the frequencies examined.

The effect of different intensities is also seen in the data of Baughn (1966). From Robinson's data are shown in Fig. 11.2 the effect on hearing at 4 kHz of occupational exposure for various durations at various values of sound level A, calculated from the relations developed in the course of the investigation (Robinson, 1970). These

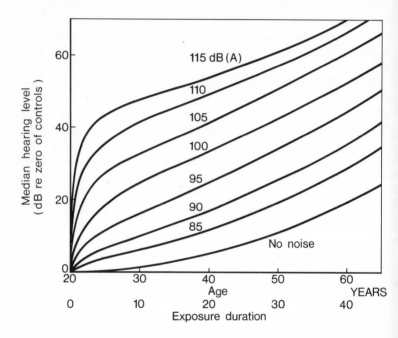

11.2   Development of hearing deterioration as a result of age and noise exposure, the latter being assumed to begin at age 20. Values are calculated from Robinson (1970). Parameter is A-weighted sound level of the occupational noise. Audiometric frequency 4 kHz. (Burns & Robinson, 1970b.)

different levels were not derived from the same sound, but the tendency is clear enough, that for a given duration, the hearing level increases with the intensity of the sound. Having established the basic tendency, it is not profitable to pursue the question of the effect of intensity on threshold shift without including other factors, and of these, the duration of exposure is essential in providing an adequate picture of these relations.

Duration of exposure

The pattern of onset of occupational hearing loss and its subsequent development are characterised by a fairly predictable sequence of events. The first sign is a small depression in the audiogram between 3000 and 6000 Hz, commonly at 4000 Hz. Identification by the audiometer is the only means of detecting the beginning, since the early change will be quite unnoticed subjectively. Fig. 11.3A shows a normal audiogram, together with a fairly well marked early occupational hearing loss (Fig. 11.3B). The localised nature of the elevation of hearing level at 4000 Hz is obvious.

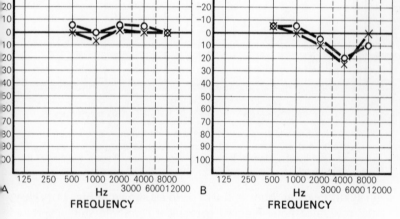

11.3 Audiograms, showing hearing level against frequency. A: An audiogram which would be regarded as within normal limits, for a young person. B: Moderate hearing loss in both ears at 4000 Hz. Subjectively this loss would be unnoticed. ○ = right ear. × = left ear. (Burns, 1958.)

If exposure to noise continues, the dip at 4000 Hz deepens, but still remains predominantly in the same frequency region. This stage is seen in Fig. 11.4A, as a result of the occupation of weaving over a period of 20 years. Continued exposure produces a definite slowing of the deterioration at the most affected frequency, about 4000 Hz, but a subsequent gradual extension to higher and lower frequencies, the precise pattern depending on the spectral characteristics of the exposure. This effect is seen in Fig. 11.4B, as a result of weaving noise over a period of 33 years. The audiograms of Figs.

11.3 and 11.4 are actual hearing levels of particular persons. They are not from the same person, for the reasons already discussed, but they illustrate typical stages of the process from data gathered in the course of an investigation into hearing and noise in the textile industry (Burns, Hinchcliffe & Littler, 1964; Burns & Littler, 1960). A different type of presentation of data on hearing of weavers

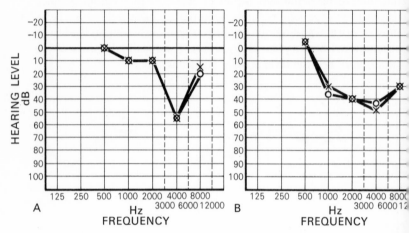

11.4   Audiograms which, in conjunction with a clinical otological examination and a case history, were classed as attributable to occupational hearing loss. A: Age 34 years, 20 years' weaving. B: Age 46 years, 33 years' weaving. (Littler, 1958.)

(Taylor, Pearson, Mair & Burns, 1965) is shown in Fig. 11.5. Here, the presumed noise-induced permanent threshold shift, expressed as median values of the different age groups are shown in an audiogram-like manner. This investigation used only women, and the age correction was from Hinchcliffe's data for presbycusis in conjunction with the British standard (BS 2497) audiometric zero of 1954. The complication of possible previous noise-induced threshold changes is thus believed to be minimal. The investigation was retrospective, and succeeded in meeting all the requirements for an investigation of this type. Thus no pathological ears were used, no other occupation was recorded, and the weaving noise had been experienced each working day from less than one year up to 52 years. It was ascertained that the noise was unlikely to have changed materially over this entire period. In all these investigations, the inclusion of temporary threshold shift must be avoided as far as

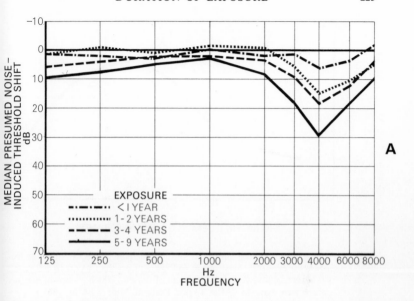

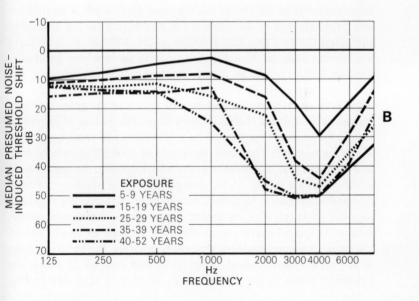

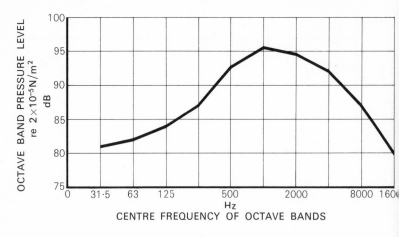

11.5   Presumed noise-induced threshold shift as median values, at different frequencies, for exposure durations as follows. A: Less than 1 year to 5–9 years. B: 5–9 years to 40–52 years. C: Spectrum of noise to which groups of persons in A and B were exposed. (After Taylor, Pearson, Mair & Burns, 1965.)

possible, by performing the audiometry before starting work in the morning, so that hearing may recover in the interval since the last noise exposure at the end of the previous day's work. This interval will probably not avoid some error due to temporary threshold shift, and the best practical solution to the difficulty is to perform audiometry only on Monday mornings in cases where Saturday–Sunday breaks occur, to secure the maximum interval since the last noise exposure. In this investigation this was done. Study of Fig. 11.5A and B gives the sequence of events in the growth of presumed noise-induced threshold shift as median values of groups of people for various exposures. The total number of ears represented in the figure is 461. The small depression at 4000 Hz in persons with less than one year of exposure to the weaving noise, the spectrum of which is shown in Fig. 11.5C, contrasts with the absence of such a depression in control subjects from the same industry, not regularly exposed to noise. The increasing depth of the depression, which retains its typical maximum at 4000 Hz through all except one of the eight exposure duration groups, is obvious. In the two groups with greatest durations of exposure (35–39 and 40–52 years) the extension down to lower frequencies, similar to that seen in Fig.

11.5B, is clear. In occupational hearing loss from other noises, similar, but not necessarily quite identical, patterns of deterioration are found. In the weaving situation of Fig. 11.5, in fact, the deterioration is rather more than expected, for reasons not yet apparent.

It is possible to study the growth at a particular frequency of presumed noise-induced threshold shift by plotting the latter against years of exposure. This has been done in earlier investigations by Nixon and Glorig (1961) and by Taylor, Pearson, Mair and Burns (1965). In the latter investigation, the initial rapid period of increase of presumed noise-induced threshold shift at 4000 Hz is succeeded by a slowing down which is marked and occurs at 10–15 years. Thereafter further deterioration attributable to noise is at a low rate, for the next 30 years. The changes at 3000 Hz were similar to those at 4000 Hz. At 2000 Hz, this saturation effect is not conspicuous, the general tendency was for an appreciable rate of deterioration to continue for a longer period. The same behaviour at 2000 Hz is also seen in data from Nixon and Glorig (1961). At 1000 Hz similarly a steady deterioration continues for the entire period of the investigation, covering exposures from less than one year to over 52 years in the case of Taylor et al. (1965). In the latter investigation, which is believed to have an adequate degree of validity, there are discontinuities in the curves, but these have always been a feature of such data, and are also seen in those of Nixon and Glorig (1961), and are presumably due to inadequate sample sizes. Nixon and Glorig (1961) believe that the deterioration at 4000 Hz ceases after about 10 years, at a level determined by the severity of the exposure. More work has been done on occupational noise exposure, by Baughn (1966) and by Burns and Robinson (1970a).

These two investigations have each made a deliberate effort to include a sufficient number of different noise conditions to build up a picture of the response of the ear to a variety of different exposures in terms of intensity and of duration. They differ, however, inasmuch as the data of the Baughn study is derived from an unselected industrial population, while Burns and Robinson's subjects were free as far as could be ascertained from aural pathology, or other sources of hearing loss except noise. The two sets of data are in a sense, complementary to one another. The effect of the inclusion of some subjects with ear pathology in their investigation has been considered by Burns and Robinson and in detail by Robinson (1971). The effectively pathology-free population of their investigation provides the logical basis for a study of the relations between noise and hearing, and is the appropriate background against which practical situations may be considered for the purpose of evolving measures for the preservation of hearing.

UK Government investigation

This investigation was carried out jointly by the British Medical Research Council and the National Physical Laboratory on behalf of the Department of Health and Social Security. It was a field study of hearing and noise in a variety of industries, in which known noise exposure was related to the state of hearing in some 1000 volunteer subjects, both noise-exposed and controls without noise exposure. Retrospective and, to a lesser extent, prospective data were obtained. The conditions for acceptance of subjects were broadly as indicated previously under the heading 'retrospective studies'. Thus the state of hearing could be associated only with the effects of age and of a noise-exposure history which was in the main known with considerable precision. Pure-tone air-conduction audiometry on the Rudmose (Chapter 6) principle was employed, with the modification of interruption of the tones (Copeland, Whittle & Saunders, 1970). All audiograms were taken (except for the purpose of measuring TTS) before starting work at the beginning of a day or shift. Hearing levels measured thus, can include a small component of temporary threshold shift which for this purpose may be ignored.

The audiometric hearing levels expressed relative to those of a young industrial control population not exposed to noise, were subjected to a simple subtraction of a decibel value determined by age, from a smoothed version of Hinchcliffe's 1959 presbycusis data, so yielding, as described above, the quantity 'presumed noise-induced threshold shift' or 'age-corrected hearing level' denoted by the symbol H. (In figures in this text showing either of these terms without the symbol H, the audiometric zero used was that given in BS 2497: 1954.)

In Burns' and Robinson's study, in which a variety of noises with different spectral characteristics and intensities could be examined, the choice of a suitable measure from the virtually infinite variety of physical derivatives of the sound stimulus is a major decision. In this case extensive trials of different measures showed that sound level A (L_A), or slightly better, a statistical derivative of it, the level in dB(A) exceeded for 2% of the time, (L_{A2}), gave excellent predictions of the effects on hearing. Although not quite optimal they were selected for the present, particularly in view of the widespread use of the A-weighting. The advantages conferred by a single-figure noise measure are immense, and for reasons which are discussed below, sound level A is regarded as a valid measure for spectra within certain limits of 'rising' and 'falling' characteristics.

NOISE IMMISSION LEVEL   It was noticeable that a particular value of H could be associated with a high noise level for a short time or a lower noise for a longer time. This is in effect the equal-energy concept, long postulated to cover these permutations. Systematic testing of the data (Robinson & Cook, 1970; Robinson, 1970) showed, moreover, that the extent of H was related to the sound energy reaching the ear, over a wide range of exposures. This sound energy was termed by Robinson *noise immission*, or in logarithmic form *noise immission level* (NIL); it is the sum of sound level and a logarithmic expression of duration of exposure. Noise exposure may thus be expressed as

$$E_A = L_A + 10 \log_{10} \frac{t}{t_0} \tag{11.1}$$

where E_A = A-weighted NIL
    t = duration in years
    t_0 = 1 year.

In some circumstances, t and t_0 may be expressed in months, or L_{A2} used instead of L_A; in the latter case E_A becomes E_{A2}. In the ensuing discussion NIL refers to the E_A version, so that NIL and E_A are here synonymous. The minimum period of exposure in the original data is 1 month or about 160 h of actual noise exposure, and the maximum 50 years.

The significance of the concept of noise immission level will be evident in many aspects of the relations between noise exposure and hearing loss; it can be derived from a virtually infinite range of combinations of sound level and duration. We recall that the sound energy will remain the same if an increment of 3 dB of sound level is accompanied by a halving of duration, or any other similar proportional adjustment. As examples, the following daily occupational exposures (a) to (e) all give the same NIL value ($E_A = 100$ dB).

(a) 1 year at 100 dB(A)
(b) 2 years at 97 dB(A)
(c) 10 years at 90 dB(A)
(d) 40 years at 84 dB(A)
(e) 1 year at 97 dB(A) followed by
        10 years at 87 dB(A)

It will be seen that the two components of (e) each give NIL = 97 dB. Addition of the two components then yields 100 dB (Appendix F). Thus any exposure, or combination of exposures, can be allotted the appropriate NIL value on the basis of Equation 11.1.

These summations, it must be noted, are made possible by the use of a single-figure noise measure. The limitations of this usage for different spectra are not serious, and are noted below under the effect of frequency. Another limitation of course exists. This is that, to be meaningful, exposure equivalent to an entire occupational lifetime (e.g. NIL = 100 to 105) cannot be expected to be compressed into excessively short durations. If it were so the corresponding sound levels could be excessively high.

The derivations of the relations between the values of NIL and the associated hearing loss are described in the original publications. The value of H is predictable as noted above, for the various audiometric frequencies. Due, however, to the large individual variation in susceptibility to noise-induced hearing loss, it is necessary to specify H in terms of a statistical distribution, for each audiometric frequency, in a noise-exposed population.

DEVELOPMENT OF OCCUPATIONAL HEARING LOSS   The type of onset and subsequent growth of the value of H is of a characteristic pattern. In terms of smoothed median curves each applicable to a particular occupational sound level (in dB(A)), the hearing levels with continued exposure are shown in Fig. 11.2 for 4 kHz. Note that this figure shows median hearing level (not H) at an assumed age of 20 years for the start of work in noise. In order to derive H, the presbycusis (bottom curve) is subtracted from the hearing level value on any desired curve. The dramatic early deterioration as a result of high sound levels contrasts with the slow onset of presbycusis, but the two together summate, in the end, to produce disability in some persons.

The detailed picture shown by Robinson's (1970) analysis, reveals

an orderly and harmonious system of relations, which he has for-
mulated quantitatively with great elegance and economy of mathematical expression. Essentially, the relation between noise immission
level and H can be represented by a curve having the properties of
the trigonometrical function known as a hyperbolic tangent. The
curve is basically S-shaped; the rise is slow at first, then steepens and
finally flattens off, eventually tending towards a horizontal direction.
It is seen in the lower half of the nomogram of Fig. 11.10. The significance of this is that for low values of NIL, little deterioration occurs;
increased NIL gives rapid deterioration at the more susceptible frequencies, as seen in Fig. 11.1 on a linear plot, and finally, with continued exposure, deterioration slows markedly at these frequencies.
The striking feature of Robinson's analysis is his demonstration
that the same curve describes the deterioration at different audiometric frequencies and for different degrees of susceptibility of the
population, by an orderly system of shifts of the co-ordinate scales
of the curve, fully described by Robinson (1970). In this way the
older, strictly descriptive approach to the pattern of noise-induced
threshold shift is seen to be reducible, within entirely acceptable
limits, to this simple and fundamental basis. The usual pattern of a
rapid increase of hearing level, the '4 kHz dip' at the 4 kHz region,
followed by slowing of the process to a low rate of deterioration, is
mathematically reproduced. So also is the slower onset of deterioration at the lower frequencies, which, however, continue to deteriorate after the saturation effect is established at the 4 kHz region.
Eventually, for the most severe exposures continued for a long
period, the noise-induced shifts at the lower frequencies climb up
towards the values attained in the 4 kHz region. All this is faithfully
represented in Robinson's formulation. An illustration, in an
audiogram-like presentation of some of these features is seen in
Fig. 11.6. It should be studied in conjunction with Fig. 11.1 and
Fig. 11.10. It shows the calculated median values of noise-induced
threshold shift H as a function of A-weighted NIL (E_A), at the
various audiometric frequencies. The contours are placed at approximately 5 dB intervals on the H scale at 4 kHz, but it will be seen that
the E_A values are compressed at the high and low ends, corresponding to the initial and final portions of the hyperbolic curves of Fig.
11.10.

Inspection of Fig. 11.6 shows the noise-induced component growing but altering in configuration; always maximal at 4 kHz, the rate

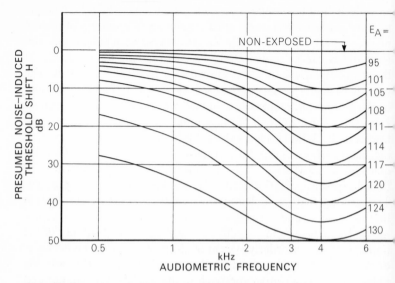

11.6 Median values of presumed noise-induced threshold shift at various frequencies calculated from Robinson (1970). The parameter is A-weighted noise
immission level (E_A). Note the compression in the lower and upper ranges of
the E_A scale, compared with the equal increments of H at 4 kHz; also the
different response at different frequencies. (After Burns & Robinson, 1970a.)

of deterioration becomes slower as the highest exposure levels are
reached. On the other hand the deterioration at the low frequencies
becomes the predominant feature when saturation is occurring at
the higher frequencies, at the higher values of E_A.

SUSCEPTIBILITY TO NOISE-INDUCED HEARING LOSS   When the
dispersion of values of hearing level rather than the median value is
examined, a more complex picture emerges. The dispersion found
in normal hearing levels of young people (Chapter 6) is much increased in the case of noise-affected ears.

   As an example, Fig. 11.7 shows the values of age-corrected hearing
level at 4 kHz, for 172 noise-exposed subjects against the logarithm
of time. The greater dispersion with longer durations of exposure,
as well as increased mean value of H, is striking. The fact that some
persons show little or no apparent ill-effects after many years of
exposure, whereas others deteriorate rapidly, is obvious. The range
of values between these extremes, for the same exposure, covers a

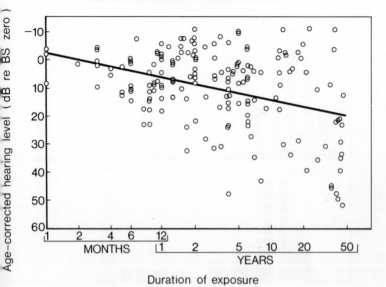

11.7   An illustration of the scatter of values of age-corrected hearing level
       for different durations of exposure. Audiometric zero, BS 2497:1954;
       frequency, 4 kHz; 172 subjects; noise level, 2%, A-weighted (L_{A2}),
       95–99 dB. (Burns & Robinson, 1970b.)

span of some 60 dB. Some variability is attributable to the audiometry itself, in which it is inherent (Appendix J), but the fact remains
that noise exposure which may seriously affect some persons, is
apparently innocuous to others.

When dispersions of this kind are depicted graphically, a convenient way of doing so is to use histograms, as in Fig. 11.8. This
shows, from an earlier study by Taylor *et al.* (1965) the degree of
presumed noise-induced threshold shift attained by different percentages of ears in four different groups of persons exposed to the
same noise for different durations, each during a period when the
noise-induced deterioration was not increasing markedly with time.
The very considerable spread of values is obvious, as well as some
tendency to the occurrence of isolated values for greater threshold
shifts. The standard deviation tends to increase for higher audiometric frequencies.

Another feature of these derived hearing levels is seen conspicuously in the histogram for 1000 Hz. It is that some values of presumed noise-induced threshold shift are negative. This, as expressed,

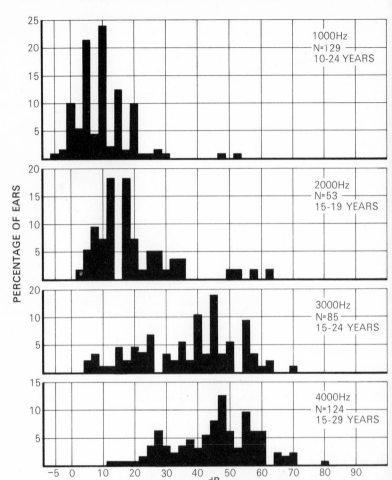

11.8   Distribution of presumed noise-induced threshold shift at 1000, 2000, 3000 and 4000 Hz at durations of exposure within which noise-induced threshold shift is not increasing markedly with time. Note the occurrence of some negative values in the histogram for 1000 Hz, making the term 'age-corrected hearing level' preferable in such cases. Audiometer steps 2·5 dB; N = ears. (Taylor, Pearson, Mair & Burns, 1965.)

implies that noise improved the hearing of some persons. This unlikely inference presents a logical difficulty in the invariable use of the term 'noise-induced threshold shift'. We may reasonably assume that this operation of deriving an age-corrected hearing level by the subtraction of a median value appropriate to age (or no subtraction at all for people 18–25 years of age) produces negative values of hearing level for those with more acute hearing, or small deterioration, or both. It is thus desirable that, in such presentations, this possibility is recognised by the use of the alternative term 'age-corrected hearing level'.

Hearing levels presented as a cumulative distribution (see p. 99) are shown in Fig. 11.9, from the data of Burns and Robinson (1970a). In this method, the distribution of age-corrected hearing level may be compared with a hypothetical normal distribution having the same standard deviation, by plotting the values as cumulative percentage distributions on so-called probability paper. This has a probability-integral scale on the vertical (ordinate) axis, while the hearing levels are plotted on a decibel scale on the abscissa. Thus in general terms the variable is shown horizontally, with the percentage of occurrence vertically. The vertical scale is compressed centrally, with 50% at the mid-point, in such a way that if the cumulative percentages of occurrence in a series conforming to a normal distribution are plotted, the points will fall upon a straight line. Thus, in a normal distribution, with the median value at 50%, the slope of the line would be determined for any particular horizontal scale of values by the size of their standard deviation. The larger the standard deviation the more will the line incline towards the horizontal. In Fig. 11.9 the cumulative percentage distributions of the values of age-corrected hearing levels of 4 groups of subjects are shown for the 4 kHz audiometric frequencies. Of these groups, one is a control, consisting of young persons not exposed to noise, no age correction being involved. The groups C, D and E represent persons who have sustained progressively greater noise exposure classified by noise immission level, which is least in C and most in E. The control group of 97 young, normally-hearing persons is seen, at 4 kHz, to have a nearly Gaussian distribution, the line through the points being almost straight. The noise-exposed groups, on the other hand, incline more towards the horizontal the greater the noise exposure, with its consequent increase of age-corrected hearing level. The scatter shown in this way progressively becomes greater the higher

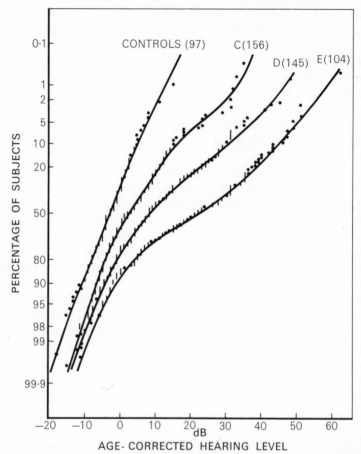

11.9 Cumulative distributions of hearing levels of four groups of subjects comprising: a control group (age 18–25), expressed as measured hearing levels; noise-exposed groups C, D and E in that order of ascending values of noise immission level, expressed as age-corrected hearing level (presumed noise-induced threshold shift). Frequency, 4 kHz; audiometric zero, BS 2497: 1954; figures in brackets are number of subjects in each group. The percentage scale shows the percentage of subjects with age-corrected hearing levels equalling or exceeding the corresponding value on the hearing level scale. (Burns & Robinson, 1970a.)

the noise immission level (i.e. the standard deviation increases with NIL) and this trend is accompanied by a progressively more marked distortion, away from the Gaussian distribution. This effect is quite systematic and takes the form of an inflexion; the curves deviate to the right at a position which is nearer the origin (i.e. lower percentages and lower values of H) the greater the NIL. Study of this figure is instructive, inasmuch as the median, standard deviation and any derived percentile values can be obtained. The increasing scatter with increased NIL in the noise-exposed groups is striking. In group E, for example, the extreme values range from over 60 dB to -10 dB, for a median value of about 22 dB. These tendencies are apparently a characteristic feature of noise-exposed populations, and have been found throughout this investigation and elsewhere (Taylor, Pearson, Mair & Burns, 1965).

The large spread of values of presumed noise-induced threshold shift obviously prompts the question of how much of this dispersion represents true variation in the degree of damage due to noise. The extent of the damage can be indicated by comparing the variation in audiometric hearing level of a similar population which has not been exposed to noise, with the variation in presumed noise-induced threshold shift of the noise-exposed population. Thus for example, if we take the 4000 Hz frequency of Fig. 11.8, with a standard deviation of 13·2 dB, strictly we should compare it with the standard deviation of a population of women of about the same age, without noise exposure, and preferably from the same environment as the noise exposed population. Data are not available in precisely this form, but for practical purposes a standard deviation of about 8 dB could be assumed, compared to about 13 dB for the values of presumed noise-induced threshold shift. When comparing such data, the variance, which is the square of the standard deviation, is used. Thus the variance in one case is 64 dB² and in the other 169 dB², the difference being 105 dB². The square root of this difference, about 10·2 dB, represents in terms of standard deviation the component of the total variability in the values of presumed noise-induced threshold shift which is attributable to the effects of the noise.

The question of dispersion of different origins, in the estimation of noise-induced hearing loss, is considered in detail by Robinson (1970, 1971) and by Robinson and Burdon (1970). These papers should be consulted for the detailed analysis.

CALCULATIONS INVOLVING NOISE IMMISSION LEVEL To return to the concept of noise immission level, we may examine how it can be made to indicate the effect on hearing of a particular exposure. The relation between exposure and hearing loss can be described by an equation developed by Robinson which encompasses audio-metric frequency and the statistical distribution of age-corrected hearing levels in an exposed population, by the incorporation of suitable constants, thus:

$$H_p = 27 \cdot 5 \left\{ 1 + \tanh \frac{E_A - \lambda_f + u_p}{15} \right\} + u_p \qquad (11.2)$$

where H_p = age-corrected hearing level equalled or exceeded by p percent of the population
E_A = A-weighted noise immission level
λ_f = a constant depending on audiometric frequency
u_p = a constant depending on the selected percentage p.

If actual hearing level equalled or exceeded by p percent of the population (H'_p) is required to be estimated, the expected median value of hearing level appropriate to the audiometric frequency and age is added to the value of H_p.

To find H'_p

$$\begin{array}{lll} & H'_p = H_p + C_f(N - 20)^2 & \text{(for } N > 20\text{)} \qquad (11.3) \\ \text{or} & H'_p = H_p & \text{(for } N \leqslant 20\text{)} \end{array}$$

where C_f = a coefficient depending on frequency
N = age in years.

However, the formula has been translated into nomographic form by Robinson, which makes very rapid estimates possible. The nomogram is given in Fig. 11.10 and as an example of its application we may take the assessment of the potential hazard to hearing of a given noise environment.

ASSESSMENT OF HEARING HAZARD   In this case, we assume a steady noise for 8 hours per day, 5 days per week, of 88 dB(A). The median effect on hearing at 4 kHz of such a spectrum, over a working life of 40 years is to be estimated. The steps are

Step 1   Determine the sound level A of the
noise   88 dB(A)
Select the desired duration   40 years
Step 2   Calculate E_A, i.e. $L_A + 10 \log 40$,
and enter the appropriate NIL curve
(interpolated if necessary) in top $88 + 10 \log_{10} 40$
half of nomogram $= 104$

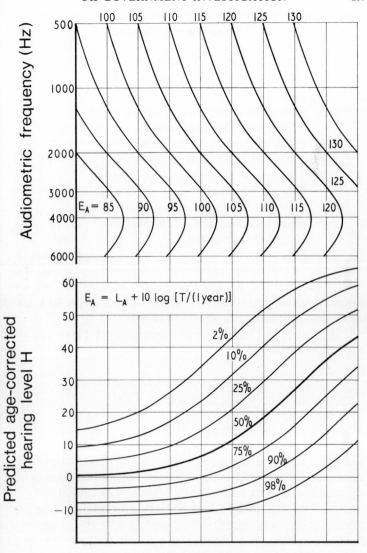

11.10 Nomographic chart relating noise exposure in A-weighted noise immission level (E_A) with the percentage of persons equalling or exceeding the indicated values of age-corrected hearing level H, for different audiometric frequencies. For use see text. (Burns & Robinson, 1970a.)

Step 3	Locate the intersection of the NIL curve (or interpolated curve) with the audiometric frequency scale	4 kHz
Step 4	Descend vertically from this intersection to the lower half of the nomogram	
Step 5	Locate the intersection of this vertical line with the median curve	50% curve
Step 6	Read out the value of the intersection on the left hand lower vertical scale. This gives the presumed noise-induced threshold shift or age-corrected hearing level (H) relative to controls, equalled or exceeded by 50% of those exposed	13·5 dB approximately

To illustrate the method of finding H', the actual hearing level, assume that the start of work is at age 18 years with 40 years of subsequent exposure, then age = 58 years.

Using Equation 11.3, the median hearing level due to age alone is found and added to H:

$$H' = 13·5 + 0·0120(58 - 20)^2$$
$$= 13·5 + 17·3$$
$$= 30·8 \text{ dB}$$

This value H' is the hearing level equalled or exceeded by 50% of persons of this age free from ear disease or any other source of hearing impairment, if they sustained an exposure of 104 dB NIL. This estimate is applicable to any value of p (the percentage of the population equalled or exceeded) in addition to the 50% value used in the illustration above. For example, to estimate the value of hearing level equalled or exceeded by 10% of the exposed population for the same conditions as the example, we use the same procedure down to step 4. At step 5, instead of the 50% curve, we find the intersection on the bottom half of the nomogram of the vertical line with the 10% curve. The value is approximately 34 dB. To find the corresponding hearing level we again add the expected median value due to age from equation 11.3:

$$H' = 34 + 17 \cdot 3$$

$$= 51 \cdot 3 \text{ dB}$$

This is the hearing level at 4 kHz equalled or exceeded by 10% of persons aged 58 years, who have been exposed to an NIL of 104 dB.

In Appendix N are given all constants needed to work equations 11.2 and 11.3, together with tabulations of the expression

$$27 \cdot 5 \left\{ 1 + \tanh \frac{E_A - \lambda_f + u_p}{15} \right\}$$

for various values of $(E_A - \lambda_f + u_p)$

OTHER APPLICATIONS   For example, it can be determined what degree of exposure is compatible with some defined degree of restriction of occupational hearing loss, in a given percentage of the exposed population. This exposure can be translated into various combinations of sound level and duration, by means of the equivalence or 'trading relation' of level and time whereby 3 dB added is equivalent to halving the duration, and vice versa.

In addition mixed exposures at different levels and durations can easily be reduced to a single value of NIL. In such cases the NIL value is ascertained for each component of the exposure separately. These NIL values are then added logarithmically (Appendix F) and the resultant value used to calculate the value of H, either by equation 11.2 or by the nomogram of Fig. 11.10.

Where the exposure, or a component of it, consists of daily durations of less than 8 hours, the contribution of the duration component in the exposure $E_A = L_A + 10 \log t/t_0$, is reduced accordingly. Thus a year's exposure at 4 h/day would be calculated as 1 year at 8 h/day and reduced by 3 dB, since it represents one half of the energy. Any other fraction can simply be derived by subtracting the appropriate value in dB, from the NIL value for 8 hour working.

This principle has been adopted in Recommendation R1999 (ISO, 1971). The calculations have been greatly simplified to enable the basic data to be used to predict the consequences of different exposures. The procedure describes how occupational noise exposure, including complex patterns of level and duration, within a period of 1 week, may be expressed quantitatively, and from this is derived an estimate of the percentage of people expected to show a specified increase of hearing level as a consequence of the noise.

Specifically, measurements are made during one week (the standard is limited in its assumption that all weeks are the same) of the level in dB(A) and duration in minutes or hours, or if the exposure is not uniform, the levels and durations of each of the separate components. From a table, each component is assigned an additive index, the *partial noise-exposure index*. These are then added arithmetically, the sum being designated the *composite* (i.e. weekly) *noise-exposure index*. The latter is then used to indicate from a table the *equivalent continuous sound level*. This is the sound level in dB(A) which if present for 40 hours in one week, would produce the same exposure as the various components, any possible effects of intermittency being ignored. Appendix N gives the necessary tables (ISO, 1971) for deriving the equivalent continuous sound level. The equivalent continuous sound level is used in R1999 to indicate from a table the percentage of persons in which a defined degree of hearing impairment will occur as a result of occupational exposure to the particular sound level. Periods of exposure of up to 45 years for normal working durations as well as the effect of age, are included. The *risk* of hearing impairment is here taken to be the difference between the percentage of people with impaired hearing in a noise-exposed group, and the percentage of people with impaired hearing in a group not exposed to noise, but otherwise equivalent. The latter percentage is based on the distribution of threshold shifts due to age and also a certain component of ear pathology. This is a basically unsatisfactory feature, inasmuch as the degree of pathological hearing loss is likely to vary between populations. However, if a pathology-free population is assumed, the results would not be typical of working, or any other, real populations. Thus in the data in this ISO Recommendation actual industrial populations have been used to arrive at a tentative working value for the basic distributions of hearing impairment due to noise, age and some ear pathology. The index of hearing impairment has been taken to be the attainment of a mean hearing level of 25 dB, from the individual hearing levels at the 500, 1000 and 2000 Hz audiometric frequencies. Both this index, and the hearing levels taken to represent a population not exposed to noise, may be reviewed in the future.

Robinson's formulations can of course be used in the estimation of quantities associated with the statistical concept of risk, which is discussed below. Suppose that we wish to estimate the percentage of persons with known exposure who may be expected to attain, as a result of exposure to a specified noise immission level, at a particular frequency or combination of frequencies a hearing level equal to or greater than a defined value. For example, using the previous example of a noise immission level of 104 dB, find the percentage of persons aged, say 60 years, whose hearing level H' equals or exceeds, at 4 kHz, some specified hearing level, e.g. 45 dB. To do this, first subtract the correction for 60 years of age at 4 kHz, which is approximately 19 dB. The resulting age-corrected hearing level is 45 − 19 = 26 dB. Now, from the upper part of the nomogram of Fig. 11.10, establish the point of intersection of the 104 dB NIL curve with the 4 kHz line, and descend vertically to the lower part of the nomogram. Where this vertical line intersects the vertical scale of age-corrected hearing level at $H = 26$, the percentage p is read by interpolation. In this case the value is approximately 20%. That is to say 20% of subjects aged 60 years with an exposure of 104 dB NIL, in the absence of any other cause of hearing loss other than noise or advancing age, may be expected to have attained, or exceeded, at 4 kHz, a hearing level of 45 dB.

This percentage will be inflated if pathological ears, such as would occur in any unselected population were to be included (Burns, Wood, Stead & Penney, 1970). The concept of risk has been considered in some detail by Robinson (1971) and will be discussed in the appropriate context in Chapter 12.

Frequency characteristics of the exposure

A wide-band noise, such as commonly encountered in industry, can be divided for the purposes of frequency analysis into bands of various widths (Chapter 4). One octave is a convenient bandwidth, and the effect on the ear of such bands can to some extent be regarded as separate. Thus a particular frequency band can be assumed to influence hearing over a particular frequency range. These conclusions stem from a number of investigations by means of temporary threshold shift (Kryter, 1963; Kylin, 1960) and accord with the 'place theory' of cochlear dynamics. It may be further stated that the amount of temporary threshold shift at the frequencies associated with individual bands of noise is about the same irrespective of whether the various noise bands are presented singly, or together as one wide spectrum noise. It has been assumed that this principle is applicable also to permanent threshold shifts. This is a fundamental concept, since it enables the effect on hearing of a wide band noise to be judged on the basis of the levels in various frequency bands, such as octave bands.

If we assume that the frequency relations between the exposure noise and the resultant threshold shift are in fact similar for temporary and permanent threshold shifts, it is possible to ascribe noise-induced changes at particular frequencies to particular bands of noise (Ward, Glorig & Sklar, 1959; Kryter, 1963). Thus noise in the octave band 300–600 Hz would affect hearing mainly about 1000 Hz; 600–1200 Hz, about 1500 Hz; 1200–2400 Hz, about 3000 Hz; and 2400–4800, 4000 Hz or 6000 Hz. The permanent threshold shifts would be expected to be spread rather widely over 2–3 octaves around the particular frequencies and the general upward trend of the threshold shift with respect to the noise exposure frequency is obvious (Fig. 10.7, Chapter 10).

The data derived from actual investigations of permanent threshold shifts and the associated noise exposure are not in fact very plentiful. Many of the present working approximations are derived from temporary threshold shift data. Taking all the evidence available, however, there is a clear tendency for the ear to be more tolerant of noise at the low than the middle and high frequencies. The evidence suggests that the ear is particularly vulnerable to frequencies in the range 2000–4000 Hz. This tendency is certainly seen in temporary threshold shift, which is greater the higher the exposure frequency, at least up to 4000 or even 6000 Hz (Ward, 1963). In considering the effect on the ear of different frequencies of exposure, it is relevant to note the particular susceptibility to threshold shift in the range 3000–6000 Hz as a result of noise. This effect was discussed at some length in connection with temporary threshold shift. The permanent noise-induced threshold shift in this frequency range, and particularly associated with the audiometric frequency of 4000 Hz, is so typically found as a result of such causes as industrial noise, gunfire noise or head injury, that it is obviously not acutely dependent on the exact pattern of the exposure spectrum. It is a conspicuous warning sign of the utmost use in anticipating occupational hearing loss.

Robinson (1970) has investigated the influence of different types of frequency spectrum of the noise in the causation of occupational hearing loss. The various spectra investigated were all those of industrial noises, and the average spectrum was of a rather flat configuration. The extremes ranged from spectra rising with increasing frequency to those falling with increasing frequency within the limits shown in Fig. 11.11. When the mean age-corrected audiograms of groups of subjects exposed to the extremes of rising or falling spectra are compared, a result of great significance emerges;

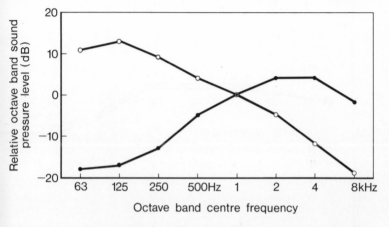

11.11   Limiting shapes of rising or falling spectra for the occupational noises used in the comparison of different weightings, of Fig. 11.12. (Burns & Robinson, 1970a.)

the 4 kHz dip in the audiogram remains the conspicuous feature in each case, and the difference in magnitude of the age-corrected hearing level is not large. Specifically, noises with substantially the same sound level expressed in dB(A), produce slightly more noise-induced hearing loss if they are of the falling spectrum type than in the case of those with rising spectra. The explanation is to be sought in the weighing used. In Fig. 11.12 (Burns & Robinson, 1970b) are shown the results of using the A-, B- and D-weighting to quantify noise exposure. It can be seen that the same exposures, as measured by the three weightings, produce different age-corrected hearing levels. In the case of A-weighting, the greater noise-induced change with falling spectra would appear to indicate that the reduction in in the contribution of the low frequencies to the A-weighted level is a little too much. The D-weighting shows the same effect, presumably associated with its enhancement of the high frequencies. The

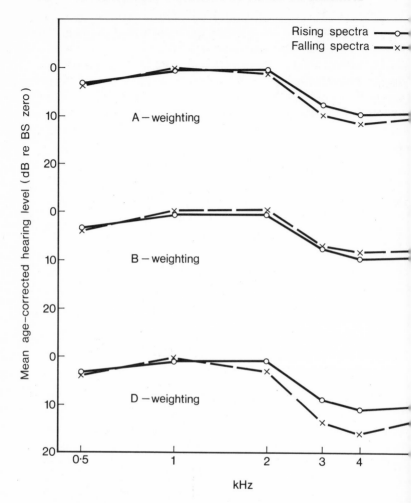

11.12   Mean age-corrected hearing levels of groups of persons with matched average exposures to noise of (a) rising or (b) falling spectra, the exposure calculated on the basis of A-, B- or D-weighting. (Burns & Robinson, 1970b.)

B-weighting is the best on the basis of this data, although it apparently goes slightly beyond the optimum, in the opposite direction to the A-weighting. Robinson (1970) found that for spectra rising or falling markedly (gradients of not more than 5 dB/octave) the use of A-weighting in estimating noise-induced threshold shifts implies an error of not more than about plus or minus 2 dB for noises with rising or falling spectra respectively, compared to the average of the various industrial spectra investigated.

Further investigation is required to provide more precise relations between the spectral characteristics of the stimulus and the pattern of noise-induced threshold shift. The above latitude in spectral slopes, however, appears to be accompanied by such modest errors, for A-weighting, that the claims of other weightings, such as B, must be assessed on the basis of further studies.

Assessment of susceptibility to noise-induced hearing loss

The very varied effect of the same noise exposure on the hearing of different individuals, previously noted, has prompted much speculation on possible ways of estimating susceptibility in a quantitative way. Great practical advantages would accrue from such an estimate. In the meantime, it is convenient to note the efforts, unfortunately not fully successful, that have been made to establish relations between temporary and permanent threshold shift, in a search for the basis of a prognostic test of susceptibility.

It has long been assumed that susceptibility to TTS is correlated with susceptibility to occupational hearing loss. This hypothesis has been difficult to prove or to refute. Jerger and Carhart (1956) appear to be the first to produce data supporting a relation; they showed low correlation coefficients, but nevertheless positive and statistically significant, between measures of permanent threshold shift and TTS. A great volume of work before and since has failed to show a correlation between temporary and permanent threshold changes in individuals, as Ward (1965) points out. Another effort to investigate these relations was made by Burns, Stead and Penney (1970). In individuals, positive and statistically significant correlations of rather low numerical value ($+0.38$ at best) were found between measures of susceptibility to TTS and permanent threshold shift (PTS). The methods and results are summarised here.

Two indices, D_T and D_P, of susceptibility to TTS and PTS respectively, were correlated. They depended on the availability of subjects who had known histories of exposure to one noise only, in which they worked daily. The derivation of the indices was as follows.

Index D_T   The TTS due to one day in the person's occupational noise is the basic quantity for this index. The threshold shift, six minutes after the end of the day's work (TTS_6), was corrected for the hearing level of the particular person, since TTS and hearing level have been found by Ward to be inversely related. This correction was made by deriving the regression line between TTS and PTS (for the same frequency or frequency combination) in a group of

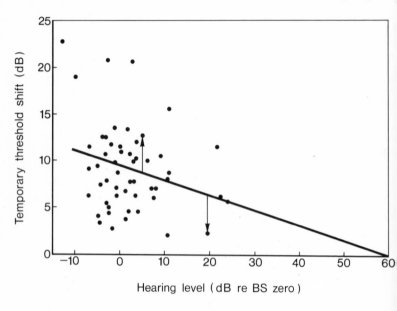

11.13   Derivation of values of D_T for a group of 53 persons exposed occupationally to sound level L_{A2} 99–104 dB. The deviations, two of which are indicated by arrows, from the regression line constitute the values of D_T. Frequency: mean of 1 and 2 kHz. (Burns & Robinson, 1970a.)

subjects subjected to the same occupational noise. The deviation in decibels (positive or negative) of the value of TTS for the particular person, from the regression line, is the value D_T. Obviously, this can be done for any audiometric frequency or frequency combination. The derivation of D_T for the mean of 1 and 2 kHz, is shown in Fig. 11.13. The arrows illustrate two individual cases, both of approximately 4 dB, one positive, one negative.

It can be seen that this procedure has the effect of expressing each TTS value relative to a mean value, adjusted for hearing level throughout the range.

Index D_P This is also a relative value, in which the hearing level of each individual is expressed relative to the expected median for his particular noise exposure history. It is, specifically, the difference in decibels between the observed age-corrected hearing level and the median age-corrected hearing level predicted from Robinson's equations (11.2 and 11.3). Absolute values, as in the case of D_T, are

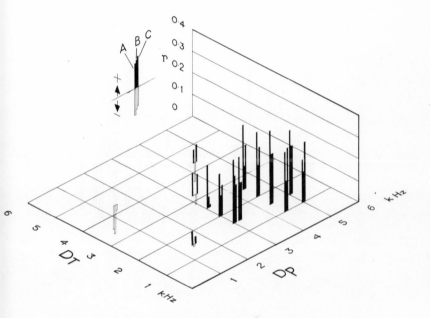

11.14 Correlation of D_T and D_P. The position of the vertical bars on the horizontal surface indicates the frequency or average of the frequencies used for deriving D_T and D_P. The height of the solid bars indicates on the vertical surface, the positive value of the correlation coefficient r for the regression of D_T on D_P. Negative values of r are shown on the same scale, as open bars pointing downwards below the horizontal surface. A, B and C indicate 3 groups of subjects habitually exposed to occupational noise of intensity diminishing from A to C. (Burns & Robinson, 1970a.)

thus avoided, and only the person's position in the range of values denoting susceptibility, is involved. Persons with different ages, noise histories and hearing levels may thus be legitimately compared.

Correlation of D_T and D_P This was done within groups classified by the level of the occupational noise. Obviously the audiometric frequencies or combinations of frequencies employed could be various; the significant combination was unexpected: low frequencies for D_T, high for D_P. This yielded a value of $+0.34$ for the product-moment correlation coefficient, which was increased to $+0.38$ by the use of the size of the 4 kHz dip rather than its absolute value. This is statistically significant, and indicates that TTS susceptibility is a valid index of PTS susceptibility. The value of the

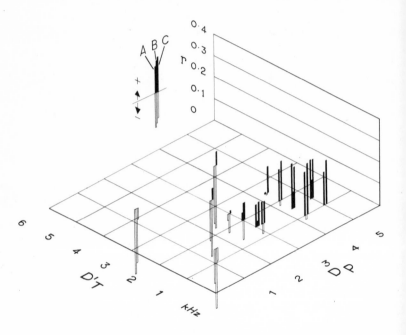

11.15   Correlation of D_T' and D_P. The plot is the same as Fig. 11.14 except for the use of D_T' instead of D_T. Note the occurrence of expected negative values of r in certain frequency combinations, but the persistence of positive values in the favourable correlations. (Burns & Robinson, 1970a.)

correlation coefficient is, unfortunately, too low as it stands for a generally useful prognostic test, say for persons proposing to enter a noisy occupation, but the results appear to merit further development.

These relations are shown in a 3-dimensional plot Fig. 11.14, which illustrates the optimum condition: D_T derived from the arithmetical average of the TTS values at 1 and 2 kHz, D_T from the mean values of 3, 4 and 6 kHz.

The validity of this correlation might be questioned on the grounds that D_P is directly related to hearing level and so also is D_T, through the methods of derivation. This link can be eliminated by deriving the TTS index, in this case designated D'_T, from the arithmetical mean of the values of TTS of the individual subjects of the group, and not from the regression line of Fig. 11.13 as in the case of D_T. The effect of this is shown in Fig. 11.15 which is identical with Fig. 11.14, except that D'_T replaces D_T. The values of the correlation co-efficients are expectedly less, but in the optimum range, they remain positive, and in the most favourable conditions reached significance level ($p = 0.05$). Improvements in the test seem possible by reduction of the variance of the audiometric basis by replication, and possibly by the introduction of a time component into the derivation of the index D_T. The former possibility receives support from the fact that, if groups of 9 or 10 individuals are used in the regression of D_T on D_P, the correlation coefficient rises to the much more encouraging value of $+0.74$.

These results afford some encouragement in a search for an effective predictive test for susceptibility to noise-induced hearing loss, and as such do not agree with such negative findings as those of Miller, Watson and Covell (1963). These workers found, in animal experiments, that maximum TTS occurred at 4 kHz for a particular noise, while permanent losses from the same noise were maximal at 2 kHz. In the meantime, the Burns, Stead and Penney approach is still inadequate for use for individuals.

12

Preservation of hearing

The usual situation requiring precautions against noise-induced hearing loss is noise at work. The basic relations between noise and its effects on hearing, which have been discussed in the previous two chapters, form the basis of the practical measures to combat occupational hearing loss. The formulation of rules for this purpose, however, is fraught with various difficulties, so that the problem is more difficult than many similar situations in occupational medicine, where people must be protected from some potentially harmful effect or agent.

Hearing preservation (also known as hearing conservation) is based on the following principles (Burns & Littler, 1960; Burns, 1965). The basic requirements are simple: noise exposure must be known and controlled, and ideally each person's hearing should be measured before employment and at intervals throughout the period of employment if a hazard is judged to exist.

Objectives of hearing preservation

One question of a fundamental nature must be answered, and in a quantitative way, before any practical code can be laid down. This concerns the degree of protection of hearing which it is desired to achieve. At first sight, it might seem self-evident that the aim should be to avoid any hearing deterioration whatsoever attributable to noise exposure sustained in the course of work, by every person so exposed. The difficulty in attaining an objective such as this is the variability in susceptibility to noise-induced permanent threshold shift, which was noted in the previous chapter. We cannot, in consequence, determine a level of exposure which will give strictly predictable effects on the individual, only that the average effect will be in a certain range, and that approximate proportions of the population will suffer less or more than the average, according to the distribution of susceptibility. Since the effects at the different audiometric frequencies are different, the situation is complicated. In addition, it is highly probable that restriction of exposure so that no deterioration whatever, at any audiometric frequency, would occur in the most susceptible individuals would over-protect those less susceptible, so causing unnecessary inconvenience to the majority, for instance in the wearing of ear protection. The restriction of noise exposure for this purpose would be fairly severe and such exposure could probably be exceeded in the course of work or even recreation, without much justification for regarding the noise environment as significantly hazardous to hearing. For these reasons, as well as others of a practical nature, it has tended to become accepted that in devising regulations for the preservation of hearing, it should be hearing for speech that should primarily be protected. Speech is carried on frequencies between 100 and 8000 Hz and of these the essential minimum range covers from 300 to 3000 Hz. This range must therefore be protected. We shall discuss how this may be achieved, or at least as far as present knowledge permits. Recalling the requirements for a practical procedure for hearing preservation, we shall review these in this sequence: (a) determination of noise exposure in physical terms; (b) assessment of exposure as a risk to hearing; (c) restriction of exposure, if found necessary in the light of (b); (d) measurement of hearing before and during employment in noisy occupations. Assessment of disability is of the highest importance. It is a clinical problem and it will be referred to briefly later. First, however, the relevant background to practical hearing preservation may be profitably discussed.

Review of rules for hearing preservation

The possible effect of a particular noise on hearing must be judged from the relations discussed in the previous chapter.

Depending on the degree of preservation of hearing aimed at, some physical description of the permissible noise exposure will be decided upon, to meet the particular requirements. These requirements, in terms of an intended degree of protection against occupational hearing loss or the acceptance of some defined amount of occupational hearing loss at various frequencies, have been called a *criterion* by Kryter, Ward, Miller and Eldredge (1966). The physical description of the noise compatible with the attainment of a particular criterion will be described as a *noise specification* (Burns, 1965). Many such specifications (some have in fact been described as criteria) have been suggested.

The basic assumption has been that continuous broadband noise, fairly evenly distributed throughout the spectrum, without prominent pure tones or intense narrow frequency bands, continued throughout the working day for about five days per week for periods of years, is the acoustic stimulus. Specifications have usually taken the form of SPL values in octave bands covering as much of the audio frequency range as knowledge justified. Should the noise be discontinuous or impulsive, or composed of pure tones, other specifications of a supplementary nature to the above have been suggested.

Historically, previous recommendations have been strongly influenced by the American National Standards Institute (ANSI) (1954) data, and most have made concessions in a compromise between the nearly impossible aim of complete protection for all those exposed, and the less exacting target of adequate protection of the speech frequencies. Following suggestions by Kryter (1950) and Rosenblith and Stevens (1953), a comprehensive code was issued by the United States Air Force (1956). The basis of the regulation was that, in any of the octaves 300–600, 600–1200, 1200–2400 and 2400–4800 Hz, should the noise attain SPL values of 85 dB, the risk to hearing was 'small', but that measures to conserve hearing were desirable. Detailed recommendation for assessment of the noise and the operation of the conservative measures were provided and the specification was based on a daily duration of exposure of about 8 hours, with higher permissible levels if the daily durations were reduced. The American Academy of Ophthalmology and Otolaryngology (1957) advised action to safeguard hearing if 85 dB SPL was attained in either of two octaves only, 300-600 and 600-1200 Hz, at the usual eight-hour daily duration. Appreciably lower values of SPL appeared to T. S. Littler and the author to be necessary, after examining the ANSI (1954) data, if damage was to be avoided in most persons during an indefinite period of normal working hours (Littler, 1958). This specification is however somewhat rigorous, and another specification, still reasonably cautious, but permitting somewhat higher SPL values, was suggested (Burns & Littler, 1960) and subsequently was re-stated, in ISO octaves (Burns, 1965), in the light of all the information then available. Glorig, Ward and Nixon (1962) on the basis of their studies into permanent and temporary threshold shifts, suggested that the proposed ISO NR 85 contour (Kosten & van Os, 1962) should be used s the specification. Kryter, Ward, Miller and Eldredge (1966) employed spectra for specifying exposure limits. Since that time, however, a marked swing of opinion, at a national and international level, has occurred, in favour of the use of A-weighted sound levels for a variety of purposes. These applications will have been noted in the contents of Chapters 8, 9 and 11 and an ISO Recommendation (ISO, 1971) has been published, using sound level A to quantify noise exposure for hearing conservation purposes. These practical applications of sound level A in specifications for the control of noise-induced hearing loss will be dealt with below.

It is relevant to return to the underlying purpose of these specifications. To recapitulate, the starting point is the degree of protection, or conversely the degree of deterioration, of hearing which is regarded as acceptable. If all persons reacted in an identical way, a sharp line of demarcation could be drawn which would separate damage from no damage and the problem would then be easily controlled. The variability of normal hearing, the variability of change to higher hearing levels with advancing age, and the variability in susceptibility to noise-induced change, makes such an idealistic approach illusory.

The consequence of these facts is that noise exposure at work would have to be at the same level as in everyday life, to enable it to be said that occupational noise had been eliminated as a possible source of hearing deterioration. This would leave the problem vague, since the noise of everyday life is variable, but certainly not negligible. In addition some recreations (shooting is a recognised hazard to hearing) may be capable of causing hearing deterioration, especially in the more susceptible individuals.

It might be possible to hazard a guess that occupational noise at or less than the octave band SPL values of the ISO NR 65 or 70 contours (corresponding to about 70 to 75 dB(A)) would be unlikely to add very much to the exposure of everyday life and thus to influence significantly the prevailing levels of hearing in a population in a technologically developed environment. This, however, is speculation, and at the present time the suggestion of a general imposition of such a level is not realistic. Let us say then that over a working day, to give no significant excess over what would be experienced by urban dwellers not engaged in noisy occupations, noise would have to be restricted to some level of exposure in the region of 70 or even 65 dB(A). Kosten and van Os (1962) actually suggest the former level for 'workshops' (Chapter 9). Any industrial plant operating at this level throughout the working day would be free from any criticism from the point of view of noise, and even the most susceptible ears could not be regarded as suffering significant change due to occupational noise, if the latter can be regarded as not contributing to an ordinary urban noise exposure. Thus, in the courtyard of a residential building the noise reached 60–70 dB(A) during the day, and inside a small saloon car 70 dB(A) would be usual at 48 km/h (30 mile/h), while a suburban electric train was measured at 76 dB(A) (Committee on the Problem of Noise, 1963).

Such maximum levels would, however, be difficult to impose, because of the practical difficulties of achieving them in many industrial situations, and the fact that the great majority of people would not agree that such restrictions were necessary, or would quite reasonably refuse to go to the trouble of wearing ear protection or adopting other precautions to reduce a noise which was not felt to be objectionable in any case. In the foregoing discussion, the fact should be remembered that the initial noise-induced change for occupational noise is almost always in the 3000–6000 Hz frequency range, and that threshold shifts at lower and higher frequencies are of lesser degree. This effect is a fortunate one in two ways; it provides the early warning of a rise in hearing level at or about 4000 Hz, and it means that appreciable elevation of hearing level at 4000 Hz is quite compatible with the retention of speech perception to a reasonable degree, provided the lower frequencies are unchanged or changed much less than at 4000 Hz. For example, the person whose audiogram is shown in Fig. 11.4A was subjectively unaware of any impairment of hearing; although speech perception must have been affected to some extent, it had not been noticed. In view of these facts it has become usual, as we noted briefly above, to attempt to ensure that minimal changes in hearing level occur in the frequencies below 4000 Hz, in order to prevent impairment of speech perception. In practice, this has involved designating a criterion which defines either permitted permanent noise-induced threshold shifts expressed as median values, for some prolonged period of years; or, alternatively, degrees of permanent noise-induced threshold shift which should not be reached. An estimate of what this means in terms of noise exposure is then made, and this is the noise exposure specification appropriate to the particular criterion. It is the yardstick with which any particular noise exposure is compared, in order to assess

ts possible danger to hearing. In the original specification of Burns
and Littler (Littler, 1958) the intention was to provide a noise
specification which would result in no loss of hearing to an unspeci-
fied majority of persons after a normal working life-time's exposure
to the noise. This specification, in certain circumstances, would in
practice be most difficult or impossible to attain, which prompted
the subsequent specifications (Burns & Littler, 1960; Committee on
the Problem of Noise, 1963; Burns, 1965). These were based on all
available information, and the criterion adopted, again on the basis
of a normal lifetime of work, at a nominal 8 hours per day for perhaps up to 40 years, was that the presumed noise-induced threshold
shift (H) should be prevented from reaching these median values:
at 4000 Hz, 15 dB; at 3000 Hz, between 10 and 15 dB; at 2000
and below, less than 10 dB. The estimate of the noise specification corresponding to the criterion consisted of the SPL values in the
various octaves as was the current practice at that time, and hearing
preservation measures were advised if the sound spectrum at any
octave centre frequency attained or exceeded the corresponding
specification values. These corresponded to a sound level of 88 dB(A).

These levels should not be regarded as acceptable but as defining
an arbitrary level of hazard.

Implications of noise specifications

Having thus surveyed the scene historically, the prescription of
acceptable noise environments for work can be considered in the
prevailing climate of opinion and practice. In the situation just
recounted, which obtained until about 1965, the industrial medical
or safety officer was offered a single noise specification, with additional guidance on the effect of reduced duration and of intermittency (Kryter, Ward, Miller & Eldredge, 1966); today, however,
wider options are open. These derive from the data mainly of Baughn
(1966) and Burns and Robinson (1970). In the latter study, described
in Chapter 11, Robinson developed the concept of noise immission
level (NIL), and with it the benefits already explained, of convenient
assessments and predictions of the effect of actual or hypothetical
noise exposures on the basis of A-weighted sound energy. The
principle having been embraced by ISO in Recommendation 1999
(ISO, 1971), it is now likely to receive widespread usage so as to
permit further refinement, if experience shows this to be necessary.

Thus, instead of the obligation to accept or reject a specific spectrum or daily sound level, the primary quantity to be decided is that of total exposure, in terms of A-weighted NIL suitably parcelled into acceptable daily exposures. Provided the noise spectrum is not excessively sloped (not exceeding 5 dB/octave in the range 63 Hz to 8000 Hz) the NIL value appropriate to a selected criterion can be determined. Such a criterion would imply a maximum permissible degree of noise-induced threshold shift over a defined period of exposure, for a defined percentage of the population. These are administrative decisions, in which informed opinion must be involved. They are of the essence of preventive medicine in the field of industrial health, and the levels selected must be related to the type of exposure; duration; intermittency; medical management of exposed populations, and other factors.

In reaching a decision on permissible total NIL in a working life time, the effect of age and of disease on hearing must be included to obtain a realistic picture of the end result in terms of hearing levels as a statistical distribution throughout a noise-exposed working population at specified ages.

We may give more realism to these considerations by taking an actual example of a specification (Burns, 1965) which has actually been used and which represents an 8-hour daily level of 88 dB(A). This, as noted above, was envisaged as covering a 40-year working life, but even this may be inadequate in some cases. The studies of Taylor, Pearson, Mair and Burns (1964), and of Burns and Robinson (1970) included subjects who had spent up to 52 and 50 years respectively, in the same noise environment. It would be realistic, therefore to consider durations of 40 and 50 years at a level of 88 dB(A); using Robinson's (1970) procedure for estimating values of H, these combinations then give A-weighted NIL (E_A, equation 11.1) values thus:

$$\text{for 40 years NIL} = 88 + 10 \log_{10} 40 = 104 \text{ dB}$$
$$\text{for 50 years NIL} = 88 + 10 \log_{10} 50 = 105 \text{ dB}$$

The implications of these noise immission levels can now be examined in the light of the predictions of Equation 11.1 and compared with the criterion values for which the specification of 88 dB(A) was intended (Table 12.1).

It can be seen that for an A-weighted noise immission level of 104 dB (i.e. 40 years at 88 dB(A)), the original criterion is easily met for 50 years (NIL = 105 dB) the values of H are still just within the

TABLE 12.1

Effect on hearing of daily exposure to 88 dB(A) for different durations

Audiometric frequency Hz	Median age-corrected hearing level (H) dB		
	Intended criterion values 40 years	Calculated values	
		40 years $E_A = 104$	50 years $E_A = 105$
500	} 'Less than 10 dB'	1·7	1·9
1000		2·6	3·0
2000		5·8	6·5
3000	'Between 10 and 15 dB'	10·9	12·1
4000	'Less than 15 dB'	13·4	14·8

limits originally stipulated. In conformity with the NIL concept, the NIL value of 105 can be attained for higher sound levels and shorter durations. For example, if 30 years is the maximum duration, a level of 90 dB(A) would be permissible (NIL = 90 + 10 $\log_{10}$ 30).

The modest values of H in Table 12.1 should not, however, be allowed to obscure the realities of the situation. These values are medians, and if the distribution is now calculated, again from Robinson's equations, an appreciably less reassuring picture emerges. These distributions should for a full appreciation be considered in the form of actual hearing levels (H') at different ages. A selection of these levels for various centiles of the population for age 70, are given in Table 12.2.

These values, instructive though they are, are less immediately comprehensible than the single figure average values of hearing level which have been used, and still are used, to provide an indication of social adequacy of hearing. Such indices have been framed on the basis that speech perception is the main determinant of the social adequacy of hearing and much effort has been expended in trying to establish a valid index using pure-tone threshold levels. Finality has by no means been reached, and the most widely-used index of this sort is that described by Davis (1962; 1970) and recommended by the Committee on Conservation of Hearing of the American Academy of Ophthalmology and Otolaryngology (AAOO). Reference to Table 12.3 shows that the index is very simply derived; it is the average of the hearing levels at 0·5, 1 and 2 kHz, and the range of

TABLE 12.2

Hearing levels at age 70, as a result of a noise immission level (A-weighted) of 105 dB.
for different centiles of the exposed population

Frequency	Hearing level (median) age only	Hearing levels, age + noise for different centiles of the population dB				
kHz	dB	50	25	10	5	1
0·5	10	11·9	17·3	22·7	26·3	34·1
1	10·7	13·7	19·8	26·0	30·2	39·3
2	15	21·5	29·3	37·8	43·4	54·4
3	20	32·1	42·2	52·0	58·1	69·2
4	30	44·8	55·5	65·6	71·7	82·5
6	35	45·9	55·6	65·2	71·3	82·5

effectively normal speech perception was taken to be up to about 15 dB on the then current American standard for normal hearing, equivalent to about 25 dB ISO. Specifically, the normal practice (Davis, 1970) is now to regard 26 dB as signifying zero handicap. The table shows the probable effect of higher mean values of the index. This system has become incorporated into many rules and has penetrated into State legislature in the USA as an indication of hearing ability (or handicap), from any cause.

This AAOO method, using 0·5, 1 and 2 kHz to derive the average, may not be ideal, and other combinations have been examined, as has speech audiometry, as indices of social disability. This has been approached by means of comparisons of the particular index with the results of questionnaires designed to elicit the desired information on the effectiveness of hearing in everyday work, recreation and social activities. Of the many possibilities, an index derived from the mean of the hearing levels at 1, 2 and 3 kHz has been considered to possess some additional merit in theory (e.g. it includes the higher frequencies significant for speech articulation) and at present, appears to warrant further investigation. Mild handicap can be considered to be established if the average at 1, 2 and 3 reaches 40 dB, since this corresponds to the lower levels of the category of 'mild handicap' Class B, of Table 12.3.

The consequences of exposure of specific NIL values can now be looked at in terms of the AAOO index, together with the 1, 2 and 3 kHz index for comparison. Using an A-weighted NIL of 105 dB,

TABLE 12.3
*Classes of hearing ability based on average value of hearing levels at
500, 1000 and 2000 Hz**

Class	Degree of handicap	Average hearing level dB	Ability to understand ordinary speech
A	Not significant	Less than 25†	No significant difficulty with faint speech
B	Mild	25 to less than 40	Difficulty only with faint speech
C	Moderate	40 to less than 55	Frequent difficulty with normal speech
D	Marked	55 to less than 70	Frequent difficulty with loud speech
			———————— Educational deafness‡
E	Severe	70 to less than 90	Shouted or amplified speech only understood
F	Extreme	90	Usually even amplified speech not understood

* These hearing levels, in dB, apply to the better ear; if the average for the poorer ear is 25 dB or more than that for the better ear, add 5 dB to the average for the latter.

† Normality is now taken to extend to 26 dB mean hearing level.

‡ Children with this degree of impairment, or worse, require special education if they are to learn to talk.

these expected average hearing level values are shown for the 50th, 25th, 15th, 10th, 5th and 1st centile levels of the hypothetical exposed population, at age 70 years, on Table 12.4A. On the basis of the original AAOO index, the expectation for these stated centiles, is that at the 25th centile level there is no handicap, but at the 15th there is incipient mild handicap. This takes no account of the fact that some persons will have sustained some hearing loss of a pathological kind in the course of their lives, but the hearing level values attributable to age and noise are the basic information. The aspect of pathology in such circumstances has been discussed by Burns, Wood, Stead and Penney (1970) and by Robinson (1971). This will be further considered in the context of risk.

The fact that, even in the absence of pathology, noise and age combined can cause hearing deterioration of the degree shown in

TABLE 12.4

A

Estimated hearing levels as mean of 0·5, 1 and 2 kHz, and of 1, 2 and 3 kHz, at age 70, A-weighted NIL = 105 dB, for different centiles of the population

Centile of the population	Mean hearing levels	
	0·5, 1 and 2 kHz dB	1, 2 and 3 kHz dB
50	15·7	22·4
25	22·1	30·4
15	26·0	35·3
10	28·8	38·6
5	33·3	43·9
1	42·6	54·3

B

As in A, except that A-weighted NIL = 100 dB

Centile of the population	Mean hearing levels	
	0·5, 1 and 2 kHz dB	1, 2 and 3 kHz dB
50	13·9	19·3
25	19·3	25·8
15	22·7	29·2
10	24·8	32·5
5	28·5	37·2
1	36·4	46·5

Table 12.4A suggests the possibility of imposing a lower lifetime noise exposure. Thus reduction from the NIL value of 105 dB (Table 12.4A) to one of 100 dB (Table 12.4B) might be considered. Study of the latter table, however, shows that a useful but not dramatic improvement occurs. In effect the AAOO 25 dB level is not quite reached at the 10th centile, instead of being just exceeded at the 15th in the case of NIL = 105 dB. The actual improvement, resulting from a reduction of 5 dB in NIL, is 3–4 dB in AAOO mean values in the 10 to 15 centiles region. Thus, in order to reduce to very small proportions the incidence of the lowest degrees of handicap due to noise in persons otologically normal in other respects, further reductions of exposure would need to be considered. Thus, if NIL = 85

dB were employed, the 1st centile (the most susceptible) of the exposed population would theoretically be expected, at age 70, to show an AAOO average hearing level of 27·7 dB. The calculations, at this level, however, begin to have less meaning, since we are then encroaching on the domain of hearing unaffected by noise, where hearing levels merely show the variance expected of normal ears. Thus, the hearing levels for the 1st centile in a young, otologically normal, unexposed population would theoretically be 15·1, 13·2 and 14·2 dB for 0·5, 1 and 2 kHz respectively (using Dadson & King's data, Table 6.1). Making the same allowance for normal presbycusis as in the noise-exposed case, the AAOO average would be 26·1 dB for age 70. This is an under-estimation of the values but even so, it is negligibly different from those with a total exposure of 85 dB NIL. To achieve the average of about 25 dB, the daily noise level would have to be not more than 70 dB(A) over a period of 30 years, which is not a meaningful result. It does not spring from any shortcomings of the data or the method of derivation, but illustrates the fact that normal variability of hearing levels can produce conditions in which social handicap may arise without any apparent pathological basis in ears not exposed to excessive noise. In fact, measured hearing levels in ears not exposed to noise may show much larger variance than this estimate (Hinchcliffe, 1959). Further, the use of centiles as low as 1 should be regarded with some caution in making such assessments. It should also be noted that the use of fractions of a decibel in Tables 12.1, 12.2 and 12.4 represent the actual result of these calculations of expected hearing levels. Such precision, as will be noted later in this chapter and in Appendix J, is not realised in the realm of practical audiometry.

The values of Tables 12.4A and B have been interpreted in a rather rigorous manner, inasmuch as the AAOO convention (Davis, 1970) is that hearing handicap is regarded as beginning at a mean hearing level of 26 dB for the average of 0·5, 1 and 2 kHz. Although this condition is just attained at the 15th centile for NIL = 105 (Table 12.4A), or between the 5th and 10th centiles for NIL = 100 (Table 12.4B), it is true that, at the age of 70 chosen in these assessments the contribution of presbycusis is considerable. Nevertheless the limit of moderate handicap (i.e. frequent difficulty with normal speech) is assumed to be 40 dB on the AAOO scale, and this is exceeded for NIL = 105 dB by the 1st centile; but for NIL = 100 dB, the 1st centile is appreciably below this value, again provided that ear pathology does not elevate the levels additionally. In practice, how-ever, it must be accepted that this will occur, and must be taken into consideration. Having thus surveyed some relevant factors in the relation between noise exposure and hearing loss, we may consider the techniques of hearing preservation at a practical level, in, for example, an industrial situation.

Procedure for hearing preservation

SUPERVISION   This is a responsibility which should be assigned by management to the industrial medical officer, or his equivalent. Under his general supervision associated services will be needed. These include the acoustician or industrial hygienist, the audio-metrician and the necessary facilities for record keeping. Where an industrial plant is small and these services are not available as part of the medical services, equivalent advice and assistance might be provided on some form of shared basis. In any case the otologist should be the final arbiter of the otological status of individuals. Management of noise-exposed populations at this level of com-petence may be easy to arrange in the case of large industrial plants, but very different and requiring much effort to achieve the same situation in the case of the small plant.

In any case, there is a limit to which general instructions can apply, since different industries or even branches of industries vary so much that a large measure of individual thought and attention is likely to be needed to devise the best variant of the basic procedure to suit individual circumstances. A general outline guide will there-fore be given covering the principles governing the supervision of noise-hazardous situations and exposed personnel.

NOISE SPECIFICATION   If the spectrum of the industrial noise lies within the limits shown in Fig. 11.11, the A-weighted NIL can con-fidently be used to define the limit of exposure beyond which hearing preservation measures should be introduced. There appears to be good reason to examine the previous recommendations (Burns & Littler, 1960; Burns, 1965) for exposure limitation which were equivalent to 104 or 105 dB NIL for 40 and 50 years respectively. An important point at once arises in comparing the previous (Burns, 1965) recommendation with the practice now in use. In the former case, the noise specification was in the form of a spectrum

(Table 12.5). If any of the octave band pressure levels of a noise attained or exceeded the specification values, it was by definition a hazard. In consequence, noise levels indicated as hazardous in this way are unlikely to attain the value of 88 dB(A) of the spectrum, and might be considerably less. Accordingly 105 dB NIL (A-weighted) should be regarded as an absolute maximum. Where this can be reduced, some benefit in terms of diminished expectation of hearing deterioration can be expected, and a value of 100 dB NIL or less should be the objective.

How this overall limitation is to be translated into daily permitted maximum sound levels will vary with the particular circumstances. The simplest situation is that involving steady noise, similar in level throughout the work area, for each working day of a normal 8 hours, five days per week, for periods of many years. This circumstance will be discussed first, and later the complicating factors of varying level, intermittency, impulse noise, or noise containing pure tones will be mentioned.

For steady noise, various recommendations on desirable maxima have been made. Where these are not in the form of dB(A), it is helpful to translate them into this measure. Some of the recommendations made since 1960 may be noted. Burns and Littler (1960) suggested a spectrum essentially similar to that of Burns (1965) equivalent to 88 dB(A); Glorig, Ward and Nixon (1962) favoured noise rating (NR) contour 85, equivalent for practical purposes to 90 dB(A); Kryter, Ward, Miller and Eldredge (1966) proposed a spectrum which is approximately equivalent to 93 dB(A); the USA Department of Labor proposed 85 dB(A) as a limit, but after a public hearing, modified it to 90 dB(A) (Van Atta, 1971); the Svenska Elektriska Kommissionen (1969) produced a standard recommending 85 dB(A); and ISO (1971) states that 'Limits for tolerable noise exposure during work may be set by competent authorities who generally demand the institution of hearing conservation programmes if the limits are exceeded. In many cases 85 to 90 dB(A) equivalent continuous sound level has been chosen.' Thus there is very widespread concurrence that a limit in the range 85 to 90 dB(A) is desirable. For continuous noise 90 dB(A) for 30 years is equivalent to 105 dB NIL and 85 dB(A) to 100 dB NIL, on the basis of normal 8-hour days. If, however, the noise does not occupy the whole day, higher levels or longer total durations are permissible but the differences are small. For example, to retain the

NIL for lifetime exposure at 105 dB, with 50 years of work duration instead of 30, a reduction of only 2 dB, to 88 dB(A), is needed; halving of the work duration will achieve a reduction in NIL of 3 dB, a somewhat unrewarding exercise but in certain circumstances valuable.

The choice of a maximum permissible level for daily exposure to steady noise (or the corresponding equivalent continuous level for fluctuating noise) is thus likely to have been made already, and it will probably be in the 85 to 90 dB(A) region. D. W. Robinson and the author (Burns & Robinson, 1970) recommended that 'in a given period, which may be taken according to convenience as a day or a working week, only a certain maximum energy is permissible. This limit can be set at a variety of levels according to the ultimate risk judged to be acceptable, and we suggest that it should not be set higher than 90 dB(A) for a normal continuous daily exposure which is likely to persist for many years.' This statement means exactly what it says, that 90 dB(A) should be regarded as the absolute upper limit of the various possible levels for continuous steady daily noise, in the absence of precautionary measures. It would appear that this value may have been interpreted as a specific recommended level, but the intention of the authors was that if a lower limit were preferred (e.g. 85 dB(A)) it would be, by implication, scientifically justifiable and beneficial in terms of the risk implied. This view is endorsed here and a maximum level of less than 90 dB(A) is recommended wherever possible, and preferably not more than 85 dB(A).

If we use as an illustration an A-weighted NIL of 105 dB, the corresponding levels for different durations of continuous noise would be, theoretically:

10 years	95 dB(A)
20 years	92 dB(A)
30 years	90 dB(A)
40 years	89 dB(A)
50 years	88 dB(A)

The references to 10 years at 95 dB(A) and to some extent also 20 years at 92 dB(A) have been included to draw attention to a possible misuse of the energy principle. If these higher sound levels and short durations were employed, an individual could have received a normal lifetime dosage at an early age. Subsequent work in noise would therefore result in the risk of unduly large noise-induced losses.

Where such higher levels are accepted in consequence of shortened duration, these proportionate adjustments must be confined to periods of one day or one week, to ensure that the dosage is prevented from exceeding acceptable rates. The small differences of one or two decibels at the upper end of the range are a little academic in the context of practical noise measurement and limitation, and are of most relevance where the noise is very steady in time and uniform in space, in the work area.

To summarise, the question of choice of a continuous daily level beyond which hearing preservation measures must be adopted simplifies down to the range 85 dB(A) to 90 dB(A), and the latter should definitely not be exceeded. However, the tendency to mixed exposure, some of which may be at lower levels, and some at higher, with ear protection, is likely to reduce the exposure; and also the tendency to shorter working hours in the future could modify these relations. For new plant, levels of about 80 dB(A) would almost remove the risk of noise-induced hearing loss; and if such reductions were impracticable, a level of 85 dB(A) would give modest, but worthwhile reductions in the expected incidence of noise-induced hearing loss.

Practical problems and palliatives will be mentioned below in the assessment of noise environments. Having indicated the problems of noise specifications we can now proceed further in the sequence.

Determination of noise exposure

The data required will vary with the situation, but a comprehensive set of measurements would be as follows:

(1) Sound levels A, B and C.
(2) In case of fluctuating noises, the equivalent continuous level or a statistical noise level distribution.
(3) Octave and possibly one-third octave band SPL values for the preferred frequency bands provided on analysers. These will usually range from centre frequencies of 31·5 to 8000 Hz; measurements of lower frequencies may sometimes be desirable.
(4) If subjectively detectable intense narrow bands or pure tones are present, or suspected to be present, a more selective analysis, such as by a 6% bandwidth analyser.
(5) Total daily duration of exposure of those working in the noise.

Noise measurement is so well described in the manuals of the principal manufacturers of acoustical measuring equipment that no description is needed here. Reference to accounts such as those of Broch (1969) and Peterson and Gross (1967) in the bibliography of Chapter 4 will be found to be very helpful.

The complete adherence to this procedure, with full quantitative description of the noise exposure, may be impossible in some circumstances. At one extreme there may be a constant noise, with only small variation in level or frequency characteristics in time and space throughout the work area. Such noise is amenable to accurate description and the exposure sustained by the people can be stated with confidence. At the other extreme is impulsive noise such as that of chipping, hammering, riveting, and similar procedures either by hand or by power tools. Noises of this type may have little or no regular time pattern, so increasing the difficulty of assessment and they must be assessed from the viewpoint of potential hearing damage by specific methods, which will be discussed below. In most cases, detailed study of a particular noise is facilitated by making a high quality tape-recording for subsequent analysis. Where such noise surveys are regularly being made, it is convenient to do so by means of a mobile acoustic laboratory, with extensive facilities for noise measurement and analysis, which can be taken to the noise situation in its entirety. Such facilities for example are used by the British National Physical Laboratory (Copeland & Saunders, 1970). Having obtained the relevant information on the physical characteristics of the noise, and its time pattern of exposure, the next step is to try to estimate the degree of hazard to hearing.

Assessment of hazard

The judgement as to whether a given noise environment is hazardous, as defined, is of variable difficulty, chiefly occasioned by the acoustic characteristics and variation with time. It is helpful to consider a number of these conditions separately.

STEADY NOISE   If the noise varies over only a small range (say 1–2 dB) with time and position throughout the work area (such as may occur in weaving, for example), and if the spectrum is within the limits of the spectral slopes of Fig. 11.11, and it lasts for the full working day its sound level (A-scale) is merely compared with whatever daily maximum is in force; this will probably be, as discussed above, in the range 85 to 90 dB(A). If the sound level is properly measured, and exceeds the selected limit there is a clear case for instituting hearing preservation measures. If it is well below this value, the situation is not in need of attention. If the level, on the other hand is shown to be just below the limit, particular attention should be given to the measurements throughout the work area, to ensure that the value arrived at was a reliable and representative one, and including a check on the daily duration of work in the noise. Where there is evidence that the case is a marginal one, it would be prudent to err on the side of safety and regard the level as hazardous. Whatever limit of a practical nature is chosen it can be seen from the previous discussions that it does not define the transition from safety to danger, but merely indicates an estimated degree of hazard. Exercise of judgement in borderline cases is therefore essential.

INTERMITTENCY AND FLUCTUATION   In some cases the noise duration in one working day is less than 8 hours, either in one continuous period or broken up by intervals of quiet or reduced noise. This is a common feature of industrial noise, an example being the noise of blast furnace operation. Where the time pattern and alterations of level are reasonably clearly defined, a meaningful measure of the exposure can be obtained over a defined period such as one day or one week. The latter duration is provided for in ISO Recommendation R1999 (already described in Chapter 11), and the measure, as noted above, is the *equivalent continuous sound level*. Reference back to Chapter 11 will recapitulate how this measure is derived from tables of duration and sound level. It provides a single value in dB(A), which if present for 40 hours per week, would produce the same exposure as a sound duration of less than 40 h/week, or alternatively as various components of different levels and durations of sound stimulus, any possible biological effects of intermittency on the hearing mechanism being ignored. The equivalent continuous sound level is then used just as if it were the continuous sound level to which it is equivalent, to judge whether the specification value is exceeded or not. Again, it is necessary to be aware of the possible danger to the ear of very short duration high-level exposures. In R1999, for example, the tables do not go below durations of 10 minutes per week or above 120 dB(A); these two parameters in combination produce an equivalent continuous sound level of 96 dB(A). This is therefore excessive relative to even the least conservative commonly recommended level, i.e. 90 dB(A) for continuous noise, and measures for preservation of hearing would therefore need to be taken. The Recommendation advises that durations of less than 10 minutes should be regarded as 10 minutes. The permitted level at a duration of 10 minutes, over a period of one week, to give an equivalent continuous sound level of 90 dB(A) is in fact, approximately 114 dB(A). Thus one week's dosage would be expended in 10 minutes at this level. It is desirable, however, that for fluctuating noises of this description, two precautions should be taken:

(a) where sound levels exceed 110 dB(A) even though the equivalent continuous sound level does not exceed 90 dB(A), the conditions should be classed as requiring hearing preservation measures;

(b) where any doubt exists about the reliability of the measurements, or if no satisfactory measurements are possible, and therefore there is no assurance that the equivalent continuous sound level is below 90 dB(A), hearing preservation measures should again be instituted if any measurement shows values above 90 dB(A).

In the actual calculation of equivalent continuous level the tables from R1999, reproduced in Appendix M are convenient, and they refer to a period of one week. One day may also be used, and it is only necessary to divide the value of the composite noise exposure index by 5, to reduce the result for a 40 hour week to that for an 8 hour day. A nomogram is provided for calculations on a daily basis (Appendix M).

IMPULSE NOISE   By this is meant any noise having a rapid variation of level with time, by reason of a succession of impulses or impacts. Again, the necessary operation is to derive a value for the equivalent continuous sound level, in dB(A), for comparison with the noise specification level. This solution has only been recently considered, in view of the demonstration of the validity of an energy measure as a predictor of hearing damage (Robinson & Burdon, 1970; Robinson, 1970). The A-weighted equivalent continuous noise level for impulse and impact noise of certain types can be derived, using three dimensions of the noise: (1) peak height of the waveform; (2) repetition rate; and (3) the decay-time constant of the waveform envelope (Atherley & Martin, 1971), whilst for more general cases it can be measured directly by suitable instrumentation. The equivalent continuous sound level is then used to assess the noise as though it were continuous, by comparison with the specification level, and if need be, to make predictions, from Robinson's formulations, of the probable effect on hearing.

USE OF NOISE DOSE METERS   Dosimetry is an attractive alternative to the use of lengthy measurement procedures followed by calculation. The requirement is to provide suitable circuitry to summate sound energy; such a circuit, coupled to the output of a sound level meter using A-weighting (or other desired characteristic), would indicate total sound immission and the equivalent continuous sound level. If the equipment could deal with impact, in addition to steady or fluctuating noise, assessments of a wide variety of sound levels would be greatly facilitated. This form of noise dose meter is becoming available and in its present form, while by no means a personal item of equipment of the compactness attained in radiation dosimetry, it is a valuable item in surveying areas, and produces data the acquisition of which would be laborious and time-consuming, or even impracticable, by less sophisticated means.

NOISE OF IRREGULAR SPECTRAL CHARACTERISTICS   All the foregoing presupposes that the spectrum is within the limits of Fig. 11.11. Where these are transgressed, particularly if the departure is gross,

TABLE 12.5

Values of sound pressure level in specific frequency bands which indicate a hazard to hearing at a duration of approximately eight hours per day

Octave band specified as centre frequency Hz	Sound pressure level dB
63	97
125	91
250	87
500	84
1000	82
2000	80
4000	79
8000	78

it is not yet known how much A-scale values would be invalidated. Where extremely unusual spectra are encountered, the spectrum of the noise is compared with that of Table 12.5. If any of the octave band pressure levels of the noise attain or exceed the corresponding values in this table, hearing preservation measures should be adopted.

Where sounds with irregular spectra are intense the question of the permissible duration may arise. It is advised that the spectrum is derived, as usual, as octave band pressure levels, and compared with the family of spectra given in Table. 12.6. These are the same spectrum as in Table 12.5, elevated by 3 dB throughout for each halving

TABLE 12.6

Values of sound pressure level in specific frequency bands which indicate a hazard to hearing at stated total daily durations

Octave band specified as centre frequency Hz	Sound pressure levels at specified durations dB						
	4 h	2 h	1 h	30 min	15 min	7 min	3½ min or less
63	100	103	106	109	112	115	118
125	94	97	100	103	106	109	112
250	90	93	96	99	102	105	108
500	87	90	93	96	99	102	105
1000	85	88	91	94	97	100	103
2000	83	86	89	92	95	98	101
4000	82	85	88	91	94	97	100
8000	81	84	87	90	93	96	99

Note. No exposure of the unprotected ears should be allowed to reach 135 dB sound pressure level overall value, and no exposure of the body as a whole, no matter how short, should be allowed to reach 150 dB sound pressure level overall value.

of duration, i.e. on an energy basis. If in one or more of the octave bands the spectrum values exceed those of the corresponding octaves in the Table, the sound is to be regarded as a hazard at that particular total daily duration, and appropriate steps should be taken. For example, if a sound has a value of 99 dB SPL in the octave centred at 1000 Hz, and so exceeds the corresponding value of 97 dB SPL for 15 minutes total duration, it would be pronounced a hazard at this duration, but not at a total duration of 7 minutes, where the hazard level is 100 dB SPL. These total daily durations may be in one exposure, or a number of individual exposures, the durations of which are added to give the total in minutes.

PURE TONES The situation where the sound energy is present entirely or largely in a small frequency range, or even as a pure tone, is not very common in industry and other practical applications. For purposes of description, a pure tone is as defined in Appendix A; the narrow bands of noise which may occur could be of various widths but for practical purposes any octave or one-third octave band which shows SPL values 10 dB more than either of the two adjacent bands can be regarded as containing a pure tone or a band narrow enough to merit consideration for its potential effect on hearing. The actual physical characteristics of such tones or sounds can of course be confirmed by a narrow band analysis. No well documented evidence of the permanent effects of exposure to pure tones appears to be available, although Ward has examined this subject in his extensive studies of temporary threshold shift, and it is included in the recommendation of Kryter, Ward, Miller and El-dredge (1966), but not in recent international recommendations (ISO, 1971). Much inconclusive discussion has centred round the effects of pure tones. This has involved the possible effect on the vibration pattern of the cochlear partition of the pure tone compared to broader bands of noise, the concept of the critical band and its influence, if any, on damage effects, and the lesser effectiveness of the aural reflex in response to pure tones compared to that due to bands of noise, manifest at frequencies below about 2 kHz. Kryter *et al.* (1966) regard pure tones as being more damaging than octave bands of noise of the same sound pressure level. Their 'damage risk contours' (i.e. noise specifications for restriction of occupational hearing loss) place the SPL value for pure tones 5 dB, for all practical purposes, less than the value for the octave having the same centre frequency as the pure tone, for 8 hours daily exposure. For exposure durations of fractions of a day they allow less compensatory increase in SPL than in the case of broad band noise, on the evidence of temporary threshold shift findings. The latter are interpreted as the result of less protection by the aural reflex generated by pure tones compared to that due to broad band noise. A simplified interpretation of these attitudes would be to consider the sound pressure level of a continuous pure tone as hazardous if it approaches to within 5 dB of any value of the spectrum of Fig. 12.5, for a normal pattern of industrial exposure. Shorter daily exposures could be covered by the normal procedure of increasing SPL by 3 dB for each halving of duration, but care must be exercised in the extent to which this procedure can be used. Some uncertainty of definition may arise where narrow bands occur in a spectrum.

The question is sometimes raised as to whether spectra which include a pure tone as well as broad bands can be accommodated in the energy system (Robinson, 1970). Robinson (personal communication) suggests that if one wishes to err on the safe side, the assessment of a noise for this purpose should be based on measurements in one-third octave bands. Thus, if the band containing a tone exceeds its neighbours not containing such a tone by 10 dB or more, the band should be deemed to contain a hazardous tone, and 5 dB should be added to the measured level of the band. The sound level A of the noise as a whole is then calculated in the usual way. Robinson and the writer stated (Burns & Robinson, 1970) that we believed that the role of pure tones would turn out to be marginal or negligible unless they dominated the spectrum, but did not feel that a firm recommendation could be made at that time. This is indeed still true and if an occupational noise is demonstrated to consist effectively of a pure tone with insignificant levels elsewhere in the spectrum, the questions of its compatibility with the use of sound level A would have to be considered individually.

In all the cases which are judged to transgress the selected level of the noise specification, and therefore by definition, imply a noise hazard, the following procedure should be adopted.

Procedure for hazardous noise

1. The noise should be reduced at source, or in the working areas.
2. If this is not possible, the number of persons exposed should be reduced to the minimum, and the following expedients employed.
3. Ear protection should be used to reduce the amount of noise entering the ear.
4. The duration of exposure may be reduced, either as a single measure or in conjunction with ear protection; in the latter case, if ear protection is inadequate alone, duration reduction may well be necessary in addition. Such conditions, however, should be avoided if at all possible by reduction of the noise itself.
5. In general, in noise situations assessed as hazardous to hearing,

routine audiometric examinations should be conducted, before and at intervals throughout the period of work in the noisy occupation.

Item 1 involves problems of an acoustic and engineering nature. The techniques of noise reduction are well understood, and do not concern us in this context. However for the benefit of those who are concerned with noise reduction in machinery, buildings and other situations, the following are some of the available sources of information. Certain books, such as Beranek (1971), Harris (1957), provide a wide variety of information on noise reduction. An excellent summary of the principles of noise suppression is given by Fleming and Copeland (1958) and Parfitt (1958) deals with analysis and control of vibration. A wide variety of topics of a practical nature is dealt with by the various contributions to a symposium held at the National Physical Laboratory on the Control of Noise (National Physical Laboratory, 1962). The reduction of the noise of internal combustion engines was the subject of a symposium (Institution of Mechanical Engineers, 1958) to which King (1958) contributed. An account of these topics is given by Waters, Lalor and Priede (1970). Noise in the iron and steel industry is reviewed by Adam and Thomas (1967). Richards (1965) has discussed noise considerations in the design of machines and factories. Bazley (1966) and the British Standards Institution (1960) deal with building applications of acoustics, as have Parkin and Humphreys (1969). The *Journal of the Acoustical Society of America* and the *Journal of Sound and Vibration* are regular sources of information on noise characteristics and noise reduction techniques.

It should however be noted that much trouble and expense could be avoided if the acoustic implications of engineering design of factories and equipment, ships and vehicles had been studied at the outset.

Item 3, ear protection, is a palliative which must often be an essential part of the procedure.

Ear protection

The entry of sound into the ear canal can be reduced by wearing some form of plugs in the ear canal, or external cups usually known as ear muffs. The theoretical basis of these devices has been extensively studied (Thiessen, 1962) and the design of commercially available articles is in consequence excellent. The most widely used ear plug is based on the type V51R, and their performance has been analysed (von Gierke, 1956). Sponge-sealed muffs are effective and comfortable; fluid-seal muffs give the best attenuation, using a fluid-filled annular tube (Shaw & Thiessen, 1958). A word of caution must be said on the subject of ear protection. The very fact that external sounds seem less loud, while the wearer's voice sounds to him different and louder, results in a tendency to talk more quietly. What is needed for conversation between persons wearing ear protection in noise is to talk loudly, and this must be emphasised. If this is done, up to the level of shouting if need be, it has been shown (Kryter, 1946) that in high noise environments the wearing of ear protection can improve intelligibility. The noise level at which it becomes beneficial to intelligibility to wear ear protection will vary with the degree of attenuation of the device used. The ear plugs used by Kryter (1946) improved intelligibility above a combined (noise plus speech) SPL of about 85 dB. For ear muffs, Thiessen (1962) places this level at about 95 dB SPL, the reason being that the ear muffs provide about 10 dB more attenuation in the frequency region which is important for communication.

The use of ear protection incurs the risk that accidents could conceivably occur if danger signals by sounds or by speech are not heard or even if audible malfunction of machinery or equipment is not noticed. This situation can easily arise, particularly in inter-mittent noise, and unless this hazard can be guarded against, ear plugs or muffs may introduce dangers far worse than those of impaired hearing. This pitfall can be aggravated if the wearer has some occupational hearing loss, or is in the older age groups, or both; because of the elevated threshold, the additional sound reduction due to the ear protection may result in very high thresholds. The effect of the change from not wearing ear protection to wearing ear protection will be increased by the fact that occupational hearing loss is a recruiting form of deafness, in which the threshold is raised but the sensation of loudness at high levels may be for practical purposes normal.

In recent years, glass down, a material similar to cotton wool in appearance, but composed of glass fibres of very small diameter instead of cotton fibres, has been marketed in a form suitable to insert in the ear as ear plugs. Acoustically it is much superior to cotton wool, which provides poor attenuation at low frequencies, and it is more comfortable than rubber or plastic ear plugs. It must however be carefully inserted in the recommended way, and there is always the possibility that parts may be left in the ear canal. Fears of irritation due to spicules of glass which might lodge in the skin of the ear canal have not apparently been justified. The fibre diameter is presumably too small, and quite different from coarser

TABLE 12.7

The attenuation provided by ear plugs or muffs in specific frequency bands

Octave band centre frequency Hz	Attenuation values dB		
	Glass down	V51R plug	Fluid seal muff
63		11	17
125		13	18
250	11	15	20
500	13	18	30
1000	16	22	38
2000	26	27	40
4000	33	32	43
8000		29	35

glass fibre commonly used for thermal or acoustic insulation. The disadvantage is that its success depends on the skill of the user, a feature which tends to limit the effectiveness of any protective device. However, it can also be argued that skill and conscientious attention are also needed to insert ear plugs, and to keep them properly inserted at all times.

Sound attenuation values for commercial ear plugs made of glass down, V51R ear plugs and fluid-seal muffs are given in Table 12.7. The values for attenuation are based on those given by Thiessen (1962) and by Piesse (1957). Where the attenuation is varying markedly with frequency, octave centre frequency attenuation values may tend to overstate the degree of attenuation over the octave. The values given are estimated, and are in certain instances about 3 dB less than the pure tone attenuations at octave centre frequencies, in order to avoid the risk of overstating the attenuation. All values of attenuation are subject to variation and standard deviations are given by Piesse (1957), for each type of protection. The variability of attenuation is least for the fluid-seal muff, the standard deviation being about 4 dB for most of the range of audiometric frequencies.

A new development, the Gundefender plug, is an amplitude-sensitive device (Forrest & Coles, 1970). They provide evidence of increasing attenuation as the SPL increases over 110 dB, while at low levels, e.g. for speech perception, little attenuation occurs. Mosko and Fletcher (1971) have corroborated these findings; they find that for communication purposes at low noise levels, ears protected by the Gundefender are equivalent to ears without plugs. Against gunfire noise, the Gundefender is equivalent in attenuation, as judged by reduction of TTS, to the V51R plug.

The techniques of ear protection cannot yet be regarded as entirely satisfactory. The admirable attenuation, reliability and comfort in many situations of fluid-seal muffs is offset by objections of discomfort in some warm and humid environments, as well as their large size and relatively high cost. Insert plugs are not particularly comfortable, and may be either quite unacceptably uncomfortable or acoustically inadequate if gross misfits are achieved; this in unskilled hands is not as improbable as it sounds. Glass down depends on competent insertion procedures for its effectiveness, and portions of the material may, with mishandling, be left in the ear. The latter need not happen, and with proper use, glass down can be reasonably acceptable. In this field, there is no substitute for good, experienced advice, based on the practical knowledge of the industrial medical officer or hygienist who has actually supervised the allocation, fitting and wearing of ear protectors.

Reduced exposure duration

Reduction of exposure duration can be used to reduce exposure below the specification equivalent continuous level in force, or alternatively to reduce exposure from levels which are great enough even with ear protection to require further reduction. The relation between intensity and duration are obtainable from Appendix M for the purpose of such calculations.

Conclusion

This completes the consideration of the various procedures needed for the preservation of hearing, except audiometry. Before concluding this section and proceeding to the consideration of other topics including audiometry in industry, it is opportune to note some more recent actual or projected codes of practice or similar recommendations of US, UK or international origin. The original *Guide for the evaluation of hearing impairment* (Am. Acad. Ophthalmol. Otolaryngol., 1959) has recently been supplemented (*ibid.*, 1969). The USA Walsh-Healey Public Contracts Act (Van Atta, 1971) embodies recommendations which are based on the assumption that halving of noise exposure allows an increase of 5 dB in sound level A; this is not in harmony with the energy concept (Robinson, 1970), which is widely accepted internationally (ISO, 1971), and by the British Occupational Hygiene Society (1971). It is understood that the British Standards Institution also is considering the question of recommendations for acceptable noise exposures. The UK Department of Employment (1972) has now issued a comprehensive Code of Practice for the control of noise exposure in industry; such a unified and authoritative document should be of great assistance to industry. This Department has also issued a helpful short guide entitled *Noise and the worker*.

The essential difficulties of the entire hearing preservation procedure will have become evident during the discussion. The conditions at different noise levels vary greatly. Thus if the noise is well below the specification equivalent continuous level, all should be well. If clearly below, but perhaps by only a few decibels, some persons will sustain some occupational hearing loss, as discussed later under the topic of risk. This is the most vulnerable condition for a noise-exposed population, and a justification for a policy of progressive lowering of occupational noise levels, towards 80 dB(A). If the noise just exceeds the specification, or is at a marginal level deliberately classed as hazardous, the situation is at once transformed if hearing protection is worn, to one in which the hazard virtually disappears, but at the cost of the nuisance of wearing ear protection. In such a case, audiometry would not be expected to reveal much more than the existence of individuals who have acquired some hearing loss of a non-occupational sort, or who were not conscientiously wearing their ear protection. If the noise levels are greater, the safety margin of ear protection will be progressively reduced as level rises, until a stage is reached when the specification level at the ear is attained even with protection. In this case either the noise must be reduced, or the exposure time shortened. In such cases, audiometry is fundamental to the procedure, and would be expected to reveal incipient changes in the hearing of the susceptible minority, who would be considered individually to find a solution. This might mean change of work environment.

The concept of risk

Reference has already been made at some length to the predictions, made possible by the use of the energy concept, of probable hearing deterioration due to specified exposures. The term risk is collectively applied to the probability of the attainment of some defined hearing level, indicating some particular degree of handicap. Robinson (1971) defines risk generally as the percentage of a population whose hearing level, as a result of a given influence, exceeds the specified value, minus that percentage whose hearing level would have exceeded the specified value in the absence of that influence, other factors remaining the same. It is thus possible to assign risk values arising from age, from noise, or from pathological conditions affecting hearing, or to combined effects. A definition, specifically directed to hearing handicap, is found in R1999 (ISO, 1971) as 'The difference between the percentage of people with impaired hearing in a noise-exposed group and the percentage of people with impaired hearing in a non-noise-exposed (but otherwise equivalent) group'. The 'impaired hearing' is usually impaired hearing for conversational speech, defined in R1999 as existing if the arithmetical average of the permanent hearing levels of the person for 500, 1000 and 2000 Hz is 25 dB or more relative to the ISO threshold defined in R389 (ISO, 1964). The significance of this is that the comparison is being made between a so far undefined population not exposed to noise and an equivalent population exposed to noise. If the undefined population is a selected one (as briefly mentioned in Chapter 11) free from pathology, only age and noise contribute to the deterioration; if an unselected industrial population is used containing potentially any or all of the causes of deterioration, the numbers suffering handicap by the attainment of a

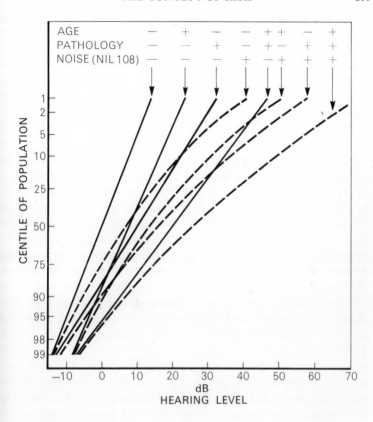

12.1   Cumulative distributions, in eight defined populations, of average hearing
       levels (0·5, 1 and 2 kHz), showing the effects of age, noise, and pathology.
       Presence (+) or absence (−) of these factors are shown as follows. Age: 20
       years, −; 60 years, +. Noise: absent, −; present, +. Pathology: absent, −;
       present, +.

       Quantitative assumptions are as follow. Young population (age 20 years)
       not exposed to noise, median hearing level = 0 dB at each frequency; Gaussian
       distribution, standard deviation = 6 dB. Older population (age 60 years) not
       exposed to noise, median average hearing level 7·5 dB; Gaussian distribution,
       standard deviation = 7 dB. Pathology population age 20 years, median
       average hearing level = 10 dB; Gaussian distribution, standard deviation
       10 dB. Pathology population age 60 years, median average hearing level =
       20 dB; Gaussian distribution, standard deviation = 11·4 dB. Populations with
       elevated hearing level due to noise: noise component calculated as in Fig.
       11.10, for NIL = 108; distribution not Gaussian, but skewed. (Robinson,
       1971.)

specified average hearing level will be increased compared to the pathology-free population. The risk as defined, however, has certain peculiarities when worked out numerically, which can produce perplexing and unexpected results if the principles are not understood fully. The interrelations involved in the statement of risk in this way depend on the assumption, usually, but not always justifiable, that threshold shifts due to age, noise and pathology are additive in decibel values. It is found that the risk values, calculated according to the definition, are not proportional to the threshold shifts. An example of the consequences of age, noise and some pathology in the hearing of a population can be quoted (Burns & Robinson, 1970). We take the original AAOO minimum criterion of handicap, viz. the attainment of a mean hearing level of 25 dB for the average of the 0·5, 1 and 2 kHz audiometric frequencies; suppose that the basic condition is a population with normal hearing and no aural pathology aged 18 years. By age 33 the proportion attaining the 25 dB mean hearing level due to age alone will still be negligible, in the absence of noise or disease. Had they been exposed to noise (95 dB(A) daily) 4% would have exceeded the 25 dB mean level. The risk as defined above would thus be 4%. Now assume that, instead of noise, the population suffered a median conductive hearing loss of 20 dB, in which 30% would have exceeded the 25 dB mean level. If we now combine these effects, assuming that they all (age, noise, pathology) acted to the degree stated, no fewer than 58% of the population would have exceeded the 25 dB mean level. The handling of such data thus requires a proper understanding of the arithmetic and the consequences; pathology is seen to elevate greatly the degree of handicap in the presence of age and noise, and proper definition of circumstances must be made as a part of any statements of risk. Robinson (1971) has discussed this topic very fully, and Fig. 12.1 clearly shows the effects, as cumulative distributions, of the three factors, age, noise and pathology, in various combinations.

The use of such predictions is in making a numeral estimate of the extent to which handicap may be expected to involve populations of different ages, noise exposures, and pathology. Quantification of the first two, as we have seen, is now on a reasonably firm footing, but the component of pathology, which exerts such a potent influence on the incidence of handicap as a whole, may be expected to vary between populations, countries, and in all likelihood, occupations. Knowledge of this point is still inadequate for any generalisations.

Audiometry for monitoring purposes

The preservation of hearing cannot, in general, be considered to have been completely provided for without the inclusion of audiometry. The vulnerable situations, particularly affecting the most susceptible minority, are where the sound level is moderate without ear protection, or high with protection; also, ear protection may not be effective, whether through mal-fitting, unserviceability, or neglect to use it as prescribed, and the only control is through audiometry. The same situation arises when, for some reason such as the need to retain direct voice communication, the wearing of ear defenders is not feasible.

The burden of routine audiometry is considerable, and the potential benefits are severely undermined by the known variability of the procedure. Pre-employment audiometry is essential to establish some sort of baseline, and is a protection for both employer and employee. It is unhelpful to suppose that the audiometry of large working populations is easy or cheap; it is neither, and requires time and care in interpretation. Nevertheless, it is impossible to avoid the conclusion that the problem should be faced and audiometry, despite its present shortcomings, regarded as an integral part of hearing preservation. The part which should be played by audiometry is not, however, universally agreed. As has been mentioned above, the usefulness of pre-employment audiometry is clear, and it should ideally be incorporated into pre-employment medical examinations. Monitoring audiometry, on the other hand, is likely to give different returns, in terms of the protection of the individual, depending on the prevailing sound level, and whether ear defenders are being worn. As an illustration, let us assume that a specification limit of 90 dB(A) is in force. Thus an equivalent sound level of 90 dB(A) would be acceptable without ear protection, but is clearly more hazardous than one of 92 dB(A) with ear protection, and so the rigid application of routine audiometry in the latter case, while not employing it in the former, contains an element of illogicality. Thus there is room for closer study of the application of routine audiometry in individual situations in order to give the industrial medical officer the best advice in this difficult aspect of hearing preservation. Such a study would involve a critical examination of the benefits of audiometry in various hypothetical conditions of sound level, duration of exposure, and degree of attenuation by different protective devices; in addition, the conditions in any specific industrial situation should be studied as an individual problem. Work is in progress on these topics.

The principles, application and reliability of audiometry are considered in Chapters 6, 11 and in Appendix J. The importance of pre-employment audiometry must be emphasised. For this purpose much time may be saved eventually, and uncertainty avoided, by taking for the initial audiogram the average of three separate audiometric examinations, removing and replacing the earphones between each occasion to randomise the error. By this means the expected departure of the mean from the person's true mean will be almost halved. A full audiogram, at least from 500 Hz upwards, is recommended. For subsequent examinations, audiograms must be taken in the morning, before work, to eliminate as far as possible temporary threshold shift. The interval between audiograms must be dependent on the exposure and general level of care in supervision of noise levels, duration of exposure, and efficiency of hearing protection. One year would be a convenient interval. In some circumstances, this first interval (i.e. pre-employment to first repeat audiogram) might have to be shorter, perhaps as short as three months, particularly in young persons with minimal previous exposure.

Finally, it must be emphasised that the assessment of hearing from audiograms is not a task which should be entrusted to non-medical staff. Pre-employment audiometry should be accompanied by a full clinical otological examination, and medical supervision of all subsequent audiograms should be maintained to avoid erroneous results or missed diagnoses. It may be found that the easiest way for the industrial medical officer to deal with this situation is to enlist the participation of an otologist on a sessional basis, who will then, at a specialist level, be able to supervise the entire procedure. He would not take the audiograms personally, but would review them at intervals and advise the industrial medical officer accordingly.

Practical aspects of monitoring audiometry

Where audiometry has to be carried out in a number of individual situations either as a routine, or for purposes of research, it is frequently found that the only practical means is to provide a mobile audiometric unit. This means at the simplest level an enclosure in the form of an audiometric room mounted in a vehicle or a trailer.

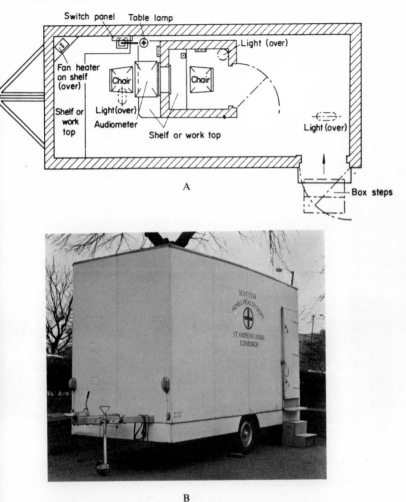

12.2 A small audiometric unit in the form of a trailer, accommodating one audiometer. A, general arrangement; B, external appearance. (Taylor, Burns & Mair, 1964.)

The audiometer, heating or cooling, ventilation and lighting can be connected to some convenient source of mains electric supply. At the other extreme, several audiometers and complete calibration facilities can be provided in a unit of high sound attenuation capable of operating in almost any sound environment, and provided with

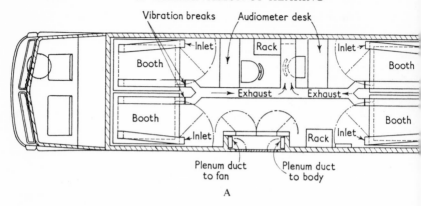

A

B

12.3   A mobile audiometric laboratory used in a research project for the UK Department of Health and Social Security, by the Medical Research Council and the National Physical Laboratory. A, general arrangement; B, external appearance. (Copeland, Whittle & Saunders, 1964.)

its own motive power. The difference in cost between these alternatives would of course be very great, but they are not expected to perform the same duties. For example, a simple moderately priced trailer vehicle was required to accommodate one audiometer, and more than one of these have been built, and have proved satis-factory. The constructional details are provided in a description of the unit and its acoustic performance (Taylor, Burns & Mair, 1964). The vehicle is illustrated in Fig. 12.2. Its small size and moderate weight are very suitable for parking and manoeuvring in confined factory spaces. A very much larger vehicle has been described by Copeland, Whittle and Saunders (1964). It is a complete audio-metric laboratory, with full calibration facilities and four small audiometric rooms with separate audiometers. The latter are modi-fied Rudmose ARJ4 instruments. It was used in the investigation of hearing and noise in industry, on behalf of the UK Department of Health and Social Security by the Medical Research Council and the National Physical Laboratory (Burns & Robinson, 1970). In virtue of the four audiometers, its capacity for carrying out audiometric examinations is large and the sound attenuation provided is such that it is capable of operating in comparatively high ambient noise. It is illustrated in Fig. 12.3. Such a vehicle, although primarily de-signed for research, is of great value where large numbers of audio-grams might have to be done in large industrial situations. The capital cost is of necessity high.

Tests of susceptibility

The description (Chapter 11) of efforts to find a test for estimating the degree of susceptibility to noise-induced hearing loss shows that this has not yet been brought to a practical level. It may be possible, by these or other means, to identify the small proportion of very susceptible subjects; if this could be done, and efforts continue to this end, much simplification would be possible in the methods of hearing preservation.

Assessment of hearing

The final judgement on the status of a person's hearing must be based on a number of factors. Assuming that the clinical picture excludes ear pathology, and that only the possible effects of noise are under consideration, the initial audiogram and any subsequent audiograms are the data on which decisions must be made. Recalling the variation in hearing levels with the best audiometric techniques and normal populations, and the effects of age, which are usually

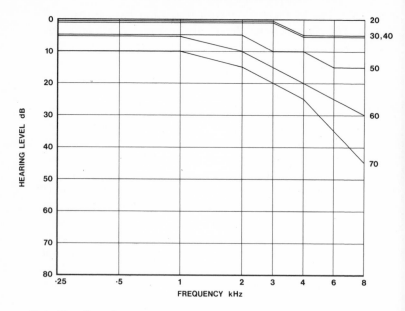

12.4   The effect of advancing age on hearing level. Curves are based on manual audiometry, one audiogram per subject, from a random-sample rural population. Values are *average* (median) hearing levels, rounded to the nearest multiple of 5 dB. (After Hinchcliffe, 1959.)

different in men and in women, due to the miscellaneous noise exposure of the latter, the designation of normality or otherwise on a single audiogram is bound to be a fairly broad generalisation. An approximate indication of the condition of hearing can be obtained by comparing the recorded hearing level with the expected hearing level (median value) for the particular age from Hinchcliffe's (1959) data. These curves apply to men or women up to the age of 54 years and for frequencies up to 2000 Hz inclusive. For higher ages and frequencies the designation of 'normal' male hearing has been somewhat variable. However, it apparently need not, in absence of noise, be worse than female hearing. In assessing the hearing of men therefore, the female presbycusis thresholds of Hinchcliffe may be used as representative of hearing of either sex minimally affected by noise.

The characteristics of presbycusis have already been noted (Fig.

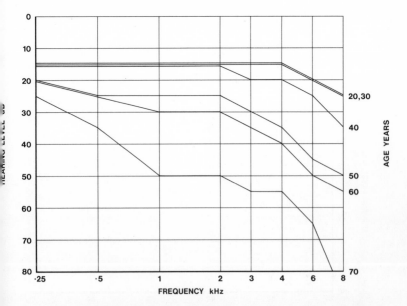

12.5 The effect of advancing age and individual variation on hearing level. Curves are based on manual audiometry, one audiogram per subject, from a random-sample rural population. Values are *probable maximum* (5th centile) hearing levels rounded to the nearest multiple of 5 dB. (After Hinchcliffe, 1959.)

Note on Figures 12.4 and 12.5

In Rudmose semi-automatic audiometry, recorded hearing levels may be expected to be some 2–3 dB less than in manual audiometry, and the 5 dB steps will also be absent.

6.3) but the additional feature of individual variability must be remembered. In addition, therefore, to the decline in sensitivity of hearing shown as median values in Fig. 6.3, the variability between individuals found by Hinchcliffe increased with age and frequency also. Hinchcliffe (1959) shows the variation of hearing level as a function of age in two ways: (1) the median values, on which Fig. 6.3 is based, and (2) the values for the hearing level which would be exceeded by only 5% of the population which he terms the probable maximum. The latter represent very considerable departures from the median, The average and probable maximum hearing levels as a function of frequency, for various ages, are shown in Figs. 12.4 and 12.5. These serve to indicate the need for caution in contemplating

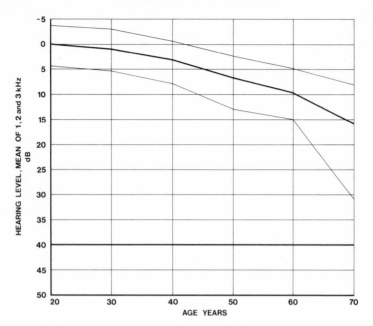

12.6   The effect of age on average hearing levels. Values are median hearing levels
in dB (heavy line), derived from the arithmetical average of the median hearing
levels at 1, 2 and 3 kHz. The thin lines indicate the 25 and 75 centile values.
The median line thus indicates a statistical trend rather than a precise statement
of any individual's deterioration. The heavy horizontal line at 40 dB average
hearing level indicates slight social hearing disability. (Data from Hinchcliffe,
1959.)

audiograms, and the impossibility of forming any opinion without
a full clinical examination and case history.

Normally, persons in noisy industries will show appreciable de-
pressions of the audiogram in the region of 3000–6000 Hz, and
especially at 4000 Hz. The first question which must be faced concerns
the person with apparently impaired hearing, who may be in danger
of reaching a stage sufficient to limit speech perception, and to constitute a social handicap. Guidance can be obtained from the pure
tone audiogram in this respect by taking the arithmetical average
of the hearing levels at 0·5, 1 and 2 kHz; if this average value
exceeds 26 dB (Davis, 1970), it is probable that some difficulty will
be encountered in the perception of faint speech. Further deterioration will give greater handicap, so that an approach to this value

should be a warning sign in any diagnostic consideration. An alternative frequency combination, 1, 2 and 3 kHz may be used. For this average, 40 dB would be in the range of mild disability. Figure 12.6 shows the values of this average in medians and quartiles from Hinchcliffe's data, plotted against age for otologically normal men or women without specifically noisy occupations, in a rural community. This diagram indicates approximately how much hearing an individual has in reserve, against the deterioration of old age. It is obvious that a noise-induced threshold shift in a young person may be unnoticed, but presbycusis may turn it into a disability in later life. These suggestions may need modification in the light of experience, but they provide a definite framework against which to consider the status of a particular case. In practice, the average in dB of the hearing levels at 1, 2 and 3 kHz is placed at the appropriate age on the diagram. By making the assumption that the rate of deterioration due to age will proceed normally, from the new level to which it has been raised by noise, an indication of the probable consequence of any particular degree of deterioration will be obtained. This assumes that no further noise-induced change will occur. If it does, the slope of deterioration will be increased. It is suggested that at age 70, the hearing level average should not exceed 40 in favourable cases. If it is seen that it will, the future of the individual, from the viewpoint of hearing can be considered as a specific problem.

Calculation of value for impairment

Davis (1970) discusses hearing handicap and medicolegal rules, based on the AAOO definitions, and this should be consulted for details. The allocation of a percentage value for impairment for purposes of compensation is achieved by attaching an arbitrary value in per cent handicap to a measured hearing level in decibels. Specifically the mean hearing level for 0·5, 1 and 2 kHz is used, 26 dB being set equal to zero handicap. For increase of hearing level thereafter, a rate of 1·5% per decibel, describes the handicap. Thus 27 dB equals 1·5% handicap, 93 dB mean hearing level equals 100% handicap. For the purpose of making allowance for unequal hearing in the two ears, the following procedure is used. The hearing levels of each ear are separately converted to a percentage handicap. The percentage in the better ear is multiplied by 5; the resulting value is added to the percentage handicap in the poor ear and the sum divided by 6. This yields the overall binaural hearing handicap value. The $1\frac{1}{2}\%$/dB rule, originally introduced as a purely nominal description of handicap for the purpose of monetary compensation has recently received the sanction of scientific vindication in terms of speech audiometry (Hood & Poole, 1971).

13

Aircraft noise

When Louis Blériot flew from France to England in his little mono-plane in July 1909, the noise from the 25-h.p. Anzani engine, heard from the ground by those fortunate enough to witness this historic event, was probably 20 to 30 dB more intense than the noise reaching the ground from a current heavy jet transport aircraft at its normal cruising altitude. This anomaly, pointed out by Hargest (1962) is of course due to the low altitude at which Blériot flew. When the transport aircraft is at these low altitudes it generates a noise field which creates a social problem of the utmost seriousness. This problem, though varying with the situation and local conditions of airports, is not confined to any particular country; it has not been neglected in the past, but efforts to contain it have not been success-ful, largely because of the growth of air transport. At the time of writing, further efforts are being made to attack the problem, and on an international scale. In all situations where personal liberty and privacy are affected, as in this problem of noise disturbance, intense feelings are apt to be aroused. The subject deserves the closest objective study if a reasonable understanding is to be achieved, and in an attempt to put the very varied facts in perspec-tive, the problem will be considered from three viewpoints. These are: the present situation and the factors which have contributed to it; the reactions of people in communities subject to aircraft noise, and its possible effects; and the remedial measures which present themselves at the present time. All this refers to the noise of aircraft in the landing and take-off phases of flight. It does not include the problem of the sonic boom, which now presents itself as the first generation supersonic transport aircraft (SST) begin to approach the operational stage. This is of an entirely different nature, separate from the low-altitude characteristics of the SST as a noise source, and will be treated separately.

Present situation of aircraft noise

DEVELOPMENT OF AIR TRANSPORT   The development of the pres-ent situation originates from the rapid growth of air transport after the end of World War II, but the evolution of the aeroplane as a factor in everyday life has been fairly continuous since the very beginning, with two main periods of accelerated development during 1914–18 and 1939 onwards to the present. The first aircraft were naturally small, and their power requirements and consequent engine and propeller noise were very modest. The rapid advances during 1914–18 contributed to the ensuing peacetime use of modified military aircraft or derivates. However, in the inter-war years, landplanes and flying boats, with up to four piston engines driving propellers, had been developed and were in general if restricted use on long-haul routes. Smaller aircraft on domestic scheduled flights were steadily increasing in numbers. Up to this time little disturbance due to noise seems to have been caused; power was moderate and numbers of aircraft small by present standards. After the advances of World War II, the large pressurised transport aircraft emerged, and in the span of about one decade the biggest and heaviest types of piston-engined propeller-driven transports evolved and were absorbed into the fleets of the airlines. The last generation of these machines (such as the Douglas DC 7C and Lockheed Super Constellation L 1649A) were sufficiently large and powerful to produce comparatively high noise intensities. The public, however, had become accustomed during the war to the sound of aircraft, including heavy bombers, and the similar noise of the four-engined transports was therefore quite familiar and in the main accepted without much comment or complaint. For example, London Heathrow Airport authorities received 87 complaints about aircraft noise during the year 1956. In that year there was a total of 109,046 transport aircraft movements entirely composed of propeller-driven aircraft. It appears that these 87 complaints came from about 60 persons, so that local disturbance was presumably at a fairly low level. However, during the year 1958, complaints had risen to 350, and aircraft movements to 117,295. To what extent this increased level of complaint stemmed from the modest increase in aircraft numbers is uncertain, since another factor was now beginning to emerge, in the form of the beginnings of jet transport operations, which possibly influenced the number of complaints in the latter part of this year in particular. To trace the course of this development we must return to the year 1949. On 27 July, 1949, the de Havilland Aircraft Company of Hatfield, England, first flew their Comet 1 jet transport aircraft. This type was operated commercially during 1952–4; it was of moderate power, but complaints of noise did occur at Heathrow, mostly concerning ground running of the engines at night. This was a vigorous community reaction against a specific noise (Bell, 1963). In America, the Boeing Company of Renton, Washington, U.S.A. on the experience of their jet bomber aircraft, produced the prototype of the 707/720 family of Boeing jet transports, which first flew in July, 1954. After vicissitudes stemming from factors in the design of pressurised fuselages, the Comet 1 was succeeded by a larger, modified version, the Comet 4, possessing increased weight, power and range. This aircraft inaugurated the first transatlantic jet services on 4 October, 1958. Shortly after, the Boeing 707-120 also took its place on the North Atlantic services. From that time the present era of international jet services can be said to have begun, and with it, a gradually worsening situation of community disturbance due to the noise of jet aircraft.

It is almost superfluous to state that this community disturbance would not occur if aircraft did not fly over populated areas at altitudes low enough to produce high noise levels on the ground. This situation occurs throughout the world where established airports exist in the vicinity of built-up areas. New airports can theoretically be built away from populated areas, but in countries with high densities of population it is difficult to arrange sites where flight paths will not interfere with some populated area. This aspect will be referred to again in the section on expedients to reduce disturbance.

The present and possible future disturbance caused by aircraft noise is now being viewed against the somewhat belated, but now world-wide, realisation of the degree to which the environment is being degraded by man-made pollutants. The disturbance due to commercial operation of transport aircraft, with the attendant noise (and in addition local air pollution by the efflux of jet engines), must be reviewed, since the scale of operations continues to grow.

This growth of air traffic, in terms of transport aircraft movements and number of passengers (*Flight International*, 1966a) has shown impressive increases. On the basis of 15 major airports throughout the world, an increase of over 72% in the number of passengers occurred between 1961 and 1965. These airports handled a total of 67 million passengers in the year 1965. This is for a small number of airports, and in the many others, large and small, similar trends may be expected to operate. The implication of these trends may be illustrated by quoting traffic data in specific airports. For example, the number of passengers using London Heathrow in one year increased from about 6 million in 1961 to about 10·5 million in 1965, while in 1969–1970 the figure was 14·68 million (*Flight International*, 1970). This itself is somewhat below the European average yearly growth, which as at September 1970 was 17·5% in passengers and 9·6% in total movements. Variations occur however; Chicago O'Hare, the leading airport in terms of air transport movements, showed a small decrease in 1969 as opposed to 1968. At the present time, a recession in air transport is occurring, but continued growth is expected. This expectation has stimulated costly and protracted investigations into the siting of new airports in a number of principal cities (Commission on the Third London Airport: Report (1971)). The decisions reached, on the basis of the various relevant factors, including the cost-effectiveness of the proposals, have been the cause of much controversy between planning authorities and conservationists. Irrespective of decisions or implementations, the Third London Airport Commission found, for example, that the third London Airport should, on its first runway, be operational by 1980. This estimate is based on the expectation that the four main airports in SE England, which handled some 17·1 million passengers in 1969, will experience a demand for about 36 million in 1975 and about 60 million in 1981. The growth at John F. Kennedy Airport, New York, from about 16·5 million passengers in 1965, has been estimated to be over 49 million in 1975 (*Flight International*, 1966a). Since the noise disturbance of aircraft is compounded of the numbers of overflights and the noise of each individual aircraft, as will be described below, reduction of numbers will reduce the disturbance. In the next 10 years some of the passenger traffic growth may be expected to be accommodated in larger aircraft, so reducing the numbers of flights relative to the numbers of passengers carried. At the time of writing the recession in the industry is achieving the same thing, but perhaps not to a significant extent, and from reputable estimates, this is a temporary setback. All in all, the problem is a formidable one, requiring the utmost effort in the technical, environmental and administrative fields. The relevant factors will be discussed below.

The jet engine as a noise source

The logical first step in considering the problem of aircraft noise as a social problem is to look at the noise itself as a physical entity.

Particularly, we should consider the characteristics of the exposure as experienced by people in the areas affected. Thus the intensity and frequency distribution of the noise, its duration and time characteristics are relevant. We shall see that the growth of community disturbance can be attributed to a number of changes in the physical nature of the exposure, and that clearly recognisable physical differences exist in the characteristics of the noise produced by jet engines as against piston engines driving propellers.

The noise of any type of aircraft engine increases with increased power. Since the take-off demands maximum power, the noise output is greatest in this flight condition. However, the noise characteristics of jet- and piston-engined aircraft are not only different at maximum power conditions, but markedly different under reduced power. Reduction from full to idling power gives a greater decrease in noise with piston than with jet engines, so that there are marked differences in both the take-off and landing phases of flight, as well as in taxiing on the ground.

The noise from the large piston-engined transport originates chiefly in the engine exhausts and the propellers. Aerodynamic noise due to the movement of the air over the fuselage, supporting and control surfaces, and mechanical engine noise are less important sources. At take-off, the noise is considerable subjectively but the high frequencies are less noticeable than in the case of the jet engine. A frequency spectrum of the noise of a piston-engined aircraft is shown in Fig. 13.1B, for comparison with the noise of a simple, or pure, turbojet aircraft (Fig. 13.1A). It will be shown that the original pure turbojet engine, by a gradual process of evolution, has now given way to the turbofan engine, with greatly reduced noise output despite much increased thrust.

The introduction of the large jet transport thus brought an unfamiliar noise, as well as a large increase in power. Any jet propulsion system for aircraft operates by ejecting a stream of gases, so generating a force analogous to the recoil of a gun, which is exerted in the opposite direction to that of the jet itself. In practical jet engines for aircraft, the production of the necessary high-velocity stream of gases depends on the burning of fuel, the oxygen being derived from the atmospheric air. Such engines are therefore dependent upon the atmospheric air for their operation. Rockets it may be noted carry the necessary materials for the evolution of large quantities of gas, and being thus independent of the surrounding air, can operate

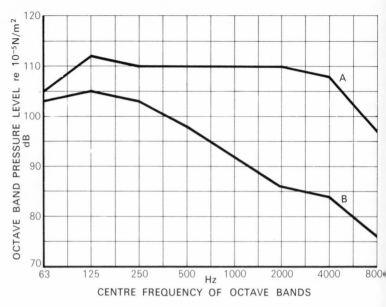

13.1   Comparative frequency spectra of aircraft types in flight. A, simple turbojet;
       B, four-engined propeller aircraft with piston engines. The actual levels have no
       particular significance. (© U K Dept. of Trade and Industry.)

away from the earth's atmosphere. The pure turbojet engine is simple
in principle. The engine as a whole can be regarded as a tube (Fig.
13.2) whose axis is approximately parallel to the fore-and-aft axis
of the aeroplane. The end of the engine pointing forward houses a
compressor which takes in air and compresses it. Following this,
fuel is burned in the combustion chamber or chambers and the
resultant hot gases are expanded through a turbine which drives the
compressor. Turbines and axial-flow compressors as normally used
in jet engines are both rotary devices with blades analogous to those
of a windmill. The turbine is rotated by the action of the gases on
its blades, in the same way as wind turns a windmill. The compressor
on the other hand when caused to rotate moves the air in a way
similar to that of a ventilating fan. After leaving the turbine the hot
gases are discharged as the jet exhaust, to the atmosphere, where they
produce part of the characteristic noise of the jet engine. The diagram
of Fig. 13.2 shows the principle simply, but it will be noted that both
compressor and turbine are unlike the simple windmill or ventilating

fan, which have only one row of blades. The jet engine normally employs a number of rows of blades, or stages, in both compressor and turbine. The actual engine of which this is a diagram, the Rolls-Royce Avon RA 29/1 turbojet engine, is illustrated in Fig. 13.3.

The noise output of jet engines comes from a number of sources. These include the noise of the jet itself, the noise of the compressor and the noise of the turbine. The characteristic roar of a jet engine is produced by the violent mixing of the exhaust gases with the air into which they are discharged. This noise is influenced by a number of factors, and is markedly dependent on the velocity of the actual jet relative to the surrounding air (Lighthill, 1952; Lloyd, 1959). The sound power may be proportional to the eighth power of the jet velocity, if the engine is restrained in a stationary position; that is, delivering its so-called static thrust. The original jet transports used the pure turbojet engine (Figs. 13.2 and 13.3) with high velocities in the jet of gases, reaching about 730 m/sec (2400 ft/sec). The power and acoustic characteristics of these engines with high jet velocities had the effect of making the aircraft subjectively noisy, and in fact these characteristics compelled the use, from the outset, of means to reduce community disturbance. These means consisted of the fitting of noise suppressors to the engines to reduce the noise of the jet itself at the source, and the creation of flight procedures designed to minimise noise at take-off. Noise suppressors fitted to jet engines take various forms, but generally they operate by modifying the

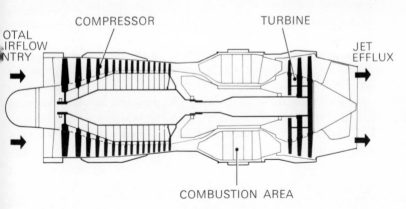

13.2 Diagrammatic section of a pure turbojet engine. (After Rolls-Royce (1971) Ltd.)

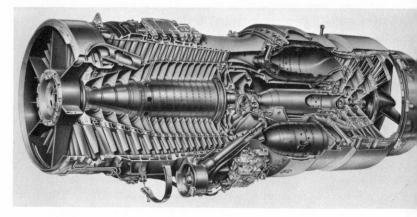

13.3   Cutaway illustration of the turbojet engine shown diagrammatically in Fig. 13.2. The engine is the Rolls-Royce Avon RA29/1 turbojet engine. (Rolls-Royce (1971) Ltd.)

flow of gases from the engine by directing them through some form of nozzle (Greatrex & Brown, 1958). The noise reduction is accompanied by a reduction in efficiency with increased operating costs of the aeroplane. It has been estimated that the total annual cost of noise control was about £1 million per year about 1960, when only about 200 heavy jet transports were operating throughout the world (Cook, 1960).

Since then, efforts have continued to reduce the noise and increase the efficiency of jet engines. The very marked influence on noise output of jet velocity relative to the surrounding air suggested that reduction of the gas velocity would be a profitable approach. Earlier developments of this principle, in the form of the low-ratio bypass engine (Hargest, 1962), enabled lower gas velocities in the jet to be employed, with consequent reduction in noise, with a compensatory increase of mass flow. The principle has been pursued still further in the turbofan type of engine, which will be referred to later, and some of its features will be described under the heading of noise reduction.

Noise associated with take-off and landing

Current jet transport aircraft require long runways, which may be up to 4000 m (about 13000 ft), for take-off and landing. The take-off run of heavily loaded aircraft will probably exceed half a minute in duration, after which the aircraft is rotated into a nose-up attitude for the climb out. For landing, a gradual descent along the Instrument Landing System glide slope of 2·5° to 3° to the horizontal can occupy as much as the final 10 miles to the runway threshold. The approach thus involves an area on the ground with the aircraft at low altitude on a fixed track; some controls on noise have at the time of writing been imposed in this flight condition. At take-off, however, airlines are required at many airports to prevent the noise on the ground from exceeding maximum values in PNdB at pre-scribed positions laid down for each direction for each runway. This is achieved by an initial steep climb to gain height as far as possible over the airport itself, after which thrust and rate of climb are re-duced to restrict the noise, followed by a continuation of the climb when populous areas are cleared or height gained sufficiently to keep the noise within permissible levels while climbing (Davies, 1971). Each airport presents its own particular conditions, and the serious problems are associated with airports near large built-up areas.

Methods of reducing community disturbance

The methods employed to minimise noise levels on the ground are numerous. Firstly, noise levels are laid down to which all operators must conform, and for London Heathrow there is a maximum permissible perceived noise level in PNdB, of 110 by day and 102 by night, at specified points for each route which are approximately four miles from the start of roll. These limits are enforced by noise measurement and the airlines are notified of infringements, and data kept on the degree of conformity with the regulations. Aircraft at different airports are subject to different noise-limiting regulations, and the principle of preferential runways is widely applied. This ensures that aircraft do not use runways involving overflying populated areas unnecessarily. For night-time operation, the lesser permissible noise level must be attained by operating the aircraft at reduced all-up weight, with subsequent fuelling at some point where the full take-off weight is permitted by less stringent noise regulations. The weight-reduction involved is very large. In Figs. 13.4 and 13.5 are shown the ground pattern of noise levels in PNdB, on approach and take-off, against heights, distances and flight paths, for day or night operation, for aircraft current at that time (1962). Quieter engines are now in course of introduction. The take-off flight paths would correspond to the worst condition at London Heathrow, where the departure is in a west to east direction, over-flying a populous area, using runways 10 Left or 10 Right.

Community reactions

In the vicinity of older established airports, sited for the requirements of piston-engined transports only, community disturbance can be acute as judged by complaints received by the authorities concerned. In the present phase of expansion of air travel it is essential to be able to judge community disturbance in an objective manner. Complaints to airport authorities, to civil aeronautics administrations, letters to newspapers and the like, while doubtless genuine manifestations of the feelings of individuals, are not sufficient evidence on which to base firm conclusions, and possible action.

One of the most serious community noise problems associated with major airports is that of London Heathrow. The situation was investigated by the Central Office of Information and the Ministry of Aviation by means of a social survey combined with a noise survey in the vicinity of the airport. The results may be found in the full report (McKennell, 1963) or in the summarised version published by the Committee on the Problem of Noise (1963) at whose request the survey was conducted. As an illustration of the methods employed in such investigations the survey and its uses will be briefly described.

Social survey in the vicinity of London Heathrow Airport

The principle of the investigation is quite simple. A representative sample of the population was asked to answer a number of questions designed to show their subjective reactions to aircraft noise, and the noise and number of aircraft were correlated with their answers.

The aspects on which information was sought included the following (Committee on the Problem of Noise, 1963):

1. The various direct effects of aircraft noise and the proportion of those exposed who experienced each effect;

2. the extent of annoyance by indirect effects;
3. the characteristics of the noise found annoying and the proportion of people annoyed;
4. the effect on annoyance of variation in any particular characteristic of the noise;
5. the social and psychological factors which might influence the degree of annoyance by the noise;
6. to what extent people could become accustomed to aircraft noise;
7. whether aircraft noise caused people to move away from the vicinity of the airport; and whether there were limits of exposure to aircraft noise beyond which movement of population away from the area would be enough to make the population remaining unrepresentative of the population as a whole;
8. any other data necessary for the understanding of the relation between noise and public reaction, in such a form as to enable predictions to be made on the effect of changes in any relevant variable.

Not all of these, and other questions, were able to be answered in decisive terms, but a most illuminating picture of the situation emerged from the results.

Methods

The survey included all the residential districts within 10 miles of London Heathrow Airport. The actual investigation was carried out in September 1961, and the adult population in the area numbered about 1,400,000.

SUBJECTIVE MEASUREMENTS The sample consisted of two categories of people. The first, numbering 1731, was a random sample of those persons whose names appeared in the electoral registers. The second category was composed of 178 persons from a list of those who had complained about the noise of aircraft. All 1909 persons interviewed answered a list of 42 questions presented as an investigation on living conditions in the various neighbourhoods. Different aspects, acceptable and unacceptable, of living in the areas, were explored, and the questions were designed to show each person's attitude to aircraft, and the effect of this attitude on their reaction to their surroundings.

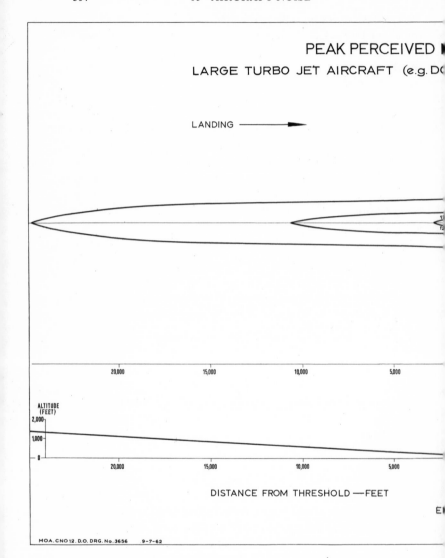

PEAK PERCEIVED

LARGE TURBO JET AIRCRAFT (e.g. D

LANDING ⟶

20,000   15,000   10,000   5,000

ALTITUDE
(FEET)
2,000⌐
1,000⌐
0⌐

20,000   15,000   10,000   5,000

DISTANCE FROM THRESHOLD —FEET

E

MOA. CNO 12. D.O. DRG. No. 3656   9-7-62

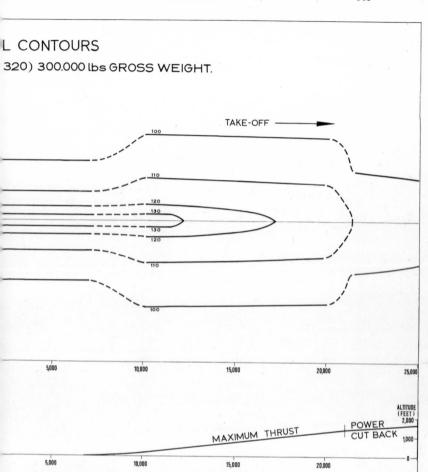

L CONTOURS

320) 300,000 lbs GROSS WEIGHT.

TAKE-OFF ⟶

DISTANCE FROM START OF ROLL — FEET

MAXIMUM THRUST

POWER
CUT BACK

ALTITUDE
(FEET)

Scale:~1:25,000
AORB. June 1962.

13.4 Noise in peak values of PNdB of heavy jet aircraft at different points in the landing and take-off phases of flight, for day-time operation. Take-off gross weight 300,000 lb (approximately 136,000 kg); Landing weight not specified. (© UK Department of Trade and Industry.)

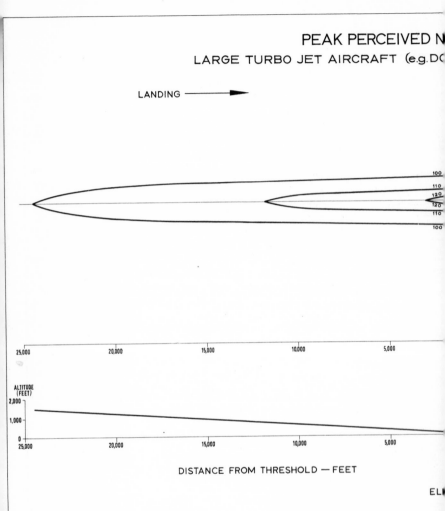

PEAK PERCEIVED N

LARGE TURBO JET AIRCRAFT (e.g. DC

LANDING

100
110
120
120
110
100

25,000   20,000   15,000   10,000   5,000

ALTITUDE
(FEET)
2,000
1,000
0
25,000   20,000   15,000   10,000   5,000

DISTANCE FROM THRESHOLD — FEET

EL

MOA. CNO 12. D.O. DRG. No. 3657.

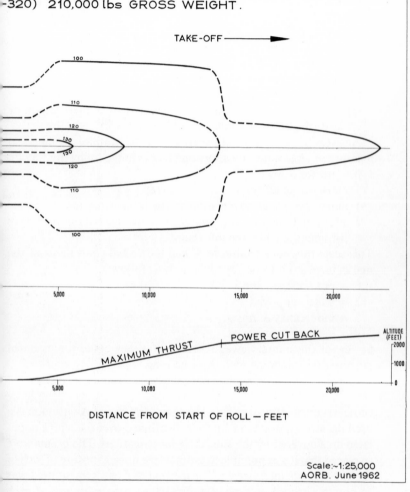

EL CONTOURS

-320) 210,000 lbs GROSS WEIGHT.

TAKE-OFF

100
110
120
130
130
120
110
100

5,000   10,000   15,000   20,000

MAXIMUM THRUST   POWER CUT BACK

ALTITUDE (FEET)
-2000
-1000
-0

5,000   10,000   15,000   20,000

DISTANCE FROM START OF ROLL — FEET

Scale:~1:25,000
AORB. June 1962

13.5   Noise in peak values of PNdB of heavy jet aircraft at different points in the
landing and take-off phases of flight, for night-time operation. Take-off gross
weight 210,000 lb (approximately 95,250 kg); Landing weight not specified.
(© UK Department of Trade and Industry.)

Having reached the subject of aircraft noise, it was necessary to ask questions which would elicit replies capable of being scored numerically, thus providing a quantitative measure of subjective reaction. The questions used occurred in question 13B of the list and were:

'Does the noise of aircraft ever

(a) wake you up
(b) interfere with listening to TV or radio
(c) make the house vibrate or shake
(d) interfere with conversation
(e) interfere or disturb any other activity, or bother, annoy or disturb you in any other way?'

Answers to (a) to (e) were scored so as to produce a numerical measure of annoyance, by allocating marks in the following way:

0 = no annoyance
1 mark for an affirmative answer to (e)
1 mark each for annoyance caused by (a), (b), (c), (d)
1 additional mark for any source of annoyance, not in the above list, mentioned by the informant.

This scale thus runs from 0 to 6, and the verbal equivalents of the marks turned out to be approximately as follows:

0 = not annoyed at all
2 = a little annoyed
3 = moderately annoyed
4 = very much annoyed.

So few of those interviewed scored 6 that they could for practical purposes be combined with those scoring 5.

The answers to the various questions are discussed as results.

OBJECTIVE MEASUREMENTS   The then Ministry of Aviation measured the aircraft noise at 85 points distributed over the whole populated area covered by the subjective investigations. The points were so placed that it was possible to estimate the noise exposure of each of the 1909 people interviewed. The method of measurement used was to obtain a record on tape of about 100 successive aircraft at each point, for subsequent analysis. Thus data were available on about 8500 aircraft, and various measures of the exposure sustained by the subjects could be obtained by subsequent analysis. These measures included peak noise levels, the total number of aircraft, the noise levels exceeded by 10% and by 50% of the aircraft, the number of seconds during which 85 PNdB and 95 PNdB were exceeded. Thus a very large number of possible measures of noise exposure could be investigated, for the purposes of correlation with the data obtained from the subjective part of the enquiry.

Analyses and results

The 42 questions produced answers from which 58 social and psychological variables could be isolated. The analyses of the noise yielded 14 variables descriptive of the noise exposure. Analyses by computer showed that the physical correlates could be reduced to two only. These were the average peak noise level, in PNdB; and the number of aircraft heard during a day-time period of about 12 hours.

This conclusion is fundamental to the measurement of community disturbance. The two factors could be combined in a variety of ways, and different solutions have been proposed and used, in different countries. The particular measure in use in Britain is known as the *noise and number index* (NNI) and was derived in the following way.

Noise and number index

The basic comparison is between annoyance as indicated on the numerical scale, and the average peak noise level of the aircraft, together with the number of aircraft heard. The significance of the use of peak noise levels is that as each aircraft passes the listener, its noise builds up to a maximum and then diminishes. It is this maximum which is meant by peak noise level.

For the purpose of comparison of the annoyance scores of the people in the random sample with the peak noise levels and total numbers of aircraft heard, the aircraft noise data were subdivided into ranges of noise and numbers of aircraft. This permitted average annoyance scores of groups of people to be compared with averages of numbers of aircraft heard, within particular noise ranges. The noise ranges used were 4 in number: 84–90 PNdB (mid-point 87); 91–96 PNdB (mid-point 93); 97–102 PNdB (mid-point 99); 103–108 PNdB (mid-point 105). The numbers of aircraft heard were divided into 3 ranges. These ranges were 1–9 aircraft heard (average 5·75); 10–39 aircraft heard (average 22·5); 40–110 aircraft heard (average 81).

We are now in a position to compare the annoyance scores of the 1731 people in the random sample with the noise level and number of aircraft heard, in the case of each person. This information is contained in Table 13.1.

TABLE 13.1

The number of people with various annoyance ratings classified by noise level and number of aircraft per day

Noise level in PNdB	Average number of aircraft per day	Annoyance score						Average annoyance score	Number of people in stratum
		0	1	2	3	4	5		
84–90	5·75	230	128	113	5	5	31	1·1	512
	22·5	45	33	26	17	12	22	1·9	155
	81	5	7	2	7	10	7	2·8	38
91–96	5·75	51	41	28	17	11	10	1·5	158
	22·5	90	64	55	45	35	32	1·9	321
	81	18	15	13	23	18	23	2·7	110
97–102	5·75	2	1	—	3	1	—	2	7
	22·5	13	9	20	16	11	13	2·5	82
	81	20	22	38	26	30	64	3·1	200
103–108	5·75	—	—	—	—	—	—	—	—
	22·5	1	—	1	5	2	2	3·2	11
	81	11	7	17	16	19	67	3·6	137

Inspection of this table shows that of the possible 12 exposure combinations, or cells, derived from the four noise ranges and three number ranges, eight yielded suitably large numbers of persons. There is also a tendency, which is probably unavoidable, for high noise to be accompanied by high numbers of aircraft heard, as well as an association between low noise and small numbers of aircraft. However, these effects are not sufficiently marked as to impair the usefulness of the comparison. The annoyance scale is seen to reflect the noise and number of aircraft in an orderly fashion. Minimal annoyance occurred at the lowest levels of noise and numbers of aircraft. Thus nearly half a group of 512 people were not annoyed at all by an average of about six aircraft per day with a peak noise in the range 84–90 PNdB. At the other end of the scale of exposure, about half the people in a group of 137 were more than 'very much annoyed' which was the highest verbal grade of disturbance in the scale, by an average of 81 aircraft per day with a peak noise in the range 103–108 PNdB. Despite these trends, highly individual reactions are seen. For example in the last group mentioned, despite the serious disturbance to the majority, 11 persons out of 137 professed no annoyance at all. Study of the table shows also that some persons are apparently seriously disturbed by aircraft giving noise levels which would be regarded by most people as acceptable for a passing motor car, and occurring about six times per day. This is the kind of idiosyncrasy which must be accepted in subjective judgements of acceptability of noise.

The relevance to disturbance of numbers of aircraft and their noise levels suggests that some relation might exist between these two factors.

This relation was investigated in the following way. Using the data available from Table 13.1 the annoyance rating is plotted vertically and the average peak noise level horizontally (Fig. 13.6). Of the 12 possible exposure cells, as previously noted, 8 contained enough persons to justify their use, the average rating being plotted against the noise and number values. It can be seen that the three highest points, in terms of annoyance values, come from the number category of 40 to 110 aircraft per day, averaged as 81 in Table 13.1. The three intermediate values of annoyance come from the 22·5 aircraft per day range, and the two lowest annoyance values from the 5·75 aircraft per day. Lines can be drawn through the points to indicate the annoyance caused by 5·75, 22·5 and 81 aircraft per day. If these lines are straight and positioned by the technique of calculation known as the method of least squares, with weight allotted to each point proportional to the number of observations on which it was based, they appear effectively as shown on Fig. 13.6. Here the lines are drawn parallel and equidistant and indicate clearly that, for a given number of aircraft heard, the average noise, in PNdB, is directly related to annoyance. The average number of aircraft represented by the lines, moreover, actually 5·75, 22·5 and 81, very nearly represent a geometrical series in which each item is one quarter of the one above it, that is the ratio is 4. If the three lines are considered to represent 5·5, 22 and 88 aircraft, which is only a small change, the slope of the lines indicates that multiplying the number of aircraft by 4, without change of noise, is equivalent to raising the noise by 9 PNdB without change in number.

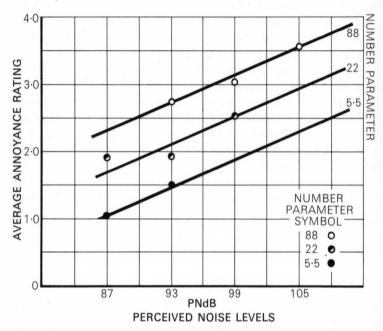

13.6 Relation between average annoyance and perceived noise levels. (Committee on the Problem of Noise, 1963.)

Also the addition of a line indicating one aircraft to Fig. 13.6 indicates that zero annoyance is about 80 PNdB. All of these findings can be incorporated in a simple relation, which gives a single figure measurement of annoyance, the *noise and number index* (NNI), thus

$$\text{NNI} = \text{average peak noise level} + 15 \log N - 80$$

where average peak noise level is the average* of the maximum

---

* The logarithmic average is normally used, thus

$$\text{logarithmic average peak noise level} = 10 \log_{10} \frac{1}{N} \sum_{1}^{N} 10^{L/10}$$

where L = peak noise level during the passage of each aircraft
        N = number of aircraft heard.

values, in PNdB, during the passage of each aircraft and N is the number of aircraft during one day or night.

The NNI, calculated for the various exposure groups and plotted against the average annoyance rating on the social survey scale (Fig. 13.7) gives a satisfactory relation, which is virtually linear.

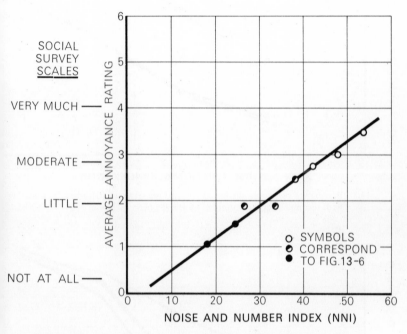

13.7 Relation between average annoyance rating (from the social survey) and noise and number index. (Committee on the Problem of Noise, 1963.)

The actual values of NNI can now be examined in relation to people's reactions to living conditions in the areas round London Heathrow Airport. These can be studied in detail in the full report of the survey (McKennell, 1963).

Specific effects

These effects include interference with relaxation, sleep, conversation, and such physical effects as house vibration and interference with television sound. Figs. 13.8 and 13.9 show how these effects

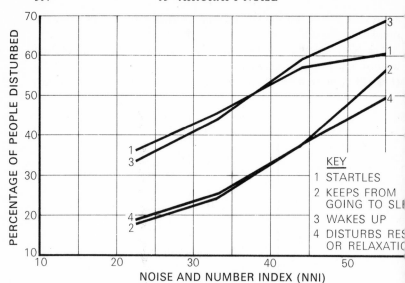

13.8 Percentage of people disturbed for various types of disturbance concerned with rest and sleep. (Committee on the Problem of Noise, 1963.)

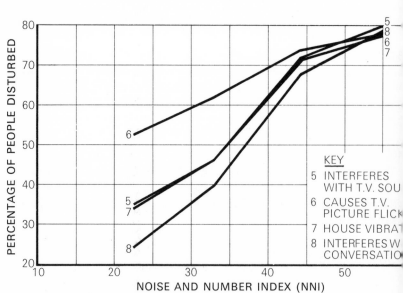

13.9 Percentage of people disturbed for various types of disturbance concerned with domestic factors. (Committee on the Problem of Noise, 1963.)

are related to NNI, in terms of the percentage of people, exposed to different NNI values, who stated that they were inconvenienced in the various ways. In Fig. 13.8 the commonest sources of complaint were of being woken up or being startled by aircraft noise. At a day-time value of about 37 NNI 50% of people stated that they were wakened and at 55 NNI nearly 70% suffered this highly undesirable effect. At 55 NNI also (Fig. 13.9), over three quarters of people suffered interference with conversation.

These are direct and specific effects. Other effects broadly connected with some form of annoyance were also investigated.

Indirect effects

The answers to the questionnaire elicited much information on the relation of people's reactions to NNI values. Some of these are shown in Figs. 13.10, 13.11, 13.12, 13.13 and 13.14. The trend of

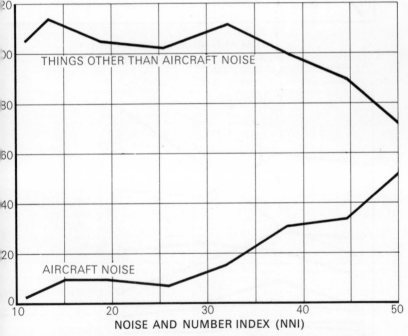

13.10   Number of times per 100 people questioned, that aircraft noise, and other things affecting local living conditions, were disliked. (Committee on the Problem of Noise, 1963.)

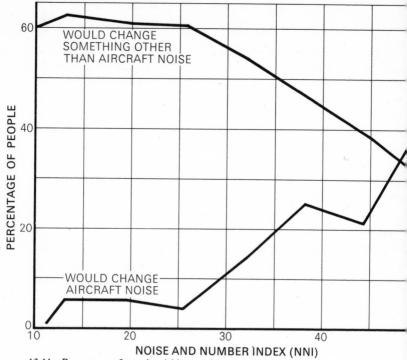

13.11 Percentage of people wishing to change their living conditions. (Committee on the Problem of Noise, 1963.)

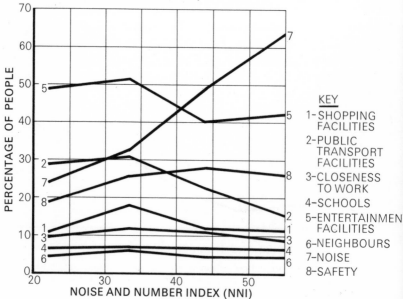

13.12 Percentage of people rating their area as poor or very poor for various reasons. (Committee on the Problem of Noise, 1963.)

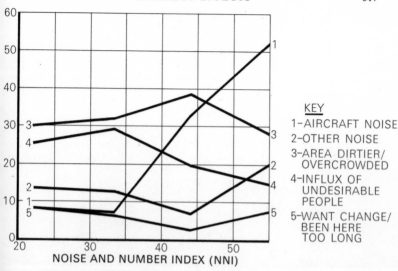

13.13 Percentage of people liking their area less now than in the past for various reasons. (Committee on the Problem of Noise, 1963.)

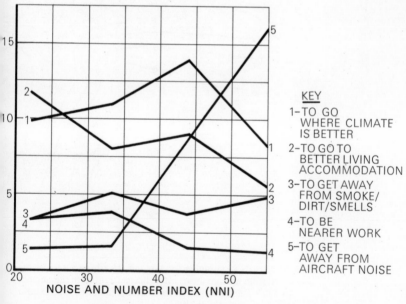

13.14 Percentage of people giving particular reasons for wanting to move. (Committee on the Problem of Noise, 1963.)

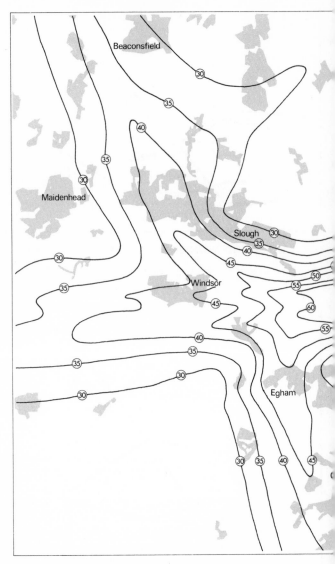

13.15 Average NNI contours for London (Heathrow) Airport derived from measurements made during day-time in the summer of 1967. These are the latest available data. (After UK Central Office of Information, 1969.)

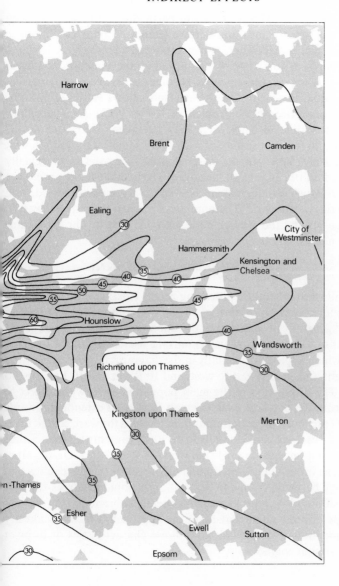

these curves is obvious. At some particular NNI value, aircraft noise becomes the most undesirable feature of living conditions in the areas concerned. There is inevitably considerable variation and uncertainty in this type of enquiry, but the conclusion is clear enough. It is that aircraft noise becomes excessive by any reasonable standards, when the noise exposure allows values of 50 to 60 NNI (Committee on the Problem of Noise, 1963). As an example, the bottom line of Table 13.1 may be taken. Here, out of 137 persons, 86 (67 + 19) were at least 'very much' annoyed, 16 were 'moderately' annoyed, 17 were 'a little' annoyed, 7 marginally annoyed and 11 'not at all' annoyed. The average annoyance score was 3·6 which is intermediate between 'very much annoyed' and 'moderately annoyed'. The number of aircraft movements was on average 81, and the average noise level 105 PNdB. The NNI value to the nearest whole number is thus

$$105 + 15 \log_{10} 81 - 80 = 54 \, \text{NNI}$$

To illustrate the noise prevailing round Heathrow in 1967 a map of the locality is given on Fig. 13.15. The contours can be seen to extend generally in an east–west direction. This is due to the direction of the prevailing winds, which entails the use mainly of the two runways which appear horizontally in the diagrams (known as runways 28 Right and 28 Left east to west, and 10 Right and 10 Left west to east). The disposition of the main populated areas with respect to these runways can be seen in the map. The areas extending towards the east are much more populous than towards the west, so that take-off in an easterly direction (10 R or 10 L) presents a more critical noise situation than towards the west (28 R or 28 L). On the other hand, the occurrence of predominantly westerly winds requires that landings be made from east to west. The populous areas towards Hounslow and Richmond are consequently subjected to the particular type of noise characteristic of the approach, which has already been mentioned in discussing the noise characteristics of turbojet and turbofan engines.

Night conditions

The NNI contours shown are those obtaining for day-time. The NNI values for night are mentioned in the ensuing section.

Second London Airport Survey

The second major survey of the London Heathrow conditions (UK Department of Trade and Industry, 1971) was undertaken six years later than the work just described, in order to check the validity of the results of the original survey and to extend its scope. A preliminary communication was given by A. E. Knowler to the 7th International Congress of Acoustics (1971) and the full details will be found in the sources cited above. The newer work consisted of three main parts: (1) a social survey in which people were interviewed, with improvements in the sampling system; (2) a comprehensive survey of the aircraft noise conditions, with knowledge of the air traffic pattern, so that the noise exposure of each person interviewed was known with some precision; (3) a comparison between social data, particularly the rating of annoyance, and the results of the physical measurements of the air traffic noise in terms of level and frequency of occurrence.

The validity of the data was increased by incorporating certain improvements, including the following. The sample size was more than doubled, 4700 persons being interviewed, with efforts to improve the distribution of the sample of persons with regard to the peak aircraft noise level and number of flights heard. Improvements were made in the derivation of the noise exposure of the respondents; these improvements included the use of radar tracking, and interpolation by means of a complicated computer programme. Lastly, night and day conditions were distinguished by the use of separate annoyance scales.

The scope of the survey was also extended by a large increase in the area covered; separate study of the landing and take-off situation; investigation of the effects on personal reaction to the aircraft noise of soundproofing of houses; and lastly, the effect of background noise level on annoyance due to aircraft.

The conclusions in part support the 1961 findings, in part do not; the following are a selection. The peak noise levels experienced had not changed significantly in the 6 years. Noise was increasingly seen as a matter for concern, and the majority thought that it was a major nuisance and likely to get worse. Against this background there was, surprisingly, little increase in awareness of, or annoyance from, aircraft noise. This was in presence of an increase of 35% in aircraft movements. Attempting to simplify conclusions from such complex data can be ill-advised, but it would appear indeed that the degree of annoyance for a given number of aircraft has fallen. It is still true that noisiness and annoyance are related in much the same way as in 1961; and it is also true that annoyance is increased by increased numbers, but the relation between noisiness and numbers of aircraft appears to have changed so that the number component has become of less importance. Obviously this conclusion prompts a re-examination of the basis of NNI. Some consideration has been given to this, but without any conclusions except to show that the NNI equation could be modified to accommodate the increased importance of loudness in relation to numbers. Many factors could operate, perhaps the most obvious speculation being that the population is merely becoming more accustomed to aircraft, and specifically to numbers of aircraft. Comparison of annoyance by day and by night showed that NNI values for zero annoyance, based on scales relevant to the particular kind of disturbance, differed by 10 NNI. This difference may not apply at other levels. The grants provided by government for soundproofing houses have been only partly successful; public awareness of the facility is limited and the average opinion rated the modifications from 'fairly' to 'very' effective.

The distribution of annoyance rating among those interviewed, for various NNI levels, can be best summarised in a table:

TABLE 13.2
Relation of annoyance to NNI

NNI	Percentage of persons expressing			
	No annoyance	Slight annoyance	Moderate annoyance	Extreme annoyance
25	25	43	24	8
35	13	36	34	17
45	6	26	42	26
55	2	16	47	35

This can be viewed against Robinson's (1970) suggestion, that 38 ± 2 NNI should be regarded as the limit for acceptability. This is equivalent to his Noise Pollution Level (L_{NP}) of 85 (using PNdB values) or 72 (using dB(A) values) for its derivation. An estimate was made of L_{NP} against annoyance in the second survey, which gave a fairly low correlation coefficient, but some uncertainty existed about background levels, which made conclusions uncertain in this respect. However, the results are consistent with the L_{NP} concept in one main respect: that annoyance due to aircraft noise is less the greater the background level, and vice versa.

No recommendations for changes in the derivation of NNI appear to be justifiable in the light of the newer results.

Possible means of reducing noise disturbance

With the present position that many major airports in different countries are admittedly giving disturbance to neighbouring populous areas, it is reasonable to expect that an international approach should be made to the solution of this problem of noise abatement. The world-wide nature and comparatively standardised procedures of civil aviation would appear to provide a favourable background for concerted action. While different countries vary in their involvement with this particular noise situation, and in their methods of trying to control it, recent events have shown great willingness, on an international scale, to improve the noise conditions associated with the operation of civil aircraft. In particular, a highly authoritative cross-section of international opinion on possible means of reducing the noise and disturbance caused by civil aircraft was obtained from a conference on the subject convened by the British Government in London in November 1966. At the time of writing the formal report has been available from HM Stationery Office since 1967, but the subject matter remains highly topical.

This was the first international conference at government level which was convened to discuss aircraft noise. The importance attached to the subject is indicated by the attendance of representatives of 26 States and 11 international organisations. The detailed consideration of separate topics was conducted by six specialist committees, and the conclusions reported in general terms by the relevant authority, the then Board of Trade, and in the technical press (*Flight International*, 1966b), represent a highly valuable assessment of the position in 1966.

The subjects discussed by the six committees were a good indication of the directions in which effort was considered to be worth expending. The different committees discussed the following topics.

Committee No. 1   The development and production of quieter aircraft and engines.

Committee No. 2   Control of noise by the siting of airports, planning their layout and limiting residential development nearby.

Committee No. 3   Operational noise-abatement procedures designed to limit the amount of disturbance caused by aircraft taking off, in flight, and landing.

Committee No. 4   Methods of specifying maximum permissible noise levels, how such levels should be determined, and methods of assessing compliance.

Committee No. 5   Methods of insulating buildings near airports against aircraft noise.

Committee No. 6   Methods of reducing noise from aircraft and engines when operated for maintenance or testing on the ground.

The conclusions reached by the committees following the presentation of papers and discussions, were naturally of a technical nature and will not be discussed in detail.

At the time of writing, however, comments are relevant on the subject matter of some of these Committees, the subjects being engine noise, airport and land usage, and noise measurement and rating. Of the various approaches to reduction of annoyance due to aircraft, the most urgent is still the reduction of engine noise. This is in no way to be interpreted as diminishing the importance of the disposition of airports and future activities in the field of land planning and similar approaches, and of possible noise-abatement procedures such as steeper landing approaches, now being studied by the National Aeronautics and Space Administration of the USA (*Flight International*, 1971). However, concrete advances have recently been made in engine noise reduction and its enforcement by certification.

Noise certification of aircraft types

The 1966 Conference on Aircraft Noise also discussed the choice of methods of specifying maximum permissible noise levels, how such levels should be determined and methods of assessing compliance. It was realised at the time that any penalties consequent upon noise reductions would have to be equally shared. This being so, no individual operator should be expected to pay an economic penalty for noise reduction, unless his competitors do the same. The highly competitive nature of air transport, which as a whole was very far from being as financially sound as it might appear (*Flight International*, 1966a) made it all the more necessary that the consequences of noise reduction should fall equally on all. It was thus agreed that it would be necessary, in order to make worthwhile progress, to ensure, by a process of certification of new aircraft, that these conformed to improved noise standards. Manufacturers would thus have realistic goals in noise reduction, and the establishment of standards would make possible more reliable predictions of community noise exposure. Questions also to be resolved were the extent of the improvements in noise characteristics and the time scale in which these might be realised. In the interval since 1966, sufficient progress has been made to achieve international agreement on procedures and to introduce actual certification for new aircraft. The details of the requirements can be found in a number of publications (UK Department of Trade and Industry, 1970; Statutory Instrument 1970/823; ICAO, 1971; Federal Aviation Administration, 1970).

The principles, as set out in Statutory Instrument 1970/823, are as follows. Aircraft noise should be measured at certain defined points relative to the flight path, in specified operating conditions. Turbojet or turbofan aircraft having a maximum total weight (in strict physical terminology maximum total mass) of more than 5700 kg must not exceed specified noise levels measured at defined points. These points are: (a) on take-off, on a line parallel to and 650 m from the extended centre-line of the runway where the noise appears to be greatest; (b) on take-off, at a point on the extended centre-line of the runway 6500 m from the start of the take-off run; (c) on the approach to landing, at a point on the extended centre-line of the runway, 120 m vertically below the $3°$ descent path. The maximum permissible noise levels at these points are stated in equivalent perceived noise level, the measured values in EPNdB (see ISO, 1970 and Appendix K) being corrected to 'the noise certification reference conditions' (Statutory Instrument 1970/823). The corrections derive from the need to standardise the noise data on the basis of aircraft position and performance, and the effect of temperature and humidity of the air on sound attenuation. The noise certification reference conditions are (1) atmospheric pressure at sea level 1013·25 millibars; (2) ambient air temperature 25°C; (3) relative humidity 70%; (4) wind velocity zero; (5) the maximum take-off and landing weights of the aeroplane are those at which certification is requested by the applicant for the certificate. The maximum permitted noise levels are dependent on the maximum permitted total weight of the aeroplane, as follows

Maximum total authorised weight of aeroplane	Maximum permissible corrected noise level in EPNdB		
	At point (a) (see text)	At point (b) (see text)	At point (c) (see text)
272,000 kg or more . .	108	108	108
34,000 kg or less . .	102	93	102

Where an aeroplane's maximum total authorised weight is between 272,000 and 34,000 kg, the noise levels not to be exceeded vary linearly according to the logarithm of the maximum total authorised weight of the particular aeroplane.

The above maxima may be exceeded at one or two of the measuring points, if compensatory reductions below the level occur at the other point or points: thus (1) the sum of the excess at two points must not exceed 4 EPNdB; (2) the excess at one measuring point must not exceed 3 EPNdB; (3) these excesses must be offset by equal or greater reductions at the other measuring point or points.

At the time of writing, more stringent temperature and altitude conditions are being proposed by the FAA (*Flight International*, 1971).

A certificate of compliance must be carried on each aircraft, and failure to do so renders an offender liable to fine or imprisonment; detention of aircraft not complying with the order is included.

This is only a brief summary, and the appropriate documents must be consulted for the details.

Reduction in engine noise

All the principal engine manufacturers have shown themselves to be fully aware of the need to reduce noise, and the modern developments of the turbofan engine have made possible significant advances in noise reduction, even in the case of engines with thrusts in the region of 40,000 lb (178 kN) or more. The advent of these so-called advanced technology turbofan engines has been particularly opportune for the new large subsonic transports now in use or approaching introduction. The engines include the General Electric CF6-6, fitted to the McDonnell-Douglas DC10 aircraft, and the CF6-50 engine of the same manufacturer, also for the DC10, and the European consortium aircraft, the A300B; the Pratt and Whitney JT9D used in the Boeing 747; and the Rolls-Royce RB211 for the Lockheed TriStar.

The development of this class of engine from the acoustical view-point, on the basis of Rolls-Royce experience, is described by Greatrex and Bridge (1967). They survey the noise output of succes-sive types of jet engines, starting with the pure jet, unsilenced, and passing through the successive phases of the pure jet with silencer, by-pass engines with progressively higher proportions of the airflow passing through the by-pass duct (higher by-pass ratio) and cul-minating in a new generation of Rolls-Royce turbofan engines. The latter have achieved appreciable noise reductions and embody means of reducing the fan noise by omission of the structural feature of previous engines known as inlet guide vanes, and having only one stage (i.e. one row of blades) in the fan itself. A diagram of such an engine is seen in Fig. 13.16 and a cutaway section of the complete engine in Fig. 13.17. The interested reader is referred to the original paper for further technical details and to a most illuminating account of the same subject up to 1970 by Smith (1970). Smith describes how the RB211 has been designed to eliminate the jet itself as a problem noise source. The remaining significant noise is therefore that originating in the fan, compressor and turbine blad-ing, and the relative disposition of these components is seen in Fig. 13.16. The main noise sources in such an engine are thus the fan, which radiates forwards and also rearwards, into the fan duct and compressor, and the turbine. Lesser sources are the compressor and the two jets labelled 'fan efflux' and 'jet efflux' in Fig. 13.16. The dominant nature of the fan and turbine is clearly heard as the aircraft flies over. Listening to the aircraft as it approaches, the fan and com-pressor noise is first heard, and increases progressively until the aircraft passes overhead, when a change occurs, the peak level of the noise thereafter being due to the rearward noise of the fan, together with the turbine noise, the jets from the two effluxes making a small

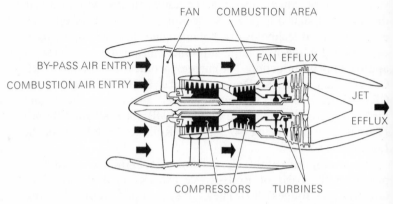

13.16   Diagrammatic section of an RB211 type turbofan engine. (After Smith, 1971.)

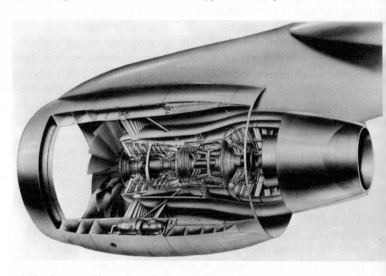

13.17   Cutaway illustration of the turbofan engine shown diagrammatically in Fig. 13.16. The engine is typical of the new Rolls-Royce type of turbofan jet engine. (Rolls-Royce (1971) Ltd.)

contribution. The relative intensities of the forward and backward radiation vary with the power setting of the engine, the situation as described is representative of an intermediate power. An indication of the origin, magnitude and direction of the various noise sources is shown in Fig. 13.18.

Thus the acoustic characteristics of transport aircraft engines at the higher power settings, has changed from one of conspicuous jet noise in the first generation pure jet engines and later in low-bypass engines, to a situation where the noise of the fan, compressor and turbines is the prominent source. Emphasis is therefore placed on further noise reduction of these components. Research into blade noise is being pursued to minimise the intrinsic noise, but in addition, such noise can be reduced by providing acoustic absorption inside the duct systems of the engine. This expedient, widely used in the past in ventilating systems and the like, has only become a

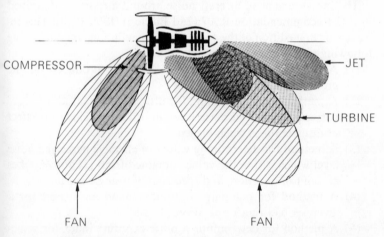

13.18   Diagrammatic indication of the noise sources of a turbofan engine. The loudness of each component in a particular direction is approximately indicated by the corresponding radial extent of the lobe. (Smith, 1971.)

profitable expedient because the jet noise is now no longer of primary importance over most of the power range of the engine. The areas which can, with advantage, be provided with sound absorbent linings are the inlet duct (Fig. 13.16) upstream of the fan, which radiates noise forwards; the fan duct, which conducts sound downstream and so outwards to the rear; and the duct leading rearwards from the turbine, where attenuation of the turbine noise in particular can be achieved.

Other facets of these engine developments involve for instance the design of the turbomachinery and lining of the absorbent structures to produce the subjectively least offensive spectrum, including the control of the 'buzz saw' phenomenon, whereby combination tones occur as a number of discordant pure tones. These considerations are too detailed to mention here. The result, however, is that substantial reductions of noise output have been achieved. This has enabled all three new large aircraft, the Boeing 747, McDonnell-Douglas DC10 and Lockheed TriStar, to be certificated, a most creditable achievement for the engine manufacturers concerned.

Measurement of aircraft noise

The measurement of aircraft noise around airports is described in ISO Recommendation R507 (2nd Edition) (ISO, 1970). This replaces the first edition of the same Recommendation, and makes the following provisions:

(1) A method of measurement of the noise produced on the ground by a given aircraft.

(2) A method for determining from noise data, values of tone-corrected perceived noise level, in order to include the effect of discrete tones when present.

(3) A method for determining values of effective perceived noise level which, using the values obtained from (2) above, takes account of duration, and spectrum of a single event.

(4) A method for mapping contours around an airport for a given set of aircraft operations.

(5) A method for determining a noise exposure index for a succession of events in a specified time interval.

These measurement techniques are more fully dealt with in Appendix K. In general terms the sound level of an aircraft may be specified as Perceived Noise Level (L_{PN}) expressed in PNdB. This may, where necessary, include a correction for the presence of irregularities of the sound spectrum (for example, pure tones); and in addition, the influence of the duration of each flyover noise episode may be provided for by a duration correction, the level being then the Effective Perceived Noise Level L_{EPN}, the value being commonly expressed as EPNdB. As will be seen from Appendix K, the proper derivation of these perceived sound levels as specified in R507 (2nd Edition) demands the calculation of the perceived noise level at a number of instants, at intervals of one half second or less. This is quite impossible without automatic data-logging equipment and impracticable without computer calculation facilities. Where an approximation by simple sound level meter readings is sufficient, L_{PN} can be derived in PNdB, by using the A-weighting of the meter and correcting by adding a value, in dB, which will range from 9 to 14 depending on the spectrum of the aircraft noise. Such a reading will not account for pure tones or duration. The same result will be obtained by using the D-weighting of the meter and adding 7 dB.

Indices of noise exposure; annoyance; land use

The need for a system of measurement of the exposure sustained by communities subjected to aircraft noise has produced an unsatisfactory situation which has been noted in Chapter 9 (p. 163). This situation is characterised by the large number of different indices broadly intended to indicate numerically the degree of annoyance, of which the NNI discussed earlier in this chapter, is one. The newer index suggested by Robinson (1970) using the Noise Pollution Level concept, has also been discussed in Chapter 9.

An admirable summary of the different national indices of aircraft noise is given by Galloway and Bishop (1970). The detailed examination of these would be tedious in this context, and some of those described and compared by Galloway and Bishop (1970) will merely be enumerated, with the appropriate sources of the original information. The object of all these indices of community response to aircraft noise can be regarded primarily as an attempt to quantify disturbance of communities.

The main indices are

Source	Index	Reference
France	Isopsophic index (N)	ICAO, 1969a
Germany	Störindex ($\overline{Q}$)	Bürck *et al.*, 1965
ICAO	Weighted Noise Exposure Level (WECPNL)	ICAO, 1969b
ISO	Aircraft Noise Exposure (L_E)	ISO, 1970
Netherlands	Total noise load (B)	Bitter, 1968
South Africa	Noisiness Index ($\overline{NI}$)	van Niekerk and Muller, 1969
United Kingdom	Noise and Number Index (NNI)	Committee on the Problem of Noise, 1963
United States of America	Composite Noise Rating (CNR)	Galloway and Bishop, 1970
United States of America	Noise Exposure Forecasts (NEF)	Galloway and Bishop, 1970

These various measures, as already noted in Chapter 9, are all essentially similar, basically on the pattern

$$\text{Index} = \bar{L} + A \log_{10} N - C$$

where $\bar{L}$ = mean of peak noise levels of each overflight, expressed in PNdB or EPNdB

A = a coefficient numerically in the range 6 to 24

N = number of noise occurrences (i.e. overflights heard)

C = a constant.

Detailed modification of some of the different formulations allow for such factors as the time of day and duration of the noise episodes. The NNI, it will have been noted, does not allow for duration, nor does the American CNR. The similarities are thus much more conspicuous than the differences between these different formulae.

Attempts to put these different scales for rating aircraft noise exposure on a comparable footing in a diagrammatic form are interesting. Galloway and Bishop (1970) have done this, by assuming daytime operations at a mean maximum noise level of 110 PNdB (or EPNdB) and an effective duration of 10 s, for various numbers of aircraft. Each national system specifies cut-off points of desirability or otherwise in different terms describing community responses or designation of permitted land use according to the level. These equivalents can be seen in Fig. 12 of Galloway and Bishop (1970). However, directly equivalent values cannot be stated in the various systems irrespective of the conditions; other relations would obtain if numbers of flyovers were held constant and the other parameters varied.

The noise exposure forecast system (NEF) (Galloway and Bishop, 1970) is a modification of the CNR formula, in which the sound level is inserted in EPNdB, instead of PNdB, so incorporating an allowance for duration, as well as other refinements. The NEF measure may in future supplant CNR in the USA.

The existence and potentialities of Robinson's (1971) Noise Pollution Level L_{NP} have already been noted in Chapter 9. Unlike the other measures, it takes account of background level and this has been confirmed recently (2nd London Airport Survey) to be a factor in the determination of disturbance. The equivalent level of annoyance to 38 NNI appears to be attained on the L_{NP} scale at about 85, using maximum perceived noise levels in the calculation.

Land usage

The problem of fitting large, or even small airports into communities has gradually evolved into major operations, and the recent history of such activities is clear proof of the massive popular forces at work in such areas. In England, the quest for a third airport for London has produced two extensive and protracted major enquiries (Cmnd. Paper 3259, 1967; Commission on the Third London Airport, 1971). The recommendations were in neither case adopted, in view of hostile public opinion. Finally a government decision was made on the basis of the least loss of amenity and environmental disturbance. Such decisions seldom if ever please everybody, and even this decision, to use a site on the Thames estuary rather remote from London, is 'regretted' by the Airport Authority (British Airports Authority, 1971).

Policy and decision on land use

Apart from the clear-cut and fundamental quest for quieter aircraft, controlled by noise certification, the whole paraphernalia of social surveys, sound measurement, and the development of indices of annoyance is merely a means to an end. The end is to insulate people from noise disturbance. In recent years the appreciation that the by-products of technological advances are in danger of irrevocably destroying the world as a place in which plants and animals can live, has slowly, and now more rapidly been gaining ground.

The concepts of maintenance of acceptable living conditions, with consideration at ethical, social and economic levels of all the factors involved, is now widely discussed and guidelines laid down. As an example of these praiseworthy activities a report by the Committee on Public Engineering Policy, National Academy of Engineering, USA (1969) may be cited. This report deals with the practicalities of 'technological assessment' by which is meant the assessment of the benefits and risks to society implied in the various possible developments of scientific and technological opportunities. Chapter 2 and Appendix B of the Report are devoted to a consideration of subsonic aircraft noise in detail, and this will repay study by those particularly concerned. The Report of the Commission on the Third London Airport (1971), deals in specific rather than general terms, with the many issues involved and emphasises the breadth of the problems entering into such decisions.

Effects of aircraft noise on hearing

The question of land utilisation is thrown into relief if one considers the situation where aircraft are at minimal height either on landing or take-off. Inevitably, noise levels, particularly on take-off, will then be high at ground level, and as many airports are at present constituted, these areas cannot be presumed to be free from people, either at work, on roadways, or even in or about dwelling houses. The extended centre line of the runway, with a band on either side, is thus the critical area, extending some distance up to the threshold and beyond the exit. The distance and sound level, as well as the numbers of aircraft movements, will be determined by the circumstances. As an example, unconnected with any particular airport or conditions, it may be assumed that it is possible (Fig. 13.4) for areas outside an airport on the extended centre line of the runway to receive sound levels of 110 dB(A) or more for a short period, of perhaps 3 s in effective duration, from some aircraft in current use. It is instructive to enquire how often this noise episode could be repeated during the day using the total energy (Chapter 11) method, without incurring undesirable noise exposure. Using 85 dB(A) as equivalent continuous noise level for occupational noise, the duration at 110 dB(A) would be about 91 s, or the equivalent of 30 take-offs in the assumed conditions. Since exposures of this order might well occur, it is clear that the previous preoccupation with the annoying or disturbing effects of aircraft noise, and the level of sideline noise, should not be allowed to obscure the need to consider the exposure of persons, however few they may be, who occupy these areas of high sound level. In making such assessments many factors must be taken into account. The utilisation of the runway, the nature of the noise reduction afforded by the buildings occupied, the proportion of time spent outdoors such as by farm or other outdoor workers, or by young children at play, or even infants in prams; the climatic conditions influencing the previous factor and the number and types of aircraft involved. Assessments of risk must therefore be made individually for each given situation. It is hardly possible to give any meaningful recommendations of a general nature, except to enjoin authorities to examine their particular situations in order to ascertain whether populations, however small in number, may in fact be otologically at risk due to aircraft. In the light of general expectations it would seem that such a possibility is of sufficient substance to warrant close examination. This has been approached by a hearing level study by Parnell, Nagel and Cohen (1972).

14

Impulse noise; intense noise; ultra, infrasound; sonic boom; vibration

There are certain problems, in the nature of special or associated aspects of noise, which are of importance as potential hazards or sources of disturbance. Four of these have been grouped in this chapter, in a summarised form. Their importance far outweighs the actual space allotted in this text, but to varying degrees, they represent advancing fields of knowledge less explored and documented than that of the effect of steady noise on hearing, or on attention and annoyance.

The topics considered are the following. Impulse noise, especially that due to gunfire and small-arms fire; the possible effects of extremely high-intensity noise; high and low frequencies; phenomena found with the sonic bang or boom, associated with the disturbance caused by bodies travelling through the air at speeds greater than the local speed of sound; and the effects of vibration. The last is often associated with intense sound, but may arise as a separate entity due to mechanical noises associated with machines, land vehicles, ships or aircraft. These environmental factors will briefly be considered in the above order.

Impulse noise

This type of noise has been mentioned in an industrial context, from such sources as riveting, hammering, chipping, stamping, and in operations such as the breaking up of masonry or concrete by power tools. Such noise is often combined with a background of other noise, or similar noise from numbers of the same type of tool operating in the same area. Impulse noise is also produced by explosions or by fireworks. For some of these sources, such as gunfire noise from small arms, their physical characteristics are quite well known, due to the widespread use of specific types of weapon, and recommendations for avoidance of hearing damage have now been evolved.

Despite the historical allusions to the effect of gunfire noise, such as descriptions of hearing damage by the noise of cannon in the eighteenth century (Chapter 1), the ill effects of such noise seem to have been ignored, or tacitly accepted, until recent years. Numerous reports of hearing damage due to gunfire have accumulated as a result of the 1939–45 war, and with the general increase of awareness of the hazard, investigations of a quantitative nature have since been made of the physical nature of the noise and the effect on the hearing of those exposed. Murray and Reid (1947) investigated the effect of gunfire on hearing in a systematic way and some aspects of this study have been referred to in Chapter 9. More recently, Coles (1962) and Coles and Knight (1965) have investigated audiometrically the effect on man of various degrees of exposure to the noise of rifle fire. Both permanent and temporary threshold shifts have been found.

The risk of hearing damage from the noise of all sizes of guns, including small arms such as service rifles and shotguns; and even of fireworks sold to the public, is now widely recognised. Some of the latter are quite dangerous to hearing at close range and cases have been published in the medical literature of hearing having been damaged permanently by fireworks.

The danger to hearing of small arms, such as rifles, machine-guns and mortars has been amply demonstrated. The usual variability in response to noise is seen here also. For example, in young men of such an age that presbycusis effects would be expected to be negligible or absent, Coles and Knight (1965) found great differences. In two such instances, both with three years of exposure to the weapons mentioned above, one had an almost normal audiogram, the highest hearing level being 15 dB in one ear at 6000 Hz; the other man showed a profound hearing loss, with hearing levels of 80 dB or more at 3000, 4000 and 6000 Hz. Ear protection was largely eschewed by both men. There is a need for a realisation of the damaging effects of small-arms fire, and for the need for ear protection. The British self loading rifle (SLR, a modification of the Belgian FN rifle) is a greater noise hazard than the older Lee Enfield ·303 in rifle, but 12-bore shotguns and other small arms are all potential hazards to hearing. Some rifle ranges may increase the hazard by their sound-reflecting form of construction. Coles and Knight (1965) emphasise these points and draw attention to the hazardous noise exposure to which many schoolboys are subjected in Cadet Forces. Ear protection is available in the form of ear plugs or muffs, which are effective against small arms fire noise. An excellent example is being set by British Services and by international teams in marksmanship competitions. Again a note of extreme

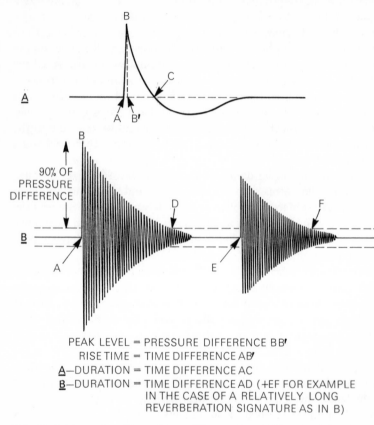

PEAK LEVEL = PRESSURE DIFFERENCE BB′
RISE TIME = TIME DIFFERENCE AB′
A—DURATION = TIME DIFFERENCE AC
B—DURATION = TIME DIFFERENCE AD (+EF FOR EXAMPLE IN THE CASE OF A RELATIVELY LONG REVERBERATION SIGNATURE AS IN B)

14.1 Evaluations of oscillographic waveforms of impulsive noise (After Coles, Garinther, Hodge & Rice, 1968.)

caution must be sounded on the possible dangers of using ear protection in an unconsidered way; there must be no possible risk of an accident due to failure to hear a warning because of the wearing of ear protection.

Coles, Garinther, Hodge and Rice (1968) and CHABA (1968) have made suggestions for a specification for avoiding damage to hearing

due to impulsive noise, based on work carried out for the British
Medical Research Council's Royal Naval Personnel Research Committee, the Institute of Sound and Vibration Research (University
of Southampton) and with the US Army Human Engineering
Laboratories. This is based on audiometric studies and on an
examination of the physical characteristics of the noise of small
arms, guns and other explosive sources. Physical factors considered
to be of primary importance are: peak level; rise time; principal
pressure-wave duration in the case of simple waveforms; and
pressure-envelope duration in the case of complex waveforms,
echoes and reverberant sound fields.

The basis of the specification (Figs. 14.1 and 14.2) is the temporary threshold shift likely to be reached or exceeded in 25% of
normally hearing persons, measured 2 min after the end of exposure
to up to 100 impulses occurring at a rate of 8 to 30 per minute. Perhaps 10 occasions per year are permitted. The permitted temporary
threshold shift is the same as the criterion of Kryter, Ward, Miller
and Eldredge (1966), namely 10 dB at or below 1000 Hz, 15 dB at
2000 Hz and 20 dB at or above 3000 Hz. The specification of Coles
et al. (1968) is used in the following way.

The wave form of the noise is the starting point. Oscillograms of
noise impulses which resemble Fig. 14.1A are classed as *A-duration*
noises, and the duration is taken as AC on Fig. 14.1A. The peak
pressure is as indicated, BB′, and is specified as dB SPL. Rise time
is not specifically included since the specification covers rise times
from less than 0·05 ms to 0·3 ms. Having thus obtained the values
for peak level in dB, and for the A-duration in ms, reference is
made to Fig. 14.2A. The quantities for peak pressure level and
duration are entered on their respective scales, and their intersection
compared with the A-duration curve. If the point lies above the
curve, a hazard, as defined by the criterion, exists.

In the case of impulses with reverberant characteristics, with
complex waveforms and prolonged decay, the procedure is as
follows. The start as before is the oscillogram of the impulse.
Referring to Fig. 14.1B, peak pressure level is, as in Fig. 14.1A, in dB
SPL. The duration to be used is the interval AD. The point D is that
at which the envelope of the waveform has decayed in amplitude to
10% of the peak value. If secondary pulses occur, their duration,
EF, is to be determined using the same 10% value of the main pulse,
and added to the duration of the latter so deriving the value for the

B-duration. Reference is again made to Fig. 14.2A, and the peak pressure level and B-duration scales are entered in the same way as before. If the intersection of the values lies above the B-duration curve, a hazard, as defined by the criterion, exists. The specification

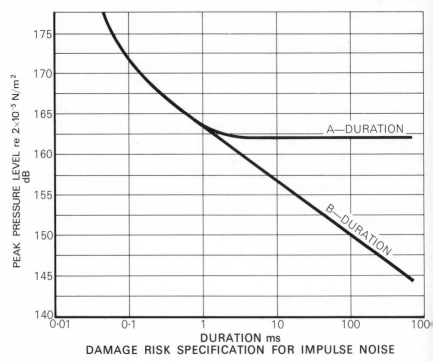

14.2A   Damage-risk specification for impulsive noise on the basis of about 100 impulses/second. (After Coles, Garinther, Hodge & Rice, 1968.)

curves are to be regarded as movable upwards or downwards according to circumstances. Thus Coles and Rice (1971) recommend that for impulses in non-reverberant fields, reaching the ear at normal incidence, that is arriving directly on the side of the head, the curve should be lowered by 5 dB. For persons known to be in the more sensitive range of noise susceptibility, the curve might have to be lowered by 5–15 dB. For occasional single impulses the curve might be raised by 10 dB. If ear protection is worn, the curve could be raised accordingly. If more than 75% of persons are to be covered (at the degree of auditory impairment noted above) the levels specified in Fig. 14.2A might have to be lowered by 5 dB to cover 90%, 10 dB for 95%, or even by 15 dB for higher percentiles. Ear plugs of V51R type would allow a rise of 20 dB, and fluid seal muffs, 30 dB on the values of Fig. 14.2A. Coles and Rice have proposed (1970, 1971) that a correction for the number of impulses may be applied for impulse repetition rates of up to 10/s, and where the wave envelope decays by at least 20 dB before the next impulse occurs (the B-duration). Fig. 14.2B shows the correction needed to the curve of Fig. 14.2A (which is for 100 impulses) for different numbers of impulses.

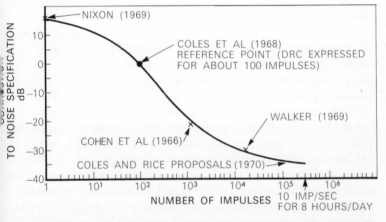

14.2B Adjustment to impulse noise specification (Fig. 14.2A) to accommodate different numbers of impulses. Proposals of other authors are entered at appropriate points on the curve. (After Coles & Rice, 1970.)

These summarised data are included to serve as an indication only and the reader is again advised to consult the original publication. However, the reader will doubtless have recalled the mention of impulse noise in Chapter 12. There the possible use of total energy as an index of the effect on hearing of impulse noise has been mentioned in the light of Robinson's formulations, and Atherley and Martin have shown that in the case of two industrial noises of impulsive character, equivalent continuous sound level is a valid indicator of the effects on hearing of such noise. Thus it should be possible to treat impulsive noises in this way, and obviously the use of instrumentation which would eliminate calculation of equivalent continuous sound level would be the method of choice.

General effects of intense noise

In Chapter 11 it was emphasised that the body should not be exposed to sound pressure levels of 150 dB and above. This presupposes that the C-weighting is used for all spectra not containing strong components below the 31·5 octave, and if such infrasonic components are believed to exist, efforts should be made to identify and include them. The value of 150 dB SPL was originally stated in USAF regulation 160-3, still appears to be a reasonable level at which to prohibit exposure unconditionally. Experience of this noise intensity is fortunately limited, and the limit must remain somewhat arbitrary.

At high levels, where the best ear protection would be used, there is a danger that the overall limit of 150 dB SPL for the body as a whole may be exceeded at levels which would be permissible at the shortest durations, for the sound actually reaching the ear. Thus, consider the use of fluid seal ear muffs (e.g. those of Shaw & Theissen, 1958) of the attenuation shown on Table 12.7, with the spectrum of Table 12.6. Here, for $3\frac{1}{2}$ min/day, 109 dB(A) would be acceptable. This, however, would be equivalent to 147·5 dB SPL, which is nearing the arbitrary upper limit for the body as a whole. Overall values of SPL on the C scale or 'linear' scale of the sound-level meter, should always be noted at these high levels; and in addition, the octave-band analysis should include the octave centred at 31·5 Hz to ensure that intense low-frequency components are not missed; if lower frequencies are likely to be present, they should also be sought. In short, extreme levels of noise should be measured with knowledge and care, and expert guidance from both the acoustic and medical viewpoint should be sought. This region, by which is meant very intense sounds, above 130 dB and including very low frequencies, is not extensively documented and the greatest care and best advice are well justified when such circumstances are encountered.

Goldman (1957) and Goldman and von Gierke (1961) have discussed some aspects of high noise and vibration and their effects on the body. Intense noise and mechanical vibration can be regarded as aspects of the same continuum. High noise fields can set up resonances in the body. Thus in very intense noise the body is shaken so that sensory receptors for touch, pressure and joint movement may be stimulated. If, as is common, mechanical vibration of surfaces on which the body is supported or with which it is in contact is present, massive sensory stimulation can occur. There are persistent statements of excessive fatigue, nausea and disorientation in the highest noise fields. Stimulation of the labyrinthine balancing mechanisms set up by vibrations in the fluids of the inner ear are probably involved in the last effect, and possibly the reports of nausea could be attributed to the same cause. These effects are very difficult to confirm in detail, but the way in which statements of such effects periodically appear, would suggest that they are genuinely experienced, but the circumstances are fortunately not common. Unpleasant sensations result from chest movements set up by intense noise and this seems to be more marked with the lower frequencies. A possible unpleasant effect in some persons is facial pain, which is presumably due to resonance in the upper respiratory tract, involving the nasal and associated cavities. The pain is felt rather deeply in the nose and cheek area, which would be consistent with such an explanation. Exposure of one ear to intense noise produces a sensation of being pushed away from the source of sound. There may be visual effects, and movements of the head may feel exaggerated or induce feelings of giddiness. These again must be due to a combination of normal rotational stimuli with abnormal stimuli to the balancing organs, presumably the semi-circular canals, conveyed through the vestibular fluids.

Ultrasonic noise

Ultrasonic sounds have sometimes been associated in the past with suggestions of serious effects on man. Parrack (1966) has written an excellent review of the effect of airborne ultrasound on humans which should be consulted for detailed information. He concludes that at the present time environmental ultrasonic fields generated by sources including turbojet aircraft engines, dental drills, and sonic cleaning systems, for which data are available do not appear to be significantly hazardous to man. He mentions some other sources such as diagnostic and therapeutic ultrasonic generators for which data are not available. He states that on the basis of experimental exposures to ultrasonic airborne sound in the range 20–37 kHz, individual environmental sounds in the ultrasonic region should be harmless to the human ear until the octave band or one-third octave band levels approach 140 dB SPL. Ultrasonic single frequency sounds gave no hearing sensation, but some temporary threshold shift occurred at half the frequency, and below, of the fundamental. Parrack also notes that 'ultrasonic sickness' as described in the period 1948–52 appears largely to be of psychosomatic origin due to apprehension about the possible effects. Non-auditory effects include sensations from vibrating hairs in the ear or nasal openings, and warming of the skin on contiguous areas of the fingers, for example. No resulting ill effects have been recorded. Acton and Carson (1967) have examined the auditory and subjective effects of industrial ultrasonic sources. Unpleasant subjective effects were experienced, including headache, nausea, tinnitus, and fatigue, by some persons. Acton and Carson suggested that the effects were actually due to noise in the high audible frequency range which also occurred in the machine noise, and not to the ultrasonic components as such.

Infrasonic noise

Infrasonic airborne sound and its effects, as mentioned above, are not so extensively documented as are the lower intensities and higher frequencies. Stephens (1969) has given a compact review of some aspects of low frequency sound. The effects on man will depend obviously on the intensity. Gavreau, Condat and Saul (1966) describe various devices for the production of intense low-frequency sounds from about 200 Hz down to frequencies as low as 7 and 3·5 Hz. The sound powers obtainable from these devices are high, reaching 10 kW of acoustic power. For comparison, a shout would be of about 0·001 W, a large orchestra about 10 W and a large jet aircraft at take-off power would yield about 100 kW of acoustic power (Beranek, 1960). Gavreau, Condat and Saul only mention briefly the biological aspects of such sound. They have experienced physiological effects similar to those of low-frequency mechanical vibration. Vertigo and nausea are attributed to excitation of the semi-circular canals, and sounds of 250, 16 and 7 Hz are mentioned as causing resonances of internal organs producing 'an intense irritation', vibration of objects in pockets of clothing, and interference with intellectual activity and visual disturbances respectively. The exact levels are not stated.

The relevance of these effects which presumably occurred at high intensity is of course whether in real situations in industry or near machines, the same sort of effects may occur.

Specific effects of intense low frequencies

Fortunately, the situation has recently been immensely clarified by a series of experimental exposures to intense low frequency air-borne vibrations, in which the subjects were highly experienced in sound and its effects on man (Mohr, Cole, Guild & von Gierke, 1965). The exposures were meticulously controlled acoustically and detailed subjective and objective records of the possible effects were kept. These courageous observers have thus produced the first authoritative account of the effects of very intense sound and infra-sound. The results provide the necessary information for proper judgements of the effects of such sound to be made. Thus the great merit of this investigation is the ability it confers to separate real and potentially dangerous effects from the vague suppositions and sinister implications which sometimes appear in non-specialist accounts.

These authors point out that in the present and future rocket motors for manned space vehicles low-frequency noise is becoming increasingly intense. The level of noise in the 1–100 Hz frequency range increases as the rocket booster increases in size and thrust, and the maximum energy tends to occur in this range. Estimates indicate that future very large boosters will produce their maximum noise energy in the infrasonic range, that is below 20 Hz. The siting of installations from which to launch space vehicles is greatly influenced by the noise hazards caused by the rockets, from the point of view of people on the ground, apart from considerations of those in the vehicles themselves. The work of Mohr *et al.* was planned to show the effect of whole body (as opposed to ear only) exposure to intense low-frequency noise, in order to try to establish the level of human tolerance to such noise. For this purpose five subjects, highly experienced in noise, exposed themselves for 2 min periods to high-intensity broad-band, narrow-band and pure-tone low-frequency noise. The effects of these exposures on cardiac rhythm, hearing threshold, visual acuity, fine motor control, spatial orientation, speech intelligibility and subjective tolerance were investigated. Exposures up to 154 dB SPL in the 1–100 Hz range indicated that for these short durations 150 dB SPL is 'well within human tolerance'. Exposures above 150 dB SPL produced responses which showed even for these highly experienced and knowledgeable people, that the limit of tolerance and reliable performance was being approached or was actually attained.

This investigation is of such fundamental importance that the full account, of what is believed to be the first systematic, controlled whole body exposure of human subjects to high-intensity very low frequency noise should be read. Summarisation cannot do justice to this work, but some of the main conclusions are as follows. The most notable responses were not auditory. Narrow-band noise in the very low frequency sonic range regularly produced chest wall vibration, gag sensations and respiratory rhythm changes. The limits of voluntary tolerance at these frequencies were not judged by these observers to have been exceeded. In the exposures to discrete frequency noise in the 50–100 Hz range, subjectively intolerable levels were reached. For example, it was decided to stop exposures at the following levels in view of the subjective effects: at 50 Hz, 153 dB SPL; at 60 Hz, 154 dB; at 73 Hz, 150 dB; at 100 Hz, 153 dB. The subjective effects were alarming, and included mild nausea, giddiness, subcostal discomfort, cutaneous flushing and tingling, at 100 Hz. At 60 and 73 Hz, coughing, severe substernal pressure, choking respiration, salivation, pain on swallowing, hypopharyngeal discomfort and giddiness were encountered. Headache in one subject occurred at 50 Hz, and headache and testicular aching occurred at 73 Hz. Impairment of vision (detectable objectively as well as subjectively) occurred in all subjects in exposures to noise at 43, 50 and 73 Hz. Marked post-exposure fatigue occurred in all subjects, but no shifts in hearing threshold measured 2 min after exposure, presumably due to the wearing of earplug and earmuff combinations. The fatigue disappeared after a night's sleep; one subject continued to cough for 20 min, and one retained the cutaneous flushing for about four hours after exposure.

Considerable individual variation occurred in the responses. The physiological effects are being studied further. To recapitulate in the words of the authors 'noise-experienced human subjects, wearing ear protectors, can safely tolerate broad band and discrete frequency noise in the 1–100 Hz range for short durations at sound pressure levels as high as 150 dB. At least for the frequency range above 40 Hz, however, such exposures are undoubtedly approaching the limiting range of subjective voluntary tolerance and of reliable performance'.

Lesser levels at low frequencies

AUDIBILITY   The equal-loudness contours published by ISO (Fig.

K.1) include 20 Hz, and Whittle (1971) is extending these still lower, some details of which are given in Appendix K.

SUBJECTIVE EFFECTS   At moderate levels, low frequency components may arise, in addition to rocket and turbojet engines, in diesel engines in rail and sea applications, electric generating plant, oil burners and ventilating systems. This type of noise presents some puzzling features, and its possible existence and effects should be taken very seriously. Where complaints of annoyance occur, certain association with low-frequency noise is not always easy. At higher levels, more often an association can be established with certainty. In practice, complaints may be received from an individual or a number of persons, couched in such vague terms, but describing a noise of perhaps continuous or pulsating nature, that it is difficult to know what to look for. Attempts at measurement may produce only normal values for the environment. In such cases low frequency noise may be involved. In other instances, with levels high enough to be clearly recognised, but not in the intense range noted above in the work of Mohr *et al.*, marked discomfort both mental and physical may be felt by persons apparently sensitive in this way. The explanation is not clearly established. In addition (Appendix K), it must be remembered that the configuration of the equal loudness contours is such that the growth of loudness at low frequencies is very much more rapid, with increase of level, than at the other frequencies. Thus, fluctuations of level are more conspicuous, which will presumably increase the distress of those sensitive to this particular type of sound. It should also be remembered that these long wavelengths are poorly attenuated in the atmosphere or by conventional sound absorbent treatments, as well as having little directional properties.

The sonic boom

Conventional artillery shells produce an audible sound in the form of a sharp crack when passing overhead. During the 1939–45 war, in addition to the actual final explosion of the charge carried by the V2 rocket used to bombard southern England, other explosive sounds were heard, in addition to the actual explosion. These sometimes gave the impression of two bangs in rapid succession. Similar sounds began to be heard when aircraft capable of exceeding the speed of sound first appeared. The effect has now become commonplace and is known as the sonic bang or boom.

Although the physical basis of the phenomenon was well known as long as 50 years ago, it was apparently not predicted in connection with supersonic flight. In the last 10 years, however, it has been subjected to intensive study.

PHYSICAL BASIS OF THE SONIC BOOM: SHOCK WAVES   The sonic boom is the audible evidence of the existence of the phenomenon of shock wave production.

It was noted in Chapter 2 that in the sound waves with which we have been dealing, the amplitude of particle displacement and thus the change in density of the air, is always small. With explosive waves, this does not obtain; the pressure and density changes are large and certain consequences ensue. In a large amplitude wave train, for example, the velocity of propagation will be appreciably greater in the regions of increased density (condensation) than in regions of reduced density (rarefaction). The crests of the pressure waves will gain on the troughs, and the waveform will therefore change as it travels. The advancing condensation phase will produce a steep wavefront which may persist for long distances from the source, and the condition is known as a shock wave. Ultimately with increasing distance the waveform assumes a rounded shape and becomes an ordinary sound wave. In such shock waves the large change in density within a very short distance in the advancing wave will produce a marked bending of light. Shock waves can thus be made visible by suitable optical techniques, or may even be seen in suitable meteorological conditions by the unaided eye. Such waves are thus rendered visible while radiating from bomb explosions. For a full discussion the reader is referred to textbooks of physics such as Stephens and Bate (1966).

SHOCK WAVES FROM MOVING BODIES   Bodies moving through the air at speeds less than the local speed of sound produce spherical sound waves propagated at the prevailing velocity of sound. If the velocity of the body exceeds that of sound, the compression of the air as a result of its motion cannot be transmitted forward, only laterally. The result of all the disturbances set up is that at speeds appreciably exceeding the local speed of sound a very thin conical shell of much increased pressure is shed from the tip of a projectile such as a bullet, and the higher the projectile air speed (indicated by the Mach number), the smaller the cone angle. At the nose of the projectile and in its immediate proximity the differences in pressure are so great that the nose wave is propagated at a velocity greater than that of sound. Subsequently normal sonic-speed propagation ensues. At the rear of the moving body a similar cone of reduced pressure is left behind. The exact pattern of shock waves of compression and rarefaction depend on the shape of the body. Even if the body as a whole is not moving at above Mach 1, the airflow over it, due to its shape, may exceed the speed of sound and small shock waves may occur. In one commercial jet aircraft, while cruising at subsonic speed, local airflow at the wing tips is apparently sufficiently rapid to create a shock wave. This can be seen from a suitable seat position, in suitable meteorological conditions, as a short thin dark line projecting upwards and backwards from a point somewhat behind the leading edge of the wing tip area. The sonic boom is a much larger and higher energy instance of the same basic occurrence, and its physical characteristics can now be considered.

THE SONIC BOOM: PRACTICAL ASPECTS   The sonic boom in a practical sense is one of the newer problems of acoustics, but far from being an isolated scientific curiosity it constitutes a problem of an immediate and urgent nature, because of the imminent introduction into scheduled services of the supersonic transport aircraft such as the British Aircraft Corporation-Sud Aviation Concorde. The acceptability and therefore the commercial future of these costly and elaborate ventures is literally dependent on the solution of the problem of the sonic booms they will produce. The peculiar difficulty is that the disturbance is not confined to the vicinity of an airport. Along the entire supersonic flight path of the aircraft a ground pattern of booms occurs in such a way that very large numbers of people may be exposed each time an aircraft overflies the area.

An authoritative assessment of the whole current sonic boom situation has been made under the auspices of the Acoustical Society of America (Acoustical Society of America, 1966, 72), and a summary of the situation was given by Hubbard and Maglieri (1967) at a symposium held in London under the sponsorship of the Institution of Electrical Engineers.

In the brief compass of this context, certain of the important features, from the viewpoint of community disturbance, of the sonic boom phenomenon will be discussed. These include (a) the physical nature of the pressure waves in relation to the aircraft and the extent and characteristics of the ground pressure patterns; (b) the effect on people exposed to these pressure patterns.

PHYSICAL NATURE OF AIRCRAFT SHOCK-WAVE PATTERNS   An aircraft travelling at a speed greater than the local speed of sound sets up multiple shock waves related to the shape of the aircraft. These travel with the aircraft, the bow shock wave leading and the tail shock wave ending the pattern. A diagrammatic sketch of the shock waves (Fig. 14.3) shows them extending outwards and rearwards from the aircraft, changing somewhat, as will be noted below, as they travel. The downwardly directed pattern eventually reaches the ground in a simplified form. The classical N-wave, shown at the bottom of Fig. 14.3 in its simplest form, is a wave literally of N-shape, with an initial portion above atmospheric pressure followed by a similar but inverted portion below atmospheric pressure. These waves travelling at the same speed as the aircraft in level flight sweep along the ground, from which they are reflected upwards (Fig. 14.3, dashed lines). The subjective impression is usually of a double bang or boom (both positive and negative phases are heard) which is audible along a track of width depending on various factors, but which may extend many miles on each side of the centre line of the ground track corresponding to the flight path of the aircraft.

NEAR AND FAR FIELD PRESSURE PATTERNS   The technique of prediction of sonic boom signatures has been reviewed by Carlson, Mack and Morris (1966). They describe the production of the sonic boom pressure pattern, which originates as noted above, as disturbances from individual parts of the aeroplane including fuselage, wing and engine nacelles. Different configurations of these components will give different pressure patterns. The latter are most marked close to the aircraft and this situation is known as the *near field* (Fig. 14.4). As the disturbances travel further from the aircraft, being left behind it in the process, they gradually coalesce, as shown in the lower altitude position of Fig. 14.4, and finally form the two-shock N-wave, which is found characteristically in the so-called *far field*. This is shown fully developed at ground level in Fig. 14.4. It

NATURE OF SONIC BOOM

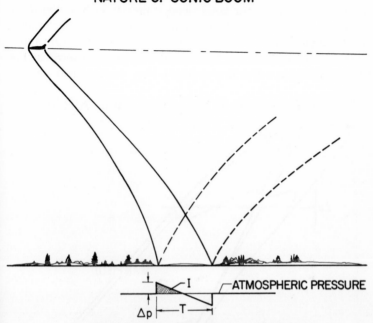

14.3   Nature of sonic boom phenomenon. (Hubbard, 1966.)

AIRCRAFT PRESSURE FIELDS

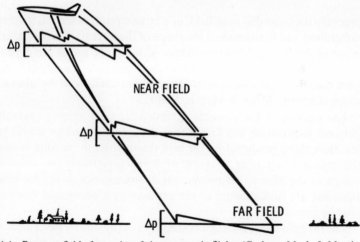

14.4   Pressure fields from aircraft in supersonic flight. (Carlson, Mack & Morris, 1966.)

is possible, however, for large slender aircraft such as the forth-coming supersonic transports, that the near field effects could extend far enough from the aircraft to reach the ground from appreciable altitudes (McLean, 1965). This situation is shown in Fig. 14.5, from McLean and Shrout (1966). Here the near field, or more

EXTENDED NEAR FIELD OF LARGE AIRCRAFT

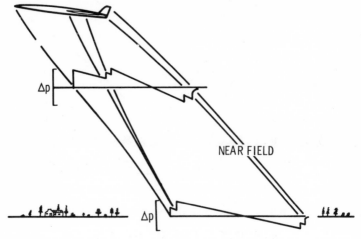

14.5   Extended near field of a large aircraft. (McLean & Shrout, 1966.)

precisely the extended near field, of a large aircraft is seen to involve the ground and to depend on the shape of the aircraft. The signature is not of the classical N-wave shape. Although N-waves are dependent on aircraft configuration for their pressure magnitude and time duration, the essential N-shape is not influenced by aircraft shape (Carlson, Mack & Morris, 1966).

The interest of the possible extended far field ground pressure patterns is two-fold: the first is that the actual pressures would be less than those predicted by far field theory, and the second is that the pressure signature may be modified favourably by suitable design of the aircraft. However, the characteristics of the far-field signature are fundamental to the problem as a whole and this will next be considered.

The far field effect is basically the N-wave previously described. A tracing of an actual sonic boom signature is shown in Fig. 14.6, from Hilton and Newman (1966). It can be seen to be composed of nearly symmetrical positive (I_{pos}) and negative (I_{neg}) components, the positive component having a peak value of overpressure (above atmospheric) Δp, amounting to 1 to 10 lb/ft^2 (about 48 to 480 N/m^2), and a short rise time τ. The total duration (of the positive and negative waves) may vary from 0·05 to 0·30 s. It is of importance, from both a physical and a biological standpoint to determine

MEASURED SONIC BOOM SIGNATURE

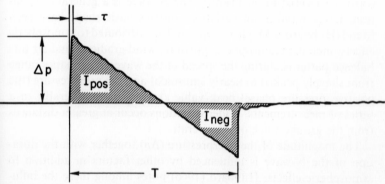

14.6   Tracing of sonic boom ground pressure signature. (Hilton & Newman, 1966.)

CHARACTERISTIC FREQUENCY SPECTRA OF N-WAVES

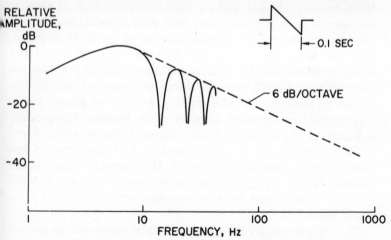

14.7   Characteristic frequency spectra of N-waves. (Hubbard & Maglieri, 1967.)

the frequency spectrum of such a waveform. If an idealised wave of duration 0·10 s is subjected to Fourier analysis the spectrum is as shown in Fig. 14.7. This spectrum is at a maximum about 10 Hz, and such spectra fall off markedly above and below this frequency. Points 40 dB below the peak occur at about 0·03 Hz and 1000 Hz (Hilton & Newman, 1966). The physical characteristics of the wave will be seen later to influence markedly their physical effects on structures and subjective effects on people.

GROUND EXPOSURE PATTERN   The N-wave is a general designation. Large numbers of variations on this basic shape have been found (Hubbard & Maglieri, 1967) and are attributed to atmospheric effects, including atmospheric pressure, wind gradients and air turbulence patterns, during the spread of the waves. The shapes range from sharply peaked to nearly sinusoidal and the overpressure (the positive pressure phase) peak value also varies considerably in a series of measurements. Wider variations occur at greater distances from the ground track of the aircraft.

The magnitude of the overpressure (Δp) together with the duration of the N-wave is influenced by other factors in addition to atmospheric effects. Hubbard (1966) notes among these the influence of aircraft design and operation. In the former category, aircraft size and weight, volume distribution and lift distribution are relevant. In the category of operation, relevant factors are aircraft Mach number, altitude, and the flight condition, such as whether the aircraft is on a straight and level path at uniform speed, or conducting manoeuvres involving change of speed or direction. The sonic boom can be diminished by reduced aircraft weight and by increased altitude. The effect of the latter is seen in Fig. 14.8, for three different aircraft types, as well as the increased overpressure caused by evolutions in pilot training. Increase in aircraft size is accompanied by increased duration of the N-wave. The pressure signature of the Concorde is now known (Warren, 1972). From 37,000 ft, the signature has not attained far-field characteristics, but from 45,000 ft, these are established. The signature is then of the classical N-wave type, with an overpressure of about 110 N/m² ($2\cdot3$ lb/ft²) and a total duration (signature interval) of 0·275 s, for about Mach 1·3.

The sonic boom ground pattern which accompanies an aircraft cruising at supersonic speed is illustrated in Fig. 14.9 from Maglieri

SONIC BOOM EXPOSURES

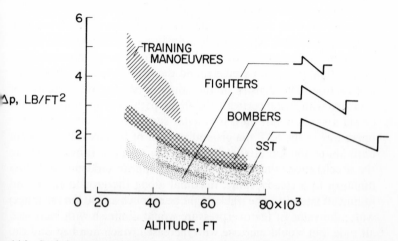

14.8 Sonic boom overpressures as a function of altitude for three different aircraft types. (Hubbard, 1966.)

GROUND PRESSURE PATTERNS

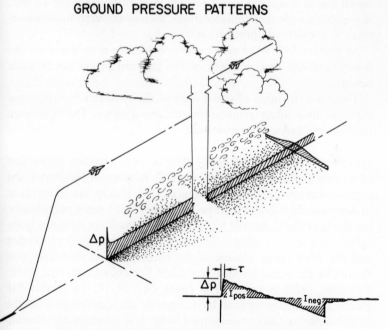

14.9 Sonic boom ground pressure patterns. (Maglieri, 1966.)

(1966). The aircraft flight path from take-off, through climb, and acceleration through the transonic range to supersonic speed is shown. Following this, level flight at two altitudes is correlated with the pattern of sonic boom production along the ground. The height of the hatched fences indicates the size of the overpressure (Δp) longitudinally and transversely, and the stippling on the ground its distribution. The density of the stippling shows the variation of pressure with distance from the flight path. The aircraft will not create a sonic boom until it reaches supersonic speed, and in the operation of the SST the ground pattern of booms will not begin until about 160 km (100 miles) from the take-off. Initially, due to the acceleration, the overpressure will appear suddenly and then diminish to a steady value, maximal along the ground track and falling off laterally. The track might be as much as 130 km (80 miles) wide; the value of the overpressure would diminish with increased altitude, but would increase with increased Mach number, and the track width would increase with increased altitude. At the end of a flight, speed would be reduced to below sonic velocity so that the boom would disappear about 160 km (100 miles) from the airport, in the absence of other considerations, such as an overriding requirement to avoid boom production.

At present ISO is known to be considering a standard on the description and measurement of the physical properties of sonic booms.

These are the main physical facts of the acoustics of supersonic flight as they affect populations on the ground. The biological implications can now be examined.

EFFECT OF THE SONIC BOOM ON PEOPLE   The situation at the time of writing (1972) is that a considerable number of incidental and organised exposures of populations to sonic booms has taken place, in America, Britain and Europe. In Britain and France particularly, test flying of the Concorde has provided the basis for data now published on subjective effects. The symposia recently held in Sweden and the USA cover aspects of the subjective and environmental factors of the sonic boom (Acoustical Society of America, 1972; Rylander, 1972). Recently (*The Times,* Oct. 29, 1971) a total of 348 claims for damage caused by Concorde on test flights over the British west coast have been settled by the UK government, totalling £11,809, for various kinds of structural damage, from loose ceiling

tiles to appreciable roof damage. The history of American studies
has been described by Nixon (1965), and in Britain by Warren (1966).
Much effort on the problem has been expended in the USA, and the
community reaction has been judged from real sonic booms made by
smaller aircraft than the SST, particularly at St Louis, Missouri and
at Oklahoma City. Training flights of the B58 aircraft have also produced community exposure (von Gierke, 1966). The St Louis experiment described by Nixon and Borsky (1966) consisted of a systematic
exposure of a populated area in the St Louis vicinity to 150 sonic
booms during a 10-month period. The ground overpressure values
occupied a range up to 3 lb/ft^2 (about 144 N/m^2). Some 2300 inter-
views, analysis of complaints and assessment of alleged damage
were related to the aircraft movements and overpressure measurements.

The sonic boom pressure changes were found to be markedly
different outside compared to indoors. The indoor pressure changes
were less, but were more complex than the outdoor signature, and
persisted for longer. Subjectively, the inside noise was more objectionable than that outside. This was attributed to its longer duration,
rattling and shaking of articles and of the house itself. As would be
expected, the conclusions (pertaining to 1961–2) were that sonic
booms over populated areas would be expected to arouse reactions.
The nature of the reaction was very variable, and no overpressure
could be defined below which sonic booms will be acceptable, nor
below which no antagonistic responses can be assured. Of the 1145
persons interviewed at St Louis, appreciably fewer expressed annoyance than admitted to interference with their normal lives and
activities. The results included shaking of houses, startle, interrupted
sleep, interruption of rest and relaxation, conversation, and radio
and television programmes. No adverse physiological effects were
reported, and this confirmed expectations that no such effects
should occur. Structural damage did occur to some extent, plaster
and glass cracks being the most numerous. A contemporary view
of estimates and observations of sonic boom effects, from Nixon
(1965) quoted by von Gierke (1966) is shown in Table 14.1.

At the time of writing, the data given in Table 14.1 are considered
by von Gierke (personal communication) to be in general still correct and representative. In addition, in the light of more recent
experience, it is probable (von Gierke & Nixon, 1972) that overland
commercial SST operations, giving sonic booms of about 1 lb/ft^2

(47·8 N/m^2) or greater would be unacceptable to a significant proportion of the population.

The loudness of sonic booms has been considered by Johnson and Robinson (1969). Their method of calculating the subjective magnitude of sonic booms involves recording of the wave form, or the use of wave forms derived from aircraft design data, from which is derived an energy spectrum by the normal procedure of Fourier analysis. This spectrum, by an adaptation of the normal loudness calculation procedure, yields a loudness value in Stevens' phons.

More recently, the annoyance value of sonic booms compared to other aircraft noise have been considered (Kryter, 1969; von Gierke & Nixon, 1972). On Kryter's equivalences, one boom per day at 1·9 lb/ft^2 would equal one aircraft flyover at 110 PNdB; this would

TABLE 14.1

Estimates and observations of the effects of exposure to sonic booms of different peak pressures

Peak overpressure (lb/ft^2)	N/m^2	Predicted and/or measured effects	
0–1	0–47·8	No damage to ground structures. No significant public reaction day or night.	
1·0–1·5	47·8–71·7	Very rare minor damage to ground structures. Probable public reaction.	Sonic booms from norm|al operational altitudes: typical community exposures (seldom abov|e 2 lb/ft^2).
1·5–2·0	71·7–95·7	Rare minor damage to ground structures; significant public reaction particularly at night.	
2·0–5·0	95·7–239	Incipient damage to structures.	
20–144	9·57 × 10^2– 6·88 × 10^3	Measured sonic booms from aircraft flying at supersonic speeds at minimu|m altitude; experienced by humans without injury.	
720	3·44 × 10^4	Estimated threshold for eardrum rupture (maximum overpressure).	
2160	1·033 × 10^5	Estimated threshold for lung damage (maximum overpressure).	

From data collected by von Gierke (1966).

be equivalent to 98 CNR, or 30 NNI. On this basis, 16 booms per day would give 110 CNR or 48 NNI, both unacceptably high. If such indices are to be used, some agreement (von Gierke & Nixon, 1972) must be established between overpressure and PNdB rating.

It is still far from easy at the present time to try to summarise briefly the various factors. There is no doubt that a factor of serious disturbance is present, and the aspect of structural damage, however modest at the levels 0–5 lb/ft^2 (0 to about 239 N/m^2), is a most undesirable feature. Ill-effects on the human body have not been shown, with the overpressures and durations so far used, but no experience with commercial SST operations has yet been obtained. The factors of safety for auditory ill-effects at present seem large.

FUTURE TRENDS   Investigation of biological responses, physiological and psychological is only at an early stage. The physics of the effect of the boom on structures is also important. Warren's (1972) work ('Recent sonic bang studies in the United Kingdom') has shown the effect of Concorde on buildings, including ancient monuments and cathedrals. The whole future of supersonic commercial operations depends on somehow making the aircraft acceptable from the viewpoint of the sonic boom. At present the situation is not clear. An obvious expedient is to limit the supersonic operation to over water, so that no booms reach inhabited land, presumably routeing supersonic operations in such a way that busy shipping lanes are avoided; some governments subscribe to this view.

Thus over these areas the SST would be no different from subsonic aircraft. This would be feasible on many routes, but long overland sections, such as the domestic networks in the USA and the shorter European routes would be excluded. The physical factors previously mentioned as relevant to reduction of sonic boom overpressures, such as efficient aerodynamic design, ability to operate at high altitudes, reduced aircraft weight, and the utilisation of the near field pressure patterns have been discussed by Baals and Foss (1966). These writers state that the estimated American limiting levels of sonic boom overpressures of 2 lb/ft^2 during acceleration and 1·5 lb/ft^2 at the start of supersonic cruise, established in 1964, cannot be said to be satisfactory from the viewpoint of community reaction. Knowledge now available does not contradict this view at the time of writing. The whole question of the use of the SST over land areas at least is thus still an open one.

Vibration

Knowledge of the physical characteristics and biological effects of vibration on the human body is still being acquired, and Goldman (1957), Goldman and von Gierke (1961), von Gierke (1964) and Guignard and Guignard (1970) should be consulted for details. The subject must be approached from a number of different aspects, depending on the method of transfer of the vibrations to the body, the position of the body, and the amplitudes and frequencies and direction of the vibrational displacements. The situation is less clearly documented than in the case of airborne noise, and at the present time no standards of an agreed nature exist to define vibration as acceptable or otherwise, as an environmental factor in given circumstances. A Committee of the British Standards Institution is at present studying these problems with a view to formulation of standards as are ISO and other national standardising bodies. This would be of great assistance to the manufacturers concerned with ships, vehicles of all kinds, machine tools, heavy production equipment, and articles such as rock drills, masonry and road drills, and small power tools of many kinds.

15

Conclusion

The preceding chapters have shown that man-made noise is wide-spread and, in excess, undesirable for many reasons. While individual susceptibility to hearing damage or disturbance in various ways is very varied, it is clear that noise as a factor in the human environment has in many situations been allowed to increase to an undesirable extent. Remedial means, however belated, must in these cases be put into effect. This in many cases is being done, with varying degrees of success.

It is satisfactory to note that, even since the first edition of this book was published in 1968, notable advances have been made in attitudes towards many of the factors which now threaten the prospects of reasonable living conditions in a large part of the world's inhabited areas. Efforts to conserve natural resources and amenities, and to restrict the harmful effect of technical by-products are now being greatly intensified in many fields, and the restriction of needless noise is one of the aims of these efforts.

The present situation can be summed up conveniently under the headings of industrial noise, local and general noise disturbance, and new or changing patterns of noise.

Industrial noise

While one determined man with a hammer can create a great deal of noise, the majority of noise problems in industry spring from the use of power. This has made possible large forces and rapidity of impact in hammering, chipping and suchlike operations, as well as the multiplication of machines under one roof. The evolution of production methods and of new prime movers has sometimes increased noise, sometimes it has reduced it, as in the case of the replacement of riveting by welding. Sometimes, however, the creation of noise is apparently accepted because the industry has been noisy in the past and the conditions seem to be established as normal and inevitable. Sometimes new machines, with perhaps higher performance than their predecessors have been produced and give appreciably more noise. This is reprehensible, and design should not be allowed to proceed without due regard to noise-producing properties. In other cases newer types of noise, especially if they are intense, or even if they are merely new, have produced a receptive climate of opinion in which not only the intense, but also moderate noises have been successfully reduced as a routine procedure. In industry some operations are intrinsically noisy and the usual measures for the preservation of hearing are the proper course. It must be expected, however, that the increased use of automation will tend to reduce the numbers of those exposed to the noise, as well as to make the protection of the remaining persons easier by such expedients as sound-reducing enclosures and the separation of noisy from relatively quiet areas.

These favourable considerations are allied to the progressively better understanding, through research, of the effect of noise on hearing, and on attention and working efficiency generally. These should lead to the general standardisation of maximum permissible levels of industrial noise. Knowledge now being gained on the distribution of susceptibility to noise-induced hearing loss in exposed populations will contribute to more precise control of noise exposure. Although good practices are already widely in use, a steadily improving situation in industry can reasonably be anticipated, as a result of improved techniques of noise control, and also because of actual or prospective legislation to control the degree of noise in industry or to compensate for hearing loss which it might cause. A recent example of such legislation is the noise aspect now included in the USA Walsh-Healey Public Contracts Act. In Britain, official consideration is known to be taking place on both prevention of and compensation for, occupational hearing loss, following the publication of the report *Hearing and Noise in Industry* by the Department of Health and Social Security, and also the *Code of practice for reducing the exposure of employed persons to noise*, by the Department of Employment. These represent useful advances in the field of occupational health.

Procedures for the control of noise exposure and thus for the preservation of hearing have developed out of all recognition in the very recent past, and even in the last two years at the time of writing (1972) new national or international recommendations have appeared or are in preparation.

Community noise

This type of noise can conveniently be separated into local or general types of disturbance. Of the local type, perusal of the newspapers will show cases where the cause of complaint may range from factories to places of amusement and even to individuals, and the noise from industrial or mechanical sources to social activities. This type of annoyance may be contained within the normal legal framework and satisfactory solutions, at least for the complainants, usually appear to be reached.

The worst general source of disturbance at present is the noise of aircraft and of road traffic. These noises are relatively out of the control of the ordinary individual, and those who are seriously disturbed have little alternative but to live in some quiet place, away from airways used at lower altitudes, or roads with heavy traffic. Transport aircraft of older types are very noisy, which is the direct result of the very high powers needed to give the performance demanded if air travel is to yield its full advantages. The reduced noise output for a given power of the new generation of jet engines, the first of which are actually in service, and backed up by obligatory noise certification requirements, is a major advance. The new large aircraft, the Boeing 747, Douglas DC10 and Lockheed TriStar will also tend to retard the growth in numbers. There appears to be no immediate prospect of any change in the pattern of flight operations, so that considerable ground areas may be expected to remain affected by the fairly extensive approach and take-off flight paths now used. The obvious solution is of course the location of airports away from centres of population. This in turn presents difficulties in densely populated countries, and demands associated very high speed surface transport, if the time gained, especially by supersonic transports, is not to be frittered away in getting to and from the airport itself. Helicopter services are used in some cases, but they also provide noise problems over urban areas. Studies of the whole socio-economic range of problems of the siting of airports have been recently made, such as by the Commission on the Third London Airport, and the subject has been very thoroughly explored.

There are considerable grounds for optimism, however, in the light of the continuing efforts of those countries which manufacture aircraft, or which regard the reduction of aircraft noise as of importance. Continuous activity is going on to reduce aircraft noise disturbance; the new engine technology has already produced power plants of increased thrust and subjectively reduced noise, and the improvement is obvious to the least observant. The noise of the Boeing 747 is less obtrusive than that of earlier aircraft of far less weight and power. Consideration is now being given internationally to the reduction of the noise of older aircraft not covered by the noise certification rules. As is usual, a cost analysis must be made of the relative economic consequences of different courses of action designed to achieve the desired noise reduction. For example, the cost of modifying or changing the engine and its installation so that the aircraft may comply with a specified standard for the remainder of its life, must be compared with the cost of complete substitution of new quieter aircraft, complying with the noise certification regulations. Proposals to make the latter embrace all the operating conditions, in terms of landing and take-off, altitude and air temperature, in which the aircraft is certificated to operate, are now being made by the Federal Aviation Administration of the USA. (At present a sea level atmospheric pressure of 1013·25 millibars, and an ambient air temperature of 25°C, 77°F, are specified.)

Other possible avenues of noise reduction are being explored. An international airline is reported to be considering the possibility of a 200-seater aircraft, powered by two RB211-type engines, for reduced take-off and landing (RTOL) from 1000 m runways. Such an aircraft could reduce, perhaps by a factor of 20, the area affected by noise in the neighbourhoods of airports. A similar approach, as discussed in the 1966 London noise conference, is the use of steeper landing approaches for noise reduction purposes. This is being studied practically and theoretically, by the USA National Aeronautics and Space Administration, in order to determine what changes in equipment and procedures would be necessary to obtain acceptance by pilot, airline and FAA for the routine use of a two-segment approach procedure, of which the first segment would be at a 6° inclination, the second at a conventional (2·5°) inclination to the touchdown.

The present situation with aircraft noise thus consists of considerable disturbance for limited numbers of people due to the placement of airports being unsuited to the type of aircraft which now use them, and a complete solution is inevitably of a long-term nature. Palliatives, in the meantime, include restriction of night operations, such as the abolition of night take-offs during the summer at London Heathrow.

The other and far more widespread source of community disturbance is road traffic. This again is a noise due to the operation of numbers of individual units, so that two noisy vehicles a day might be of little importance, but 200 might be annoying and 2000 intolerable. Road traffic noise is a nuisance because of the persistence of roads in areas where people work and live, and of their use by more and noisier vehicles. The noise produced by road vehicles is being restricted by regulations either projected or operative, in various countries and again a great expansion of work in Britain, on the fitting of roads into urban and rural areas, has been seen. Allied to this, it would appear that some restriction on the noise inside the driving compartments of motor vehicles should also be considered, in keeping with the practice already established for railway locomotives.

The increasing segregation of roads from living and working areas is beneficial and logical, but a watch needs to be kept on the tendency for the external noise of vehicles to increase, mostly due to the high-speed diesel engine.

To this end, the various regulations and prospective regulations at national level, seek to reduce excessive external noise. Opinions differ as to what constitutes a reasonable rate of obligatory progressive decrease in noise, but in Britain the opinion is held that, within about a decade, 80 dB(A) (in the prescribed measuring conditions) should be achievable for even heavy road vehicles. This might well be achieved technically, provided that the necessary international legislative backing exists to enable such vehicles to be put into production. Already it is becoming apparent that the road vehicle engineer must be conditioned in his design by the twin limitations of noise and of exhaust emission, and this is likely to have a profound influence on internal combustion engine design from now on.

Future trends

Wherever there are rapidly expanding technologies, new noise sources tend to appear. The hovercraft or air-cushion vehicle is such a source. Having great possibilities, these devices are more akin to aircraft than ships and their propulsion systems based on airscrews have so far been intrinsically noisy. Some effort has been expended on the use of submerged propellers, and this would be expected to be a quieter means of propulsion, but the air cushion has still to be provided. Obviously, their acceptability would be increased if excessive noise could be avoided.

Helicopters, again vehicles with unique advantages, nevertheless pose a problem of noise in urban areas, and their noise-producing properties are a continuing object of investigation. Aircraft using lift jets for support at low or zero speed have not been used commercially so far, but again, their characteristics as noise sources are of the utmost importance. The obvious place for STOL (short take-off and landing) or VTOL (vertical take-off and landing) aircraft in short-haul operations can only be fully realised if attention is paid to the noise aspect. At present the easier approach is by way of fixed-wing STOL aircraft and this would seem to offer real promise of reasonably rapid progress, as does the RTOL concept already noted. The whole problem of the sonic boom and its acceptability or otherwise is only at its initial stages, and it is apparently unacceptable in many quarters, over inhabited areas.

There are numbers of possibilities for different power units for road vehicles in various degrees of readiness. The gas turbine road vehicle is a reality but is not so far in quantity production. The Wankel engine which does not use reciprocating principles, is already in use in passenger cars. The less immediate prospect of fuel cells and electric drive is receiving attention. All of these types of propulsion have noise-producing characteristics quite different from the current type of diesel engine, and this must surely be taken into account in assessing their place on the roads of the future.

Thus, at the time of writing, it must be concluded that in many situations noise has been allowed to persist or actually to increase to an unnecessary extent, so that now it is yet another by-product of technical advance which has not been adequately controlled. Efforts towards its restriction or elimination are proper and justified, but a sense of proportion in dealing with the various undesirable factors of our environment should surely be maintained. The elimination of noise is only a part of the present task of reducing or abolishing these factors. It should be viewed as a part of the practice of preventive and industrial medicine and environmental preservation, such as the provision of clean water supplies, and of the continued attention to such problems as atmospheric pollution, the chemical contamination of foods, traffic and industrial accidents, and the ill-effects of tobacco smoking.

Appendices

Glossary of acoustical terms

The following definitions have been selected from the British Standard Glossary of Acoustical Terms, BS 661. The definitions are reproduced by permission of the British Standards Institution, 2 Park Street, London W.1, from whom copies are obtainable. For full information, the complete standard should be consulted.

ACOUSTICS
a) The science of sound.
b) Of a room or auditorium. Those factors that determine its character with respect to the quality of the received sound.

AMPLITUDE
Of a simple sinusoidal quantity. The peak value.

ARTIFICIAL EAR
A device for loading an earphone with an acoustic impedance which simulates that of an average ear, used for measuring the sound pressure developed.

ARTIFICIAL MASTOID
A device for loading a bone vibrator with a mechanical impedance which simulates that of an average mastoid, used for measuring the vibrations produced.

AUDIOGRAM
A chart or table relating hearing level for pure tones to frequency.

AUDIOMETER
An instrument for measuring hearing acuity.

BEL
A scale unit used in the comparison of the magnitudes of powers. The number of bels, expressing the relative magnitudes of two powers, is the logarithm to the base 10 of the ratio of the powers.

BINAURAL HEARING

1. Normal perception of sounds and/or of their directions of arrival with both ears.
2. By extension, the perception of sound when the two ears are connected to separate electro-acoustic transmission channels.

BONE-CONDUCTION HEARING LEVEL

A measured threshold of hearing by bone-conduction excitation, expressed in decibels relative to that specified as normal.

BONE VIBRATOR

Bone-conduction Receiver
An electromechanical transducer intended to produce the sensation of hearing by vibrating the bones of the head.

COCKTAIL PARTY EFFECT

The faculty of selecting one stream of information out of a number of voices speaking at the same time.

CONTINUOUS SPECTRUM

The spectrum of a wave the components of which are continuously distributed over a frequency region.

COUPLER

A cavity of specified shape and volume which is used for the pressure calibration of microphones of the reversible type. The term is also used loosely to describe certain types of artificial ear.

CRITICAL BAND

1. A band of frequencies centred round a nominal frequency which produces the same masking effect as a wider band of equal spectrum level.
2. A band of frequencies centred round a nominal frequency such that the loudness is equal to that of a narrower band of the same sound pressure level, but less than that of a wider band of the same sound pressure level.

CYCLE

Of a periodic quantity. The sequence of changes which takes place during the period of a recurring variable quantity.

DECIBEL

One-tenth of a bel. Abbreviation dB.

NOTE 1: Two powers P_1, and P_2 are said to be separated by an interval of n bels (or $10n$ decibels) when $n = \log_{10} (P_1/P_2)$.

NOTE 2: When the conditions are such that the ratios of sound particle velocities and ratios of sound pressures (or analogous quantities such as electric currents or voltages) are the square roots of the corresponding power ratios the number of decibels by which the corresponding powers differ is expressed by the following formulae:

$$n = 20 \log_{10} (u_1/u_2) \text{ dB}$$

$$n = 20 \log_{10} (p_1/p_2) \text{ dB}$$

where u_1/u_2 and p_1/p_2 are the given sound particle velocity and sound pressure ratios respectively.

By extension a single magnitude of any of these quantities may be expressed in decibels relative to a stated reference magnitude (it being understood that no impedance change is concerned). Thus for example the sensitivity of a microphone may be expressed either as:

$$x \text{ volts per N/m}^2$$

or as $\qquad y$ dB relative to 1 volt per N/m^2

where $y = 20 \log_{10} x$.

EARPHONE

Telephone receiver
An electro-acoustic transducer operating from an electrical system to an acoustical system and designed to be applied to the ear.

ECHO

Sound which has been reflected and arrives with such a magnitude and time interval after the direct sound as to be distinguishable as a repetition of it.
NOTE: In common usage the term is limited to reflection distinguishable by the ear.

FREE PROGRESSIVE WAVE

A wave propagated under conditions equivalent to those in an infinite homogeneous (but not necessarily isotropic) medium.

FREQUENCY

Of a periodic quantity. The rate of repetition of the cycles. The reciprocal of the period. The unit is the hertz. Symbol f.
NOTE: Frequency may be expressed in hertz (Hz), kilohertz (kHz) or megahertz (MHz).

HARMONIC

Of a non-sinusoidal periodic quantity. A sinusoidal component of the periodic quantity having a frequency which is an integral multiple of the fundamental frequency.

NOTE: In physics and electrical engineering the nth harmonic implies a frequency equal to n times the fundamental frequency: in music the nth harmonic usually implies a frequency equal to $(n + 1)$ times the fundamental frequency.

HEADPHONE
Head receiver
An earphone attached to a head band by which it is held to the ear.

HEARING LEVEL
Hearing threshold level
A measured threshold of hearing, expressed in decibels relative to a specified standard of normal hearing.

HEARING LOSS
The increase of individuals' hearing level above the specified standard of normal hearing when this can be ascribed to a specific cause such as advancing age, conductive deafness, perceptive defects or noise exposure.

IMPEDANCE
Acoustic impedance
Of a specified configuration or medium. The complex ratio of the sound pressure to the volume velocity through a chosen surface. Symbol Z_a.

Characteristic impedance
Of a medium in which sound waves are propagated. The specific acoustic impedance at a point in a progressive plane wave propagated in the medium.
    NOTE: In the case of a non-dissipative medium the characteristic impedance is equal to ρc.

Specific acoustic impedance
Unit area impedance
At a point in a sound field. The complex ratio of the sound pressure to the particle velocity. Symbol Z_s.

INSERT EARPHONE
Insert receiver
An earphone of small dimensions associated with a fitting for insertion into the auditory meatus.

LEVEL
Of a quantity related to power. The ratio, expressed in decibels, of the magnitude of the quantity to a specified reference magnitude.

LEVEL RECORDER
An instrument for registering, usually logarithmically, the variation with time of the magnitude of an electrical signal.
    NOTE: The input is frequently derived from a microphone.

LINE SPECTRUM

A spectrum whose components occur at a number of discrete frequencies.

LONGITUDINAL WAVE

Compressional wave
Irrotational wave
A sound wave in which the particle displacement at each point of the medium is parallel to the wave normal.

> NOTE 1: In acoustics, a particle of a medium is a volume of the medium having dimensions small compared with the wavelength of the sound but large compared with the molecular dimensions.
> NOTE 2: The term is also applied to the type of wave which may be propagated in thin solid rods or plates, in which the displacement is mainly parallel to the direction of propagation and the small component of the displacement normal to this direction is of opposite sense on opposite sides of the axis of the rod or median plane of the plate.

LOUDNESS

An observer's auditory impression of the strength of a sound.

LOUDNESS LEVEL

The loudness level of a sound is measured by the sound pressure level of a standard pure tone of specified frequency which is assessed by normal observers as being equally loud.

MAGNETIC RECORDER

Apparatus incorporating an electro-magnetic transducer and means for moving a ferromagnetic recording medium relative to the transducer for recording electric signals as magnetic variations in the medium.

> NOTE 1: The recording medium can be tape, wire, disk, etc.
> NOTE 2: The terms 'magnetic recorder' and 'tape recorder' are commonly applied to an instrument which has not only facilities for recording electrical signals as magnetic variations, but also for converting such magnetic variations back into electrical variations.

MAGNETIC TAPE

Recording medium in the form of tape which may be either of homogeneous ferromagnetic material, or of non-magnetic material (e.g. cellulose acetate, polyvinyl chloride, polyester, etc.) coated or impregnated with magnetisable powders.

MAGNETIC WIRE

Magnetic recording medium in the form of wire which may be homogeneous or have a core of non-magnetic material.

Masking

a) The process by which the threshold of hearing of one sound is raised due to the presence of another.

b) The increase, expressed in decibels, of the threshold of hearing of the masked sound due to the presence of the masking sound.

Mel

A unit of pitch. The pitch of any sound judged by listeners to be n times that of a 1 mel tone is n mels. 1000 mels is the pitch of a 1000 Hz tone at a sensation level of 40 decibels.

Microphone

An electro-acoustic transducer operating from an acoustical system to an electrical system.

Monaural Hearing

The perception of sound by stimulation of a single ear.

Noise

Sound which is undesired by the recipient. Undesired electrical disturbances in a transmission channel or device may also be termed 'noise', in which case the qualification 'electrical' should be included unless it is self-evident.

Noise Rating Curves

An agreed set of empirical curves relating octave band sound pressure level to the centre frequency of the octave bands, each of which is characterised by a 'noise rating' (NR), which is numerically equal to the sound pressure level at the intersection with the ordinate at 1000 Hz. The 'noise rating' of a given noise is found by plotting the octave band spectrum on the same diagram and selecting the highest noise rating curve to which the spectrum is tangent.

Normal Threshold of Hearing

The modal value of the thresholds of hearing of a large number of otologically normal observers between 18 and 25 years of age.

Noy

A unit of noisiness related to the perceived noise level in PNdB by the formula

$$PNdB = 40 + 10 \log_2 (Noy)$$

NOTE: This relation is analogous to the relation between phon and sone.

Occlusion Effect

The lowering of an individual's bone conduction hearing level when the normal air transmission along the meatus is impeded.

OCTAVE

a) A pitch interval of 2:1.
b) The tone whose frequency is twice that of the given tone.
c) The interval of an octave, together with the tones included in that interval.

PARTICLE VELOCITY

At a point in a sound field. The alternating component of the total velocity of movement of the medium at the point, i.e. the total velocity minus the velocity (if any) which is not due to sound. The unit is the metre per second; symbol u.

NOTE: The term 'particle velocity' may be qualified by the words 'instantaneous', 'maximum', 'r.m.s. [root mean square]', etc. The root mean square particle velocity is frequently understood by the unqualified term.

PEAK VALUE

Of a varying quantity in a specific time interval. The maximum numerical value attained whether positive or negative.

PERCEIVED NOISE LEVEL

The perceived noise level of a sound is measured by the sound pressure level of a reference sound which is assessed by normal observers as being equally noisy. The reference sound consists of a band of random noise between one-third and one octave wide centred on 1000 Hz.

PERMANENT THRESHOLD SHIFT

The component of threshold shift which shows no progressive reduction with the passage of time when the apparent cause has been removed.

PHASE VELOCITY

Of a sinusoidal plane progressive wave. The velocity of a point of constant phase in the direction of the wave normal. The point is geometrically defined, and its displacement does not necessarily involve a corresponding flow of energy.

PHON

The unit of loudness level (see under) when

a) the standard pure tone is produced by a sensibly plane sinusoidal progressive sound wave coming from directly in front of the observer and having a frequency of 1000 Hz;
b) the sound pressure level in the free progressive wave is expressed in decibels above 2×10^{-5} N/m^2;
c) the observer is listening binaurally.

PITCH

That attribute of auditory sensation in terms of which sound may be ordered on a scale related primarily to frequency.

PLANE WAVE

A wave in which successive wave fronts are parallel planes.

PNdB

The unit of perceived noise level. The numerical value of perceived noise level of a noise is equal to the sound pressure level of the reference sound expressed in dB above 2×10^{-5} N/m^2.

PRESBYACUSIS

Hearing loss mainly for high tones due to advancing age.

PROBE MICROPHONE

A microphone, or device incorporating a microphone, for measuring sound pressure at a point in a sound field without significantly altering by its presence the sound field in the neighbourhood of the point.

PURE TONE

A sound in which the sound pressure varies sinusoidally with time. The waveform may be represented by $a \sin \omega t$.

PURE TONE AUDIOMETER

An instrument for measuring hearing acuity to pure tones by determination of hearing level.

RANDOM NOISE

Noise due to the aggregate of a large number of elementary disturbances with random occurrence in time.

RECRUITMENT

An aspect of certain forms of perceptive deafness, whereby the growth of loudness of a sound with its sound pressure level is greater than it is for normal subjects.

ROOT MEAN SQUARE VALUE

(r.m.s. value)[1] Effective value—Of a varying quantity. The square root of the mean value of the squares of the instantaneous values of the quantity. In the case of a periodic variation the mean is taken over one period.

SENSATION LEVEL

Of a specified sound. The sound pressure level when the reference sound pressure corresponds to the threshold of hearing for the sound.

[1] Printed RMS in other parts of this text.

SONE

The unit of loudness on a scale designed to give scale numbers approximately proportional to the loudness. For practical purposes, the scale is precisely defined by its relation to the phon scale being given by the formula

$$Phon = 40 + 10 \log_2 (Sone)$$

SOUND

a) Mechanical disturbance, propagated in an elastic medium, of such character as to be capable of exciting the sensation of hearing.

By extension, the term 'sound' is sometimes applied to any disturbance, irrespective of frequency, which may be propagated as a wave motion in an elastic medium.

Disturbances of frequency too high to be capable of exciting the sensation of hearing are described as ultrasonic. Hypersonics is the name given to ultrasonic disturbances in a medium, whose wavelength is comparable with the intermolecular spacing.

Disturbances of frequency too low to be capable of exciting the sensation of hearing are described as infrasonic.

b) The sensation of hearing excited by mechanical disturbance.

SOUND ABSORPTION

a) Damping of a sound wave on passing through a medium or striking a surface.

b) The property possessed by materials, objects or media of absorbing sound energy.

SOUND ABSORPTION COEFFICIENT

Of a surface or material at a given frequency and under specified conditions. The complement of the sound energy reflection coefficient under those conditions, i.e. it is equal to 1 minus the sound energy reflection coefficient of the surface or material.

SOUND ANALYSER

Sound spectrometer

Frequency analyser

An instrument for measuring the band pressure level of a sound at various frequencies.

SOUND ENERGY REFLECTION COEFFICIENT

Of a surface or material at a given frequency and under specified conditions. The ratio which the sound energy reflected from the surface or material bears to that incident upon it under those conditions.

SOUND INSULATION

a) Means taken to reduce the transmission of sound, usually by enclosure.
b) Of a partition. The property that opposes the transmission of sound from one side to the other.

SOUND INTENSITY

In a specified direction. The sound energy flux through unit area, normal to that direction; symbol I.

> NOTE: It is customary to define the area parallel to the wave front, in which case $I = p^2/\rho c$ for a plane or spherical free progressive wave, where ρ is the density. (p = sound pressure; c = velocity of sound.)

SOUND LEVEL

A weighted value of the sound pressure level as determined by a sound level meter.

SOUND LEVEL METER

An instrument designed to measure a frequency-weighted value of the sound pressure level. It consists of a microphone, amplifier and indicating instrument having a declared performance in respect of directivity, frequency response, rectification characteristic, and ballistic response.

SOUND POWER

Of a source. The total sound energy radiated per unit time. The unit is the watt; symbol W.

SOUND POWER LEVEL

The sound power level (abbreviation SWL) of a source, in decibels, is equal to 10 times the logarithm to the base 10 of the ratio of the sound power of the source to the reference sound power. In cases of doubt, the reference sound power should be explicitly stated.

> NOTE 1: In the absence of any statement to the contrary, the reference sound power in air is taken to be 10^{-12} W (1 pW).

> NOTE 2: The use of 10^{-13} W as the reference sound power is deprecated.

SOUND PRESSURE

At a point in a sound field. The alternating component of the pressure at the point. The unit is the newton/square metre (N/m^2).

> NOTE: The term sound pressure may be qualified by the terms 'instantaneous', 'peak', 'maximum', r.m.s., etc. The r.m.s. sound pressure is frequently understood by the unqualified term sound pressure; symbol p.

SOUND PRESSURE LEVEL

The sound pressure level[1] of a sound, in decibels, is equal to 20 times the logarithm to the base 10 of the ratio of the r.m.s. sound pressure to the

---

[1] Abbreviation used in this book: SPL.

reference sound pressure. In case of doubt, the reference sound pressure should be stated.

In the absence of any statement to the contrary, the reference sound pressure in air is taken to be 2×10^{-5} N/m² (equals 2×10^{-4} dyn/cm²) and in water $0 \cdot 1$ N/m² (equals 1 dyn/cm²).

SOUND PROPAGATION

The wave process whereby sound energy is transferred from one part of a medium to another.

SOUND TRANSMISSION

The transfer of sound energy from one medium to another.

SOUND WAVE

A disturbance whereby energy is transmitted in a medium by virtue of the inertial, elastic and other dynamical properties of the medium. Usually the passage of a wave involves only temporary departure of the state of the medium from its equilibrium state.

SPECTRUM PRESSURE LEVEL

Of sound, of a continuous spectrum and at a specified frequency. The band pressure level for a bandwidth of 1 Hz centred at the specified frequency.

SPEECH AUDIOMETER

An instrument for the measurement of hearing by means of live or recorded speech signals.

SPEECH INTERFERENCE LEVEL

The average of the octave band sound pressure levels of a noise, centred on the frequencies 425, 850 and 1700 Hz together with the frequency 212 if the speech interference level in this band exceeds the others by 10 dB or more.[1]

STATIC PRESSURE

At a point in a medium. The pressure that would normally exist at that point in the absence of sound waves. The unit is the newton per square metre. Symbol p_0.

---

[1] Other frequency bands may be used in calculating speech interference level. See Chapter 9.

Temporary Threshold Shift

The component of threshold shift which shows progressive reduction with the passage of time when the apparent cause has been removed.

Threshold of Hearing

Threshold of audibility—Of a continuous sound. The maximum r.m.s. value of the sound pressure which excites the sensation of hearing.

Threshold Shift

The deviation, in decibels, of a measured hearing level from one previously established.

Tinnitus

A subjective sense of 'noises in the head' or 'ringing in the ears' of which there is no observable cause.

Tone

a) A sound giving a definite pitch sensation.
b) Sometimes, also, the physical stimulus giving rise to the sensation.

Transducer

A device designed to receive oscillatory energy from one system and to supply related oscillatory energy to another.

   NOTE: In acoustics one system is usually electrical, the other usually acoustical or mechanical.

Volume Velocity

The rate of alternating flow of the medium through a specified surface due to a sound wave. Symbol U.

Waveform

The shape of the graph representing the successive values of a varying quantity, usually plotted in a rectangular co-ordinate system.

Wavelength

Of a sinusoidal plane progressive wave. The perpendicular distance between two wavefronts in which the phases differ by one complete period. Symbol λ.

   NOTE: The wavelength is equal to the phase velocity divided by the frequency.

APPENDIX B

Units and equivalents

The system of units used in this book is the Système International d'Unités (abbreviation SI). The six basic SI units are the metre (m); kilogramme (kg); second (s); ampere (A); degree Kelvin (°K); and the candela (cd), for luminous intensity. The mol is likely to be accepted as the unit for amount of substance, so constituting a seventh basic SI unit. The SI is a so-called *coherent* system, in which the product or quotient of any 2 unit quantities in the system is the unit of the resulting quantity. Thus unit length (in metres) divided by unit time (in seconds) results in unit velocity (metres per second). The SI is coming into international use and the British Standards Institution (BSI) will use it in future in its standards. The SI is described in a BSI booklet No. PD 5686 (BSI, 1967a). For a summary of the basic inter-relations of physical quantities shorter physical tables, such as those of Aitken and Connor (1965) may be consulted, or for full details the reader is referred to the various publications of the BSI (1959, 1962, 1964, 1967b).

In addition to the SI system, another metric system, based on the centimetre (cm) gramme (g) and the second (s) is in use (abbreviated to cgs), and the British foot (ft), pound (lb), second (s) system is also used, with variants. The Table (pp. 380, 381) shows some of the more important units with equivalents, in the three systems. The meteorological convention for the measurement of pressure in millibars (mb) is also included.

Numerical values are all in conformity with those in the BSI publications cited, which can all be obtained from the British Standards Institution, 2 Park Street, London, W.1.

TABLE OF UNITS AND EQUIVALENTS

Physical property	Systems of units		
	SI	cgs	British
Mass	1 kilogramme (kg)	= 1000 grammes (g)	= 2·20462 pounds (lb)
Length	1 metre (m) 1 kilometre (km)	= 100 centimetres (cm)	= 3·28084 feet (ft) = 0·621371 mile (mile)
Time	1 second (s) (a defined fraction of a solar day) 3600 s = 1 hour (h)	second (s)	second (s)
Area	1 square metre (m²)	= 10⁴ square centimetres (cm²)	= 1550 square inches (in²) = 10·7639 square feet (ft²)
Volume	1 cubic metre (m³) (see note 1)	= 10⁶ cubic centimetres (cm³)	= 35·3147 cubic feet (ft³)
Velocity	1 metre per second (m/s) or 3.6 kilometres per hour (km/h)	= 100 centimetres per second (cm/s)	= 3·28084 feet per second (ft/s) or 2·23694 miles per hour (mile/h)
Density	1 kilogramme per cubic metre (kg/m³)	= 0·001 gramme per cubic centimetre (g/cm³)	= 0·0624280 pound per cubic foot (lb/ft³)
Momentum (mass × velocity)	1 kilogramme-metre per second (kg m/s)	= 10⁵ gramme-centimetres per second (g cm/s)	= 7·23301 pound feet per second (lb ft/s)
Force (mass × acceleration)	1 newton (N) (see note 2)	= 10⁵ dynes (dyn) (see note 3)	= 0·244809 pound force (lbf) (see note 4)

Acceleration due to gravity	9·80665 metres per second per second (m/s²)	=	980·665 centimetres per second per second (cm/s²)	= 32·1740 feet per second per second (ft/s²)
Pressure (force per unit area)	1 newton per square metre (N/m²)	=	10 dynes per square centimetre (dyn/cm²) or 1000 millibars (mb) or 10⁶ microbars (μb) or 10⁶ dynes per square centimetre (dyn/cm²)	= 0·0208854 pound force per square foot (lbf/ft²)
	1 bar or 10⁵ newton per square metre (N/m²) (see note 5)	=		= 29·53 inches of mercury (in Hg) or 14·5038 pound force per square inch (lbf/in²)
Energy Work (= force × distance) or Heat equivalent	1 joule (J) (newton-metre) or 2·38846 × 10⁻⁴ kilocalorie (kcal) (see note 6)	=	10⁷ ergs (1 erg = 1 dyne-centimetre)	= 0·737562 foot pound force (ft lbf) or 9·47817 × 10⁻⁴ British thermal units (Btu)
Power (rate of doing work)	1 watt (W) (1 joule per second) or 0·101972 kilogramme-force metre per second (kgf m/s)	=	10⁷ ergs per second (erg/s)	= 0·737562 foot pound force per second (ft lbf/s) or 1·34102 × 10⁻³ horse power (hp)

Note 1: 1 litre is the volume (1000·028 cm³) occupied by a mass of 1 kg of pure water at its temperature of maximum density and under a pressure of one standard atmosphere (760 mm Hg). The word litre is, however, now recognised as a special name for the cubic decimetre, but is not used to express high precision measurements.

Note 2: 1 newton is the force which will accelerate a mass of 1 kilogramme at a rate of 1 metre per second per second (kg m/s²).

Note 3: 1 dyne is the force which will accelerate a mass of 1 gramme at a rate of 1 centimetre per second per second (g cm/s²).

Note 4: In general and technological usage, 1 kg force or 1 lb force are used as units of force: that is the force exerted by standard gravity on a mass of 1 kg or 1 lb (standard gravity being such that the acceleration due to it is 9·80665 m/s²).

Note 5: In meteorology and acoustics pressure measurements are normally based on these units (the newton per square metre, millibar and microbar) which are consistent with an assumed atmosphere of 1 bar which is equal to 750·062 millimetres of mercury (mm Hg); 1 standard atmosphere is equal to 760 mm Hg or 1013·250 mb, or 14·6959 lbf/in². The above barometric heights are for Hg at 0°C, acted upon by standard gravity.

Note 6: 1 kilowatt hour = 3·6 × 10⁶ joules.

APPENDIX C

Standards and recommendations on acoustics

Many international and national standardising authorities issue publications on acoustics. The publications of the International Organization for Standardization (ISO) and the International Electrotechnical Commission (IEC) are distributed by the national standardising bodies. These, especially the member bodies of ISO, act as national representatives of each other, as sales agents and sources of information. A list of these names and addresses is available, for example from the American National Standards Institute, Inc. (ANSI), 1430 Broadway, New York, N.Y. 10018. Space does not permit of a complete list of national standards but the publications current at the time of preparation of this book, by ISO, IEC, ANSI and BSI, are as follows.

ISO recommendations

ISO R131   Expression of the physical and subjective magnitudes of sound or noise.

ISO R140   Field and laboratory measurements of airborne and impact sound transmission.

ISO R226   Normal equal-loudness contours for pure tones and normal threshold of hearing under free field listening conditions.

ISO R266   Preferred frequencies for acoustical measurements.

ISO R354   Measurement of absorption coefficients in a reverberation room.

ISO R357   Expression of the power and intensity levels of sound or noise.

ISO R362   Measurement of noise emitted by vehicles.

ISO R389   Standard Reference Zero for the calibration of pure-tone audiometers.

ISO R454   Relation between sound pressure levels of narrow bands of noise in a diffuse field and in frontally incident free field for equal loudness.

ISO R495   General requirements for the preparation of test codes for measuring the noise emitted by machines.

ISO R507   Procedure for describing aircraft noise around an airport. 2nd Ed. 1970.

ISO R512   Sound signalling devices on motor vehicles. Acoustic standards and technical specifications.

ISO R532   Method for calculating loudness level.

ISO R1996 Assessment of noise with respect to community response.

ISO R1999 Assessment of occupational noise exposure for hearing conservation purposes.

NOTE: Copies of these publications may be obtained in Britain through the British Standards Institution, 2 Park Street, London, W.1.

IEC recommendations

IEC 50(08) Electro-acoustics.

IEC 123 Recommendations for sound level meters.

IEC 126 IEC reference coupler for the measurement of hearing aids using earphones coupled to the ear by means of ear inserts.

IEC 177 Pure-tone audiometers for general diagnostic purposes.

IEC 178 Pure-tone screening audiometers.

IEC 179 Precision sound level meters.

IEC 303 Provisional reference coupler for the calibration of earphones used in audiometry.

IEC 318 An IEC artificial ear of the wide-band type for the calibration of earphones used in audiometry.

ANSI standards

S1.1-1960 Acoustical Terminology (including Mechanical Shock and Vibration) Revision and Consolidation of Z24.1-1951 and Z24.1a (*ISO R16, R131, and IEC 50(08)*).

S1.2-1962 Physical Measurement of Sound, Method for (Revision of Z24.7-1950).

S1.4-1961 General-Purpose Sound Level Meters, Specification for (Revision of Z24.3-1944) (*IEC 123*).

S1.6-1967 Preferred Frequencies and Band Numbers for Acoustical Measurements. (*Agrees with ISO R266*).

S1.8-1969 Preferred Reference Quantities for Acoustical Levels.

S1.10-1966 Calibration of Microphones, Method for the (Revision and Consolidation of Z24.4-1949 and Z24.11-1954).

S1.11-1966 Octave, Half-Octave, and Third-Octave Band Filter Sets, Specification for (Revision and Redesignation of Z24.10-1953) (*IEC 225*).

S1.12-1967 Laboratory Standard Microphones, Specifications for (Revision and Redesignation of Z24.8-1949).

S3.1-1960 Background Noise in Audiometer Rooms, Criteria for.

S3.2-1960 Monosyllabic Word Intelligibility, Method for Measurement of.

S3.3-1960 Electroacoustical Characteristics of Hearing Aids, Methods for Measurement of (*IEC 118 and 126*).

S3.4-1968 Procedure for the Computation of Loudness of Noise (*ISO 532*).

S3.5-1969 Methods for the Calculation of the Articulation Index.

S3.6-1969 Specifications for Audiometers (Revision and Redesignation of Z24.5-1951, Z24.12-1952, and Z24.13-1953).

S3.8-1967 Expressing Hearing Aid Performance, Method of.

S3-W-39 The Effects of Shock and Vibration on Man.

The following will be redesignated as S standards as they are revised or reaffirmed:

Z24.5-1951 Revised and Redesignated as S3.6-1969.

Z24.9-1949 Coupler Calibration of Earphones, Method for the.

Z24.12-1952 Revised and Redesignated as S3.6-1969.
Z24.13-1953 Revised and Redesignated as S3.6-1969.
Z24.22-1957 Measurement of the real-ear attenuation of ear protectors at threshold, Method for the.
Z24.24-1957 Calibration of Electroacoustic Transducers (Particularly Those for Use in Water), Procedures for
Z24-X-2 The Relations of Hearing Loss to Noise Exposure.

IEC or ISO numbers in brackets indicate related recommendations.
Identical content is shown by the words 'agrees with'.

BSI standards

'Amd.' signifies that an Amendment is issued with this Standard.

BS 415: 1967 Safety requirements for mains-operated domestic sound and vision equipment.

BS 661: 1969 Glossary of acoustical terms.

BS 880: 1950 Musical pitch.

BS 1568: 1960 Magnetic tape sound recording and reproduction.

BS 1927: 1953 Dimensions of circular cone diaphragm loudspeakers.

BS 1928: 1965 Processed disk records and reproducing equipment.

BS 1988: 1953 Measurement of frequency variation in sound recording and reproduction.

BS 2042: 1953 An artificial ear for the calibration of earphones of the external type. *Amd. Jan.*, 1954.

BS 2475: 1964 Octave and one-third octave band-pass filters. *Amd. May*, 1965.

BS 2497: ——— Reference zero for the calibration of pure-tone audiometers.
2497: Part 1: 1968 Data for earphone coupler combinations maintained at certain standardizing laboratories.
2497: Part 2: 1969 Data for certain earphones used in commercial practice.

BS 2498: 1954 Recommendations for ascertaining and expressing the performance of loudspeakers by objective measurements.

BS 2750: 1956 Recommendations for field and laboratory measurement of airborne and impact sound transmission in buildings. *Amd. Oct.*, 1963.

BS 2813: 1957 Polarized and non-polarized plugs for use on hearing aids.

BS 2829: 1957 35 mm magnetic sound recording azimuth alignment test films. *Amd. April*, 1957.

BS 2980: 1958 Pure-tone audiometers.

BS 2981: 1958 The dimensional features of magnetic sound recording on perforated film. *Amd. Jan.*, 1963, *July and Dec.*, 1963.

BS 3015: 1958 Glossary of terms used in vibration and shock testing.

BS 3045: 1958 The relation between the sone scale of loudness and the phon scale of loudness level.

BS 3154: 1959 Frequency characteristics for magnetic sound recording on film.

BS 3171: 1968 Methods of test of air-conduction hearing aids.

BS 3383: 1961	Normal equal-loudness contours for pure tones and normal threshold of hearing under free-field listening conditions.
BS 3425: 1966	Method for the measurement of noise emitted by motor vehicles. *Amd. Feb.*, 1967.
BS 3489: 1962	Sound level meters (industrial grade). *Amd. Feb.*, 1963.
BS 3539: 1962	Sound level meters for the measurement of noise emitted by motor vehicles.
BS 3593: 1963	Recommendation on preferred frequencies for acoustical measurements.
BS 3638: 1963	Method for the measurement of sound absorption coefficients (ISO) in a reverberation room. *Amd. Jan.*, 1964.
BS 3675: 1963	Performance of 16 mm portable sound-and-picture cinematograph projectors.
BS 3860: 1965	Methods for measuring and expressing the performance of audio-frequency amplifiers for domestic, public address and similar applications.
BS 3942: 1965	Marking of control settings on hearing aids.
BS 4009: 1966	An artificial mastoid for the calibration of bone vibrators used in hearing aids and audiometers.
BS 4142: 1967	Method of rating industrial noise affecting mixed residential and industrial areas.
BS 4196: 1967	Guide to the selection of methods of measuring noise emitted by machinery.
BS 4197: 1967	A precision sound level meter.
BS 4198: 1967	Method for calculating loudness.
BS 4297: 1968	Characteristics and performance of a peak programme meter.
CP 3. Chapter III: 1960	Sound insulation and noise reduction.

APPENDIX D[1]

Decibel conversion tables[2]

Table D.1 and Table D.2 have been prepared to facilitate making conversions in either direction between the number of *decibels* and the corresponding power and pressure ratios.

TO FIND VALUES OUTSIDE THE RANGE OF THE CONVERSION TABLES

Values outside the range of either Table D.1 or Table D.2 on the following pages can be readily found with the help of the following simple rules:

Table D.1: Decibels to pressure and power ratios

Number of decibels positive (+): Subtract +20 decibels successively from the given number of decibels until the remainder falls within range of Table D.1. *To find the pressure ratio*, multiply the corresponding value from the right-hand voltage-ratio column by 10 for each time you subtracted

---

[1] Refers to Chapter 4.
[2] By courtesy of General Radio Company (UK) Ltd.

20 dB. *To find the power ratio*, multiply the corresponding value from the right-hand power-ratio column by 100 for each time you subtracted 20 dB.

Example—Given: 49·2 dB

$$49\cdot2 \text{ dB} - 20 \text{ dB} - 20 \text{ dB} = 9\cdot2 \text{ dB}$$

Pressure ratio: 9·2 dB $\rightarrow$

$$2\cdot884 \times 10 \times 10 = 288\cdot4$$

Power ratio: 9·2 dB $\rightarrow$

$$8\cdot318 \times 100 \times 100 = 83180$$

Number of decibels negative $(-)$: Add $+20$ decibels successively to the given number of decibels until the sum falls within the range of Table D.1. *For the pressure ratio*, divide the value from the left-hand pressure-ratio column by 10 for each time you added 20 dB. *For the power ratio*, divide the value from the left-hand power-ratio column by 100 for each time you added 20 dB.

Example—Given: $-49\cdot2$ dB

$$-49\cdot2 \text{ dB} + 20 \text{ dB} + 20 \text{ dB} = -9\cdot2 \text{ dB}$$

Pressure ratio: $-9\cdot2$ dB $\rightarrow$

$$0\cdot3467 \times 1/10 \times 1/10 = 0\cdot003467$$

Power ratio: $-9\cdot2$ dB $\rightarrow$

$$0\cdot1202 \times 1/100 \times 1/100 = 0\cdot00001202$$

Table D.2: Pressure ratios to decibels

For ratios smaller than those in table, multiply the given ratio by 10 successively until the product can be found in the table. From the number of decibels thus found, subtract $+20$ decibels for each time you multiplied by 10.

Example—Given: Pressure ratio $= 0\cdot0131$

$$0\cdot0131 \times 10 \times 10 = 1\cdot31$$

From Table D.2, 1·31 $\rightarrow$

$$2\cdot345 \text{ dB} - 20 \text{ dB} - 20 \text{ dB} = -37\cdot655 \text{ dB}$$

For ratios greater than those in table, divide the given ratio by 10 successively until the quotient can be found in the table. To the number of decibels thus found, add $+20$ dB for each time you divided by 10.

Example—Given: Pressure ratio $= 712$

$$712 \times 1/10 \times 1/10 = 7\cdot12$$

From Table D.2, 7·12 $\rightarrow$

$$17\cdot05 \text{ dB} + 20 \text{ dB} + 20 \text{ dB} = 57\cdot05 \text{ dB}$$

TABLE D.1

GIVEN: *Decibels*   TO FIND: *Power and pressure ratios*

To account for the sign of the decibel

For positive (+) values of the decibel—Both pressure and power ratios are greater than unity. Use the two right-hand columns.

For negative (−) values of the decibel—Both pressure and power ratios are less than unity. Use the two left-hand columns.

Example—*Given:* ±9·1 dB. *Find:*

	Power ratio	Pressure ratio
+9·1 dB	8·128	2·851
−9·1 dB	0·1230	0·3508

← −dB+ →   ← −dB+ →

Pressure ratio	Power ratio	dB	Pressure ratio	Power ratio	Pressure ratio	Power ratio	dB	Pressure ratio	Power ratio
1·000	**1·000**	**0**	**1·000**	**1·000**	0·6310	0·3981	**4·0**	1·585	2·512
0·9886	0·9772	0·1	1·012	1·023	0·6237	0·3890	4·1	1·603	2·570
0·9772	0·9550	0·2	1·023	1·047	0·6166	0·3802	4·2	1·622	2·630
0·9661	0·9333	0·3	1·035	1·072	0·6095	0·3715	4·3	1·641	2·692
0·9550	0·9120	0·4	1·047	1·096	0·6026	0·3631	4·4	1·660	2·754
0·9441	0·8913	0·5	1·059	1·122	0·5957	0·3548	4·5	1·679	2·818
0·9333	0·8710	0·6	1·072	1·148	0·5888	0·3467	4·6	1·698	2·884
0·9226	0·8511	0·7	1·084	1·175	0·5821	0·3388	4·7	1·718	2·951
0·9120	0·8318	0·8	1·096	1·202	0·5754	0·3311	4·8	1·738	3·020
0·9016	0·8128	0·9	1·109	1·230	0·5689	0·3236	4·9	1·758	3·090
0·8913	**0·7943**	**1·0**	**1·122**	**1·259**	**0·5623**	**0·3162**	**5·0**	**1·778**	**3·162**
0·8810	0·7762	1·1	1·135	1·288	0·5559	0·3090	5·1	1·799	3·236
0·8710	0·7586	1·2	1·148	1·318	0·5495	0·3020	5·2	1·820	3·311
0·8610	0·7413	1·3	1·161	1·349	0·5433	0·2951	5·3	1·841	3·388
0·8511	0·7244	1·4	1·175	1·380	0·5370	0·2884	5·4	1·862	3·467
0·8414	0·7079	1·5	1·189	1·413	0·5309	0·2818	5·5	1·884	3·548
0·8318	0·6918	1·6	1·202	1·445	0·5248	0·2754	5·6	1·905	3·631
0·8222	0·6761	1·7	1·216	1·479	0·5188	0·2692	5·7	1·928	3·715
0·8128	0·6607	1·8	1·230	1·514	0·5129	0·2630	5·8	1·950	3·802
0·8035	0·6457	1·9	1·245	1·549	0·5070	0·2570	5·9	1·972	3·890
0·7943	**0·6310**	**2·0**	**1·259**	**1·585**	**0·5012**	**0·2512**	**6·0**	**1·995**	**3·981**
0·7852	0·6166	2·1	1·274	1·622	0·4955	0·2455	6·1	2·018	4·074
0·7762	0·6026	2·2	1·288	1·660	0·4898	0·2399	6·2	2·042	4·169
0·7674	0·5888	2·3	1·303	1·698	0·4842	0·2344	6·3	2·065	4·266
0·7586	0·5754	2·4	1·318	1·738	0·4786	0·2291	6·4	2·089	4·365
0·7499	0·5623	2·5	1·334	1·778	0·4732	0·2239	6·5	2·113	4·467
0·7413	0·5495	2·6	1·349	1·820	0·4677	0·2188	6·6	2·138	4·571
0·7328	0·5370	2·7	1·365	1·862	0·4624	0·2138	6·7	2·163	4·677
0·7244	0·5248	2·8	1·380	1·905	0·4571	0·2089	6·8	2·188	4·786
0·7161	0·5129	2·9	1·396	1·950	0·4519	0·2042	6·9	2·213	4·898
0·7079	**0·5012**	**3·0**	**1·413**	**1·995**	**0·4467**	**0·1995**	**7·0**	**2·239**	**5·012**
0·6998	0·4898	3·1	1·429	2·042	0·4416	0·1950	7·1	2·265	5·129
0·6918	0·4786	3·2	1·445	2·089	0·4365	0·1905	7·2	2·291	5·248
0·6839	0·4677	3·3	1·462	2·138	0·4315	0·1862	7·3	2·317	5·370
0·6761	0·4571	3·4	1·479	2·188	0·4266	0·1820	7·4	2·344	5·495
0·6683	0·4467	3·5	1·496	2·239	0·4217	0·1778	7·5	2·371	5·623
0·6607	0·4365	3·6	1·514	2·291	0·4169	0·1738	7·6	2·399	5·754
0·6531	0·4266	3·7	1·531	2·344	0·4121	0·1698	7·7	2·427	5·888
0·6457	0·4169	3·8	1·549	2·399	0·4074	0·1660	7·8	2·455	6·026
0·6383	0·4074	3·9	1·567	2·455	0·4027	0·1622	7·9	2·483	6·166

		−dB+					−dB+		
← →					← →				
Pressure ratio	Power ratio	dB	Pressure ratio	Power ratio	Pressure ratio	Power ratio	dB	Pressure ratio	Power ratio
0·3981	**0·1585**	**8·0**	**2·512**	**6·310**	**0·2239**	**0·05012**	**13·0**	**4·467**	**19·95**
0·3936	0·1549	8·1	2·541	6·457	0·2213	0·04898	13·1	4·519	20·42
0·3890	0·1514	8·2	2·570	6·607	0·2188	0·04786	13·2	4·571	20·89
0·3846	0·1479	8·3	2·600	6·761	0·2163	0·04677	13·3	4·624	21·38
0·3802	0·1445	8·4	2·630	6·918	0·2138	0·04571	13·4	4·677	21·88
0·3758	0·1413	8·5	2·661	7·079	0·2113	0·04467	13·5	4·732	22·39
0·3715	0·1380	8·6	2·692	7·244	0·2089	0·04365	13·6	4·786	22·91
0·3673	0·1349	8·7	2·723	7·413	0·2065	0·04266	13·7	4·842	23·44
0·3631	0·1318	8·8	2·754	7·586	0·2042	0·04169	13·8	4·898	23·99
0·3589	0·1288	8·9	2·786	7·762	0·2018	0·04074	13·9	4·955	24·55
0·3548	**0·1259**	**9·0**	**2·818**	**7·943**	**0·1995**	**0·03981**	**14·0**	**5·012**	**25·12**
0·3508	0·1230	9·1	2·851	8·128	0·1972	0·03890	14·1	5·070	25·70
0·3467	0·1202	9·2	2·884	8·318	0·1950	0·03802	14·2	5·129	26·30
0·3428	0·1175	9·3	2·917	8·511	0·1928	0·03715	14·3	5·188	26·92
0·3388	0·1148	9·4	2·951	8·710	0·1905	0·03631	14·4	5·248	27·54
0·3350	0·1122	9·5	2·985	8·913	0·1884	0·03548	14·5	5·309	28·18
0·3311	0·1096	9·6	3·020	9·120	0·1862	0·03467	14·6	5·370	28·84
0·3273	0·1072	9·7	3·055	9·333	0·1841	0·03388	14·7	5·433	29·51
0·3236	0·1047	9·8	3·090	9·550	0·1820	0·03311	14·8	5·495	30·20
0·3199	0·1023	9·9	3·126	9·772	0·1799	0·03236	14·9	5·559	30·90
0·3162	**0·10000**	**10·0**	**3·162**	**10·00**	**0·1778**	**0·03162**	**15·0**	**5·623**	**31·62**
0·3126	0·09772	10·1	3·199	10·23	0·1758	0·03090	15·1	5·689	32·36
0·3090	0·09550	10·2	3·236	10·47	0·1738	0·03020	15·2	5·754	33·11
0·3055	0·09333	10·3	3·273	10·72	0·1718	0·02951	15·3	5·821	33·88
0·3020	0·09120	10·4	3·311	10·96	0·1698	0·02884	15·4	5·888	34·67
0·2985	0·08913	10·5	3·350	11·22	0·1679	0·02818	15·5	5·957	35·48
0·2951	0·08710	10·6	3·388	11·48	0·1660	0·02754	15·6	6·026	36·31
0·2917	0·08511	10·7	3·428	11·75	0·1641	0·02692	15·7	6·095	37·15
0·2884	0·08318	10·8	3·467	12·02	0·1622	0·02630	15·8	6·166	38·02
0·2851	0·08128	10·9	3·508	12·30	0·1603	0·02570	15·9	6·237	38·90
0·2818	**0·07943**	**11·0**	**3·548**	**12·59**	**0·1585**	**0·02512**	**16·0**	**6·310**	**39·81**
0·2786	0·07762	11·1	3·589	12·88	0·1567	0·02455	16·1	6·383	40·74
0·2754	0·07586	11·2	3·631	13·18	0·1549	0·02399	16·2	6·457	41·69
0·2723	0·07413	11·3	3·673	13·49	0·1531	0·02344	16·3	6·531	42·66
0·2692	0·07244	11·4	3·715	13·80	0·1514	0·02291	16·4	6·607	43·65
0·2661	0·07079	11·5	3·758	14·13	0·1496	0·02239	16·5	6·683	44·67
0·2630	0·06918	11·6	3·802	14·45	0·1479	0·02188	16·6	6·761	45·71
0·2600	0·06761	11·7	3·846	14·79	0·1462	0·02138	16·7	6·839	46·77
0·2570	0·06607	11·8	3·890	15·14	0·1445	0·02089	16·8	6·918	47·86
0·2541	0·06457	11·9	3·936	15·49	0·1429	0·02042	16·9	6·998	48·98
0·2512	**0·06310**	**12·0**	**3·981**	**15·85**	**0·1413**	**0·01995**	**17·0**	**7·079**	**50·12**
0·2483	0·06166	12·1	4·027	16·22	0·1396	0·01950	17·1	7·161	51·29
0·2455	0·06026	12·2	4·074	16·60	0·1380	0·01905	17·2	7·244	52·48
0·2427	0·05888	12·3	4·121	16·98	0·1365	0·01862	17·3	7·328	53·70
0·2399	0·05754	12·4	4·169	17·38	0·1349	0·01820	17·4	7·413	54·95
0·2371	0·05623	12·5	4·217	17·78	0·1334	0·01778	17·5	7·499	56·23
0·2344	0·05495	12·6	4·266	18·20	0·1318	0·01738	17·6	7·586	57·54
0·2317	0·05370	12·7	4·315	18·62	0·1303	0·01698	17·7	7·674	58·88
0·2291	0·05248	12·8	4·365	19·05	0·1288	0·01660	17·8	7·762	60·26
0·2265	0·05129	12·9	4·416	19·50	0·1274	0·01622	17·9	7·852	61·66

		−dB+ ← →		
Pressure ratio	Power ratio	dB	Pressure ratio	Power ratio
0·1259	**0·01585**	**18·0**	**7·943**	**63·10**
0·1245	0·01549	18·1	8·035	64·57
0·1230	0·01514	18·2	8·128	66·07
0·1216	0·01479	18·3	8·222	67·61
0·1202	0·01445	18·4	8·318	69·18
0·1189	0·01413	18·5	8·414	70·79
0·1175	0·01380	18·6	8·511	72·44
0·1161	0·01349	18·7	8·610	74·13
0·1148	0·01318	18·8	8·710	75·86
0·1135	0·01288	18·9	8·810	77·62
0·1122	**0·01259**	**19·0**	**8·913**	**79·43**
0·1109	0·01230	19·1	9·016	81·28
0·1096	0·01202	19·2	9·120	83·18
0·1084	0·01175	19·3	9·226	85·11
0·1072	0·01148	19·4	9·333	87·10
0·1059	0·01122	19·5	9·441	89·13
0·1047	0·01096	19·6	9·550	91·20
0·1035	0·01072	19·7	9·661	93·33
0·1023	0·01047	19·8	9·772	95·50
0·1012	0·01023	19·9	9·886	97·72
0·1000	**0·01000**	**20·0**	**10·000**	**100·00**

		−dB+ ← →		
Pressure ratio	Power ratio	dB	Pressure ratio	Power ratio
$3·162 \times 10^{-1}$	**10^{-1}**	**10**	**3·162**	**10**
10^{-1}	10^{-2}	20	10	10^2
$3·162 \times 10^{-2}$	10^{-3}	30	$3·162 \times 10$	10^3
10^{-2}	10^{-4}	40	10^2	10^4
$3·162 \times 10^{-3}$	10^{-5}	50	$3·162 \times 10^2$	10^5
10^{-3}	10^{-6}	60	10^3	10^6
$3·162 \times 10^{-4}$	10^{-7}	70	$3·162 \times 10^3$	10^7
10^{-4}	10^{-8}	80	10^4	10^8
$3·162 \times 10^{-5}$	10^{-9}	90	$3·162 \times 10^4$	10^9
10^{-5}	10^{-10}	**100**	10^5	10^{10}

TABLE D.2

GIVEN: (*Pressure*) *Ratio*   TO FIND: *Decibels*

Power ratios

To find the number of decibels corresponding to a given power ratio—Assume the given power ratio to be a pressure ratio and find the corresponding number of decibels from the table. The desired result is exactly one-half of the number of decibels thus found.

Example—*Given:* a power ratio of 3·41. *Find:* 3·41 in the table:

$$3·41 \rightarrow 10·655 \text{ dB} \times \tfrac{1}{2} = 5·328 \text{ dB}$$

Pressure ratio	0·00	0·01	0·02	0·03	0·04	0·05	0·06	0·07	0·08	0·09
1·0	**0·000**	**0·086**	**0·172**	**0·257**	**0·341**	**0·424**	**0·506**	**0·588**	**0·668**	**0·749**
1·1	0·828	0·906	0·984	1·062	1·138	1·214	1·289	1·364	1·438	1·511
1·2	1·584	1·656	1·727	1·798	1·868	1·938	2·007	2·076	2·144	2·212
1·3	2·279	2·345	2·411	2·477	2·542	2·607	2·671	2·734	2·798	2·860
1·4	2·923	2·984	3·046	3·107	3·167	3·227	3·287	3·346	3·405	3·464
1·5	3·522	3·580	3·637	3·694	3·750	3·807	3·862	3·918	3·973	4·028
1·6	4·082	4·137	4·190	4·244	4·297	4·350	4·402	4·454	4·506	4·558
1·7	4·609	4·660	4·711	4·761	4·811	4·861	4·910	4·959	5·008	5·057
1·8	5·105	5·154	5·201	5·249	5·296	5·343	5·390	5·437	5·483	5·529
1·9	5·575	5·621	5·666	5·711	5·756	5·801	5·845	5·889	5·933	5·977
2·0	**6·021**	**6·064**	**6·107**	**6·150**	**6·193**	**6·235**	**6·277**	**6·319**	**6·361**	**6·403**
2·1	6·444	6·486	6·527	6·568	6·608	6·649	6·689	6·729	6·769	6·809
2·2	6·848	6·888	6·927	6·966	7·005	7·044	7·082	7·121	7·159	7·197
2·3	7·235	7·272	7·310	7·347	7·384	7·421	7·458	7·495	7·532	7·568
2·4	7·604	7·640	7·676	7·712	7·748	7·783	7·819	7·854	7·889	7·924
2·5	7·959	7·993	8·028	8·062	8·097	8·131	8·165	8·199	8·232	8·266
2·6	8·299	8·333	8·366	8·399	8·432	8·465	8·498	8·530	8·563	8·595
2·7	8·627	8·659	8·691	8·723	8·755	8·787	8·818	8·850	8·881	8·912
2·8	8·943	8·974	9·005	9·036	9·066	9·097	9·127	9·158	9·188	9·218
2·9	9·248	9·278	9·308	9·337	9·367	9·396	9·426	9·455	9·484	9·513
3·0	**9·542**	**9·571**	**9·600**	**9·629**	**9·657**	**9·686**	**9·714**	**9·743**	**9·771**	**9·799**
3·1	9·827	9·855	9·883	9·911	9·939	9·966	9·994	10·021	10·049	10·076
3·2	10·103	10.130	10·157	10·184	10·211	10·238	10·264	10·291	10·317	10·344
3·3	10·370	10·397	10·423	10·449	10·475	10·501	10·527	10·553	10·578	10·604
3·4	10·630	10·655	10·681	10·706	10·731	10·756	10·782	10·807	10·832	10·857
3·5	10·881	10·906	10·931	10·955	10·980	11·005	11·029	11·053	11·078	11·102
3·6	11·126	11·150	11·174	11·198	11·222	11·246	11·270	11·293	11·317	11·341
3·7	11·364	11·387	11·411	11·434	11·457	11·481	11·504	11·527	11·550	11·573
3·8	11·596	11·618	11·641	11·664	11·687	11·709	11·732	11·754	11·777	11·799
3·9	11·821	11·844	11·866	11·888	11·910	11·932	11·954	11·976	11·998	12·019
4·0	**12·041**	**12·063**	**12·085**	**12·106**	**12·128**	**12·149**	**12·171**	**12·192**	**12·213**	**12·234**
4·1	12·256	12·277	12·298	12·319	12·340	12·361	12·382	12·403	12·424	12·444
4·2	12·465	12·486	12·506	12·527	12·547	12·568	12·588	12·609	12·629	12·649
4·3	12·669	12·690	12·710	12·730	12·750	12·770	12·790	12·810	12·829	12·849
4·4	12·869	12·889	12·908	12·928	12·948	12·967	12·987	13·006	13·026	13·045
4·5	13·064	13·084	13·103	13·122	13·141	13·160	13·179	13·198	13·217	13·236
4·6	13·255	13·274	13·293	13·312	13·330	13·349	13·368	13·386	13·405	13·423
4·7	13·442	13·460	13·479	13·497	13·516	13·534	13·552	13·570	13·589	13·607
4·8	13·625	13·643	13·661	13·679	13·697	13·715	13·733	13·751	13·768	13·786
4·9	13·804	13·822	13·839	13·857	13·875	13·892	13·910	13·927	13·945	13·962
5·0	**13·979**	**13·997**	**14·014**	**14·031**	**14·049**	**14·066**	**14·083**	**14·100**	**14·117**	**14·134**
5·1	14·151	14·168	14·185	14·202	14·219	14·236	14·253	14·270	14·287	14·303
5·2	14·320	14·337	14·353	14·370	14·387	14·403	14·420	14·436	14·453	14·469
5·3	14·486	14·502	14·518	14·535	14·551	14·567	14·583	14·599	14·616	14·632
5·4	14·648	14·664	14·680	14·696	14·712	14·728	14·744	14·760	14·776	14·791
5·5	14·807	14·823	14·839	14·855	14·870	14·886	14·902	14·917	14·933	14·948
5·6	14·964	14·979	14·995	15·010	15·026	15·041	15·056	15·072	15·087	15·102
5·7	15·117	15·133	15·148	15·163	15·178	15·193	15·208	15·224	15·239	15·254
5·8	15·269	15·284	15·298	15·313	15·328	15·343	15·358	15·373	15·388	15·402
5·9	15·417	15·432	15·446	15·461	15·476	15·490	15·505	15·519	15·534	15·549

_{TABLE D.2 (continued)}

Pressure ratio	0·00	0·01	0·02	0·03	0·04	0·05	0·06	0·07	0·08	0·09
6·0	**15·563**	**15·577**	**15·592**	**15·606**	**15·621**	**15·635**	**15·649**	**15·664**	**15·678**	**15·692**
6·1	15·707	15·721	15·735	15·749	15·763	15·778	15·792	15·806	15·820	15·834
6·2	15·848	15·862	15·876	15·890	15·904	15·918	15·931	15·945	15·959	15·973
6·3	15·987	16·001	16·014	16·028	16·042	16·055	16·069	16·083	16·096	16·110
6·4	16·124	16·137	16·151	16·164	16·178	16·191	16·205	16·218	16·232	16·245
6·5	16·258	16·272	16·285	16·298	16·312	16·325	16·338	16·351	16·365	16·378
6·6	16·391	16·404	16·417	16·430	16·443	16·456	16·469	16·483	16·496	16·509
6·7	16·521	16·534	16·547	16·560	16·573	16·586	16·599	16·612	16·625	16·637
6·8	16·650	16·663	16·676	16·688	16·701	16·714	16·726	16·739	16·752	16·764
6·9	16·777	16·790	16·802	16·815	16·827	16·840	16·852	16·865	16·877	16·890
7·0	**16·902**	**16·914**	**16·927**	**16·939**	**16·951**	**16·964**	**16·976**	**16·988**	**17·001**	**17·013**
7·1	17·025	17·037	17·050	17·062	17·074	17·086	17·098	17·110	17·122	17·135
7·2	17·147	17·159	17·171	17·183	17·195	17·207	17·219	17·231	17·243	17·255
7·3	17·266	17·278	17·290	17·302	17·314	17·326	17·338	17·349	17·361	17·373
7·4	17·385	17·396	17·408	17·420	17·431	17·443	17·455	17·466	17·478	17·490
7·5	17·501	17·513	17·524	17·536	17·547	17·559	17·570	17·582	17·593	17·605
7·6	17·616	17·628	17·639	17·650	17·662	17·673	17·685	17·696	17·707	17·719
7·7	17·730	17·741	17·752	17·764	17·775	17·786	17·797	17·808	17·820	17·831
7·8	17·842	17·853	17·864	17·875	17·886	17·897	17·908	17·919	17·931	17·942
7·9	17·953	17·964	17·975	17·985	17·996	18·007	18·018	18·029	18·040	18·051
8·0	**18·062**	**18·073**	**18·083**	**18·094**	**18·105**	**18·116**	**18·127**	**18·137**	**18·148**	**18·159**
8·1	18·170	18·180	18·191	18·202	18·212	18·223	18·234	18·244	18·255	18·266
8·2	18·276	18·287	18·297	18·308	18·319	18·329	18·340	18·350	18·361	18·371
8·3	18·382	18·392	18·402	18·413	18·423	18·434	18·444	18·455	18·465	18·475
8·4	18·486	18·496	18·506	18·517	18·527	18·537	18·547	18·558	18·568	18·578
8·5	18·588	18·599	18·609	18·619	18·629	18·639	18·649	18·660	18·670	18·680
8·6	18·690	18·700	18·710	18·720	18·730	18·740	18·750	18·760	18·770	18·780
8·7	18·790	18·800	18·810	18·820	18·830	18·840	18·850	18·860	18·870	18·880
8·8	18·890	18·900	18·909	18·919	18·929	18·939	18·949	18·958	18·968	18·978
8·9	18·988	18·998	19·007	19·017	19·027	19·036	19·046	19·056	19·066	19·075
9·0	**19·085**	**19·094**	**19·104**	**19·114**	**19·123**	**19·133**	**19·143**	**19·152**	**19·162**	**19·171**
9·1	19·181	19·190	19·200	19·209	19·219	19·228	19·238	19·247	19·257	19·266
9·2	19·276	19·285	19·295	19·304	19·313	19·323	19·332	19·342	19·351	19·360
9·3	19·370	19·379	19·388	19·398	19·407	19·416	19·426	19·435	19·444	19·453
9·4	19·463	19·472	19·481	19·490	19·499	19·509	19·518	19·527	19·536	19·545
9·5	19·554	19·564	19·573	19·582	19·591	19·600	19·609	19·618	19·627	19·636
9·6	19·645	19·654	19·664	19·673	19·682	19·691	19·700	19·709	19·718	19·726
9·7	19·735	19·744	19·753	19·762	19·771	19·780	19·789	19·798	19·807	19·816
9·8	19·825	19·833	19·842	19·851	19·860	19·869	19·878	19·886	19·895	19·904
9·9	19·913	19·921	19·930	19·939	19·948	19·956	19·965	19·974	19·983	19·991

Pressure ratio	0	1	2	3	4	5	6	7	8	9
10	**20·000**	**20·828**	**21·584**	**22·279**	**22·923**	**23·522**	**24·082**	**24·609**	**25·105**	**25·575**
20	26·021	26·444	26·848	27·235	27·604	27·959	28·299	28·627	28·943	29·248
30	29·542	29·827	30·103	30·370	30·630	30·881	31·126	31·364	31·596	31·821
40	32·041	32·256	32·465	32·669	32·869	33·064	33·255	33·442	33·625	33·804
50	33·979	34·151	34·320	34·486	34·648	34·807	34·964	35·117	35·269	35·417
60	35·563	35·707	35·848	35·987	36·124	36·258	36·391	36·521	36·650	36·777
70	36·902	37·025	37·147	37·266	37·385	37·501	37·616	37·730	37·842	37·953
80	38·062	38·170	38·276	38·382	38·486	38·588	38·690	38·790	38·890	38·988
90	39·085	39·181	39·276	39·370	39·463	39·554	39·645	39·735	39·825	39·913
100	**40·000**	—	—	—	—	—	—	—	—	—

Calculations involving sound pressures and intensities

Example 1. The RMS pressure of a sound is $2 \cdot 0$ N/m^2. What is the sound pressure level? (Reference pressure $0 \cdot 00002$ N/m^2.)

From Equation 4.3 (p. 43).

$$SPL = 20 \log_{10} \frac{2}{0 \cdot 00002}$$

$$= 20 \log_{10} 10^5$$

$$= 20 \times 5$$

$$= 100 \text{ dB}$$

Example 2. The same pressure in cgs units would be 20 dyn/cm^2 and the reference pressure $0 \cdot 0002$ dyn/cm^2.

Again
$$SPL = 20 \log_{10} \frac{20}{0 \cdot 0002}$$

$$= 20 \log_{10} 10^5$$

$$= 20 \times 5$$

$$= 100 \text{ dB}$$

Example 3. The intensity of a sound is $0 \cdot 01$ W/m^2. What is the intensity level? (Reference intensity 10^{-12} W/m^2.)

From Equation 4.1 (p. 42).

$$IL = 10 \log_{10} \frac{0 \cdot 01}{10^{-12}}$$

$$= 10 \log_{10} 10^{10}$$

$$= 100 \text{ dB}$$

Example 4. The RMS pressure of a sound is $2 \cdot 0$ N/m^2. What is its intensity level? (Reference intensity 10^{-12} W/m^2.)

From Equation 2.4 (p. 27).

$$I = \frac{p^2}{\rho c}$$

Taking $\rho c = 409$ mks rayls

$$I = \frac{4}{409}$$

$$= 0 \cdot 00978 \text{ W/m}^2$$

[1] Refers to Chapter 4.

To find the intensity level we use Equation 4.1 (p. 42)

$$IL = 10 \log_{10} \frac{I}{I_{ref}}$$

$$= 10 \log_{10} \frac{0 \cdot 00978}{10^{-12}}$$

$$= 99 \cdot 9 \text{ dB}$$

This compares with the SPL value of 100 dB for the same sound pressure from Example 1. We thus see that, using these reference values for sound pressure and intensity, together with a value of ρc of 409 we obtain approximately the same values for SPL and IL for the same sound. Identical values for SPL and IL for the reference values of 2×10^{-5} N/m^2 and 10^{-12} W/m^2 respectively will result only if $\rho c = 400$.

APPENDIX F[1]

Combination or subtraction of decibel values

For many purposes these manipulations are necessary. Illustrations like the following, and practical applications will be found in Beranek (1960, 71) in the references of Chapter 4, p. 423.

For purposes of illustration, let us suppose that a sound source of 93 dB SPL has to be combined with another source of 95 dB SPL (we assume that each source has a different frequency, since sounds of the same frequency introduce additional factors). Alternatively, the conditions are as stated in Chapter 4, p. 45 under 'contribution of individual sound sources to a sound field'.

The essential feature of combination and subtraction of sounds is that it is on an energy basis in the cases we are considering. That is, we may manipulate sound intensities, or their equivalent, the squares of the corresponding sound pressures, in a simple additive or subtractive manner. Expressed in symbols:

$$(p_t)^2 = (p_1)^2 + (p_2)^2$$

When p_t = total sound pressure then p_1 and p_2 = sound pressures of individual components as RMS values.

This may also be written

$$p_t = \sqrt{(p_1^2 + p_2^2)}$$

Thus to find the total dB value for the addition of two sounds we could find the actual sound pressures in each case, square them, add and extract the square root. The resultant pressure would then be expressed again as SPL in dB. It is also possible to use the squares of the ratios only, add them and convert back to SPL, without the necessity of finding the actual sound pressures. Reverting to the numerical example, we shall use the latter operation.

[1] Refers to Chapter 4.

Example. What is the total SPL resulting from the combination of two separate sounds of (a) 93 dB and (b) 95 dB SPL?

Square of pressure ratio of sound (a) = antilog $\frac{93}{10}$

$$= 1\cdot99 \times 10^9$$

Square of pressure ratio of sound (b) = antilog $\frac{95}{10}$

$$= 3\cdot16 \times 10^9$$

Sum of squares of pressure ratios = $5\cdot15 \times 10^9$

$$SPL = 10 \log 5\cdot15 \times 10^9$$

$$= 97\cdot1 \text{ dB}$$

Thus the sum of two sounds of 93 dB SPL and 95 dB SPL is 97·1 dB SPL, or 2·1 dB greater than the higher value. When two sounds of the same SPL are added, the total is found to be nearly 3 dB greater than either. For subtraction, the same applies, removal of one of two sound sources of the same SPL value will reduce the level by very nearly 3 dB. These calculations need not be performed in detail since tables have been constructed for the purpose. Alternatively, charts such as the following, may be used. To derive an overall level from a number of levels such as SPL in octave bands, in practice, two values are added at a time by the use of the chart, and so on with the summed values until the overall level is found.

CHART FOR COMBINING OR SUBTRACTING DECIBELS[1]

To combine decibels

Enter the chart with the NUMERICAL DIFFERENCE BETWEEN TWO LEVELS BEING ADDED. Follow the line corresponding to this value to its intersection with the curved line, then left to read the NUMERICAL DIFFERENCE BETWEEN TOTAL AND LARGER LEVEL. Add this value to the larger level to determine the total.

Example. Combine 75 dB and 80 dB. The difference is 5 dB. The 5-dB line intersects the curved line at 1·2 dB on the vertical scale. Thus the total value is 80 + 1·2 or 81·2 dB.

To subtract decibels

Enter the chart with the NUMERICAL DIFFERENCE BETWEEN TOTAL AND LARGER LEVELS if this value is less than 3 dB. Enter the chart with the NUMERICAL DIFFERENCE BETWEEN TOTAL AND SMALLER LEVELS if this value is between 3 and 14 dB. Follow the line corresponding to this value to its intersection with the curved line, then either left or down to read the NUMERICAL DIFFERENCE BETWEEN TOTAL AND LARGER (SMALLER) LEVELS. Subtract this value from the total level to determine the unknown level.

Example. Subtract 81 dB from 90 dB. The difference is 9 dB. The 9-dB vertical line intersects the curved line at 0·6 dB on the vertical scale. Thus the unknown level is 90 − 0·6 or 89·4 dB.

[1] This chart is based on one developed by R. Musa.

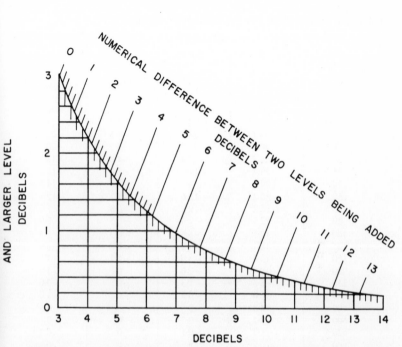

NUMERICAL DIFFERENCE BETWEEN TOTAL AND SMALLER LEVELS

APPENDIX G[1]

Relative responses for sound level meters

TABLE G

Responses, in dB, are expressed relative to the response at 1000 Hz

Frequency Hz	Curve A dB	Curve B dB	Curve C dB	Curve D* dB
10	− 70·4	− 38·2	− 14·3	− 26·5
12·5	− 63·4	− 33·2	− 11·2	− 24·5
16	− 56·7	− 28·5	− 8·5	− 22·5
20	− 50·5	− 24·2	− 6·2	− 20·5
25	− 44·7	− 20·4	− 4·4	− 18·5
31·5	− 39·4	− 17·1	− 3·0	− 16·5
40	− 34·6	− 14·2	− 2·0	− 14·5
50	− 30·2	− 11·6	− 1·3	− 12·5
63	− 26·2	− 9·3	− 0·8	− 11·0
80	− 22·5	− 7·4	− 0·5	− 9·0
100	− 19·1	− 5·6	− 0·3	− 7·5
125	− 16·1	− 4·2	− 0·2	− 6·0
160	− 13·4	− 3·0	− 0·1	− 4·5
200	− 10·9	− 2·0	0	− 3·0
250	− 8·6	− 1·3	0	− 2·0
315	− 6·6	− 0·8	0	− 1·0
400	− 4·8	− 0·5	0	− 0·5
500	− 3·2	− 0·3	0	0
630	− 1·9	− 0·1	0	0
800	− 0·8	0	0	0
1000	0	0	0	0
1250	0·6	0	0	2·0
1600	1·0	0	− 0·1	5·5
2000	1·2	− 0·1	− 0·2	8·0
2500	1·3	− 0·2	− 0·3	10·0
3150	1·2	− 0·4	− 0·5	11·0
4000	1·0	− 0·7	− 0·8	11·0
5000	0·5	− 1·2	− 1·3	10·0
6300	− 0·1	− 1·9	− 2·0	8·5
8000	− 1·1	− 2·9	− 3·0	6·0
10000	− 2·5	− 4·3	− 4·4	3·0
12500	− 4·3	− 6·1	− 6·2	0
16000	− 6·6	− 8·4	− 8·5	− 4·0
20000	− 9·3	− 11·1	− 11·2	− 7·5

* This weighting should be considered to be provisional and still under discussion. The values given are listed to the nearest half decibel.

International Electrotechnical Commission (1965), Publication 179, 1st Ed., *Precision sound level meters*. Geneva, IEC.

[1] Refers to Chapter 4.

APPENDIX H[1]

Preferred frequencies for acoustical measurements

TABLE H

Preferred frequencies in Hz for acoustical measurements and for geometric centre frequencies of filter pass bands

Preferred frequencies	1/1 octave	1/2 octave	1/3 octave	Preferred frequencies	1/1 octave	1/2 octave	1/3 octave	Preferred frequencies	1/1 octave	1/2 octave	1/3 octave
16	*	*	*	160			*	1600			*
18				*180*		*		*1800*			
20			*	200			*	**2000**	*	*	*
22·4		*		*224*				*2240*			
25			*	**250**	*	*	*	2500			*
28				*280*				*2800*		*	
31·5	*	*	*	315			*	3150			*
35·5				*355*		*		*3550*			
40			*	400			*	**4000**	*	*	*
45		*		*450*				*4500*			
50			*	**500**	*	*	*	5000			*
56				*560*				*5600*		*	
63	*	*	*	630			*	6300			*
71				*710*		*		*7100*			
80			*	800			*	**8000**	*	*	*
90		*		*900*				*9000*			
100			*	**1000**	*	*	*	10000			*
112				*1120*				*11200*		*	
125	*	*	*	1250			*	12500			*
140				*1400*		*		*14000*			
160			*	1600			*	**16000**	*	*	*

The star symbol indicates when the preferred frequency is the centre frequency of a particular ⅓ octave, ½ octave, or octave frequency band.
Bold face type: octave band centre frequencies.
BSI (1963), BS 3593.
Recommendation on preferred frequencies for acoustical measurements. London BSI.

[1] Refers to Chapter 4.

APPENDIX I[1]

Effects of background noise on audiometry

Excessive noise is not compatible with accurate audiometry. Such noise will interfere with the perception of the test tones, so that they must be of a higher intensity, to be heard compared to quiet conditions. This elevation of threshold in the presence of noise is known as masking. The quantitative aspects of this phenomenon have been described by various authors (Fletcher, 1940; French & Steinberg, 1947). Earphones normally used for audiometric purposes are not designed to exclude noise, although sound excluding telephones are now available for use where an environment sufficiently quiet for normal audiometry is impossible to obtain. It is possible to calculate how much noise is permissible for the measurement, without masking or with any defined degree of masking, of the audiometric tones, using data from Hawkins and Stevens (1950), Zwicker, Flottorp and Stevens (1957) and Scharf (1959). Examples of the derivation of sound spectra giving maximum permissible values of SPL for audiometric rooms, in both cases in mobile sound insulated vehicles, have been given by Copeland, Whittle and Saunders (1964) and by Taylor, Burns and Mair (1964). The principle of such estimates is that the prevailing, or ambient, noise in the audiometric room should be sufficiently low to avoid masking each of the various frequencies of the test tones. This does not mean that there must be no noise, which is effectively unattainable in any case, but that, according to the laws of masking, each test frequency shall not have its threshold elevated by the ambient noise. The factors influencing this condition, in addition to the ambient noise, are the values of the auditory threshold, which are in turn, defined by the audiometric zero being used; the hearing level which it is desired to measure, for example 0 dB, or −10 dB; finally, the degree to which the telephones themselves are able to exclude ambient noise. The results of the calculation as described by Taylor, Burns and Mair (1964), for the British Standard for normal hearing, using a particular type of earphone (Telephonics TDH 39 with MX 41/AR rubber cushions) are shown in Table I.

The starting point in this calculation is the SPL value, measured at the entrance to the external auditory meatus, corresponding to 0 dB hearing level for the particular frequency. This value is obtained from BS 2497[2] and is specified in column 2 of Table I. This is the actual sound pressure inside the earphone, at the entrance to the ear canal, which would just be audible by a person with a hearing level of 0 dB (BS 2497). In order that this threshold value shall not be affected by any background sound, the critical band concept of Fletcher (1940) is invoked. Fletcher showed that in the presence of a noise a pure tone is masked mainly by the components of the noise in a limited band of frequencies centred round the frequency of the pure tone. This band is known as the critical band. Provided that the background sound is of continuous spectrum type, and the critical band

[1] Refers to Chapter 6.
[2] BSI (1954).

TABLE I

Maximum permissible noise in audiometric enclosure for measurement of 0 dB hearing level

Frequency, or centre frequency of critical or octave band Hz Column 1	Threshold SPL BS 2497 dB Column 2	Maximum permissible SPL per critical band to measure 0 dB hearing level dB Column 3	Maximum permissible SPL per octave band to measure 0 dB hearing level dB Column 4	Attenuation of MX 41/AR cushions on TDH 39 dB Column 5	Maximum permissible SPL per octave in audiometric enclosure dB Column 6
125	30	20	20	2	22
250	19	9	12	4	16
500	12	2	7	11	18
1000	9	−1	5	21	26
2000	11	1	7	29	36
3000	7·5	−2·5	3·5	36	39·5
4000	9·5	−0·5	5·5	33	38·5
6000	14	4	10	30	40
8000	18·5	8·5	13·5	21	34·5

These maximum permissible values of SPL are calculated for TDH 39 telephones with MX 41/AR cushions. Different telephones would involve some modification. The SPL values given would result in an error of +3 to +4 dB due to masking noise when measuring thresholds of −10 dB. For accurate measurement of the latter, the permissible values of SPL would have to be 10 dB less, in each case, than those specified.

centred round the audiometric frequency has a SPL value at least 10 dB below that of the threshold audiometric tone (Hawkins & Stevens, 1950), the latter can be measured without error due to masking. We must thus subtract 10 dB from column 2 to yield the maximum permissible sound pressure level in the appropriate critical band to avoid masking of the tone (column 3). This value may now be converted from sound pressure level in a critical band to octave sound pressure level, with acceptable limits of accuracy, from the data on critical bandwidths of Scharf (1959) and Zwicker, Flottorp and Stevens (1957) (column 4). The maximum permissible sound pressure to avoid masking, specified in the octave band, is finally increased by the amount by which external noise is excluded by the telephone earpiece at that frequency (column 5). In this case the telephones are type TDH 39, fitted with MX 41/AR rubber cushions. The attenuations used for this combination at the various frequencies are those determined by the Royal Air Force Central Medical Establishment Acoustics Laboratory (1963). The final column (column 6) shows the permissible internal noise in the audiometric room for measurement of 0 dB hearing level at the various frequencies. The maximum permissible values of SPL in the various octaves centred at the frequencies stated will permit measurement without error of hearing levels of 0 dB corresponding to those of the British Standard BS 2497. At −10 dB hearing level, a small error of +3 to +4 dB will occur. It should be noted that the smallest values of SPL in the table, 16 and 18 dB, are at or beyond the limit of simple measurement by conventional sound level meter and octave filter combinations. Special techniques would need to be used to measure these levels with accuracy. However, if the total sound attenuation of the audiometric enclosure at various frequencies is estimated by the normal techniques, or found by actual measurement, it is possible to predict the internal sound pressure levels at the various frequencies, for given external ambient sound conditions. It will be obvious from Table I that there is little difficulty in providing adequately quiet conditions in audiometric enclosures at frequencies from about 1000 Hz and upwards, but it is below this frequency that greater difficulties arise. These stem from the greater problems of providing adequate sound attenuation for the lower frequencies, and from the relative ineffectiveness in the exclusion of outside sound by the earphones themselves at these frequencies.

APPENDIX J[1]

Sources of error in audiometry

Robinson's (1960) study of sources of variability in the measurement of hearing by the method of air conduction with conventional earphone listening draws attention to the degree of accuracy which may be reasonably expected from this technique. The sources of variability may in broad terms be divided into two categories. One is the true variability of hearing threshold, as would be indicated by the threshold sound pressure measured close to the tympanic membrane. The other category limits the accuracy with

[1] Refers to Chapters 6, 11 and 12.

which the true variability may be measured, and is compounded of factors comprising the variability due to the size of the ear canal, the particular position and pressure on the ear of the earphone, and the person's judgement of his auditory threshold on the occasion of the particular measurement.

The technique of free-field threshold measurement offers an interesting comparison with measurements of threshold by earphone listening. In the former method the subject is placed in a variable sound field and the threshold is thus found without the acoustical complications such as the effect of earphone position and interaction with the ear characteristics as determined by the size of the ear canal. Such determinations have been carried out from time to time, with the object of determining the auditory thresholds of normal people at various ages, and usually also for the purpose of study of the way in which the sensation of loudness is affected by the frequency and intensity of sounds. Robinson and Dadson (1956) have made such a study, and the effect on variation between individuals of wearing of telephones can be seen by comparing their results with those of Dadson and King (1952), where telephone listening was used (Table J).

TABLE J

Standard deviations of auditory threshold values

Frequency Hz	Standard deviation of MAF values (Robinson & Dadson) dB	Standard deviation of MAP values (Dadson & King) dB
50	6·5	—
80	—	8·0
100	5·0	—
125	—	6·8
200	4·5	—
250	—	7·3
500	4·5	6·5
1000	4·5	5·7
2000	5·0	6·1

This table shows the standard deviations obtained in these free-field threshold measurements, usually known as minimal audible field (MAF), of Robinson and Dadson, compared with the standard deviations of the thresholds found by Dadson and King for earphone listening, usually termed minimum audible pressure (MAP). The data compared are for frequencies up to 2000 Hz. In the case of the MAP, individual ears are measured and the thresholds averaged; for the MAF data, two-ear listening is used, so that the two measurements are not perfectly comparable. Nevertheless at these frequencies the tendency for the MAF measurements to show smaller standard deviations, possibly connected with the absence of the factors normally associated with the use of earphones, is reasonable. Robinson's (1960) examination of the sources of variability assesses quantitatively the extent to which the true state of hearing is obscured in normal earphone audiometry by the operation of the variables. The uncertainty due to ear canal size, uncertainty in the placement of the earphone, and the uncertainty of the subjective judgement of what constitutes the threshold, all combine to limit the accuracy of a given single determination of hearing level as an indication of the inherent sensitivity of the ear.

These particular data indicate that the standard deviation due to the combined uncertainties of earphone placement, ear canal size, and judgement of the threshold (i.e. excluding true differences between the thresholds of different people) is about 8, 6 and 10 dB for low, middle and high frequencies respectively. The true hearing level, on this basis, could thus depart from the measured hearing level by as much as 12 dB at middle frequencies, in 5% of cases, for a single audiometric measurement. However, study of Table J from Dadson and King (1952) shows that the standard deviations are either of the same order, or smaller than those derived from Robinson's (1960) data. When it is recalled that Robinson's standard deviations do not include the variability between individuals due to true difference in auditory threshold, such as would be given by a measurement of threshold sound pressure at the ear drum, and that those of Table J include all sources of variability, it is clear that, at least for young people, Robinson's standard deviations, probably due to the nature of the data, must be regarded as somewhat larger than would be commonly encountered.

Nevertheless the uncertainty of single audiometric measurements is considerable, but is doubtless influenced by the skill and practice of the listener. Where greater accuracy is required there is no alternative but to repeat the measurements, removing the earphone and replacing it each time to randomise the error due to placement. The fact that one person may show a higher hearing level because the ear canal is of large volume is inherent, and in ordinary circumstances must be accepted. When the same person is measured repeatedly, however, the ear canal factor is eliminated as a source of variability.

Various estimates have been made of the accuracy with which an audiogram may be reproduced, including studies on persons representative of those unconnected with practical psychoacoustics, and who are unfamiliar with the technique of audiometry. This information is highly necessary where changes are expected to be small between audiometric examinations such as in studies of hearing in relation to noise exposure, or in industrial monitoring audiometry for the purposes of hearing preservation, or in clinical audiometry where the progress of a condition is being followed. In the studies which have been made, a number of different circumstances have been investigated, but usually on small numbers of subjects. In order to assess the probable degree of variability the measurements should be on individuals, and at intervals such that a true change in threshold would not be expected to occur, so that factors connected with real changes in hearing should be excluded as far as possible. Brown (1948) investigated the variability of audiograms obtained at intervals of one hour, on a group of 30 subjects of which four were of normal hearing. The standard deviations of the changes of hearing level were in the region of 3 to 4 dB for the audiometric frequencies 125–4000 Hz, and 5 dB for 8000 Hz. Robinson (1960) measured the hearing of normal listeners at intervals of one or two months, and obtained comparable results. Knight (1962) using Rudmose ARJ 4 audiometry, obtained two audiograms at an interval of a few minutes, each day for five days on 20 subjects; the standard deviations of the variations between occasions, were again similar, being in the region of 4 dB at 500 Hz to 4000 Hz, and over 6 dB at 6000 Hz. Burns, Hinchcliffe and Littler (1964) measured the hearing of 18 normally hearing persons at an interval of three years during which there was no known exposure to noisy environments. Normal manual pure tone audiometry was used. The standard deviations in the range 500 Hz to 4000 Hz were in the region of 5–6 dB, and about 9 dB at 6000 Hz. High and Glorig (1962) after testing subjects not exposed to noise, at six-monthly intervals, by a Rudmose ARJ 4 audiometer, concluded that the reliability of the differences between successive audiograms was adequate for the purposes of individual audiometry, on the basis of standard deviations of from 3·7–5·4 dB over the range 500–6000 Hz. Test–retest reliability coefficients in this study were acceptably high, about 0·8, at all frequencies except 6000 Hz. Burns and Hinchcliffe (1957) using an audiometer (Burns & Morris, 1955) in which the same oscillator and attenuator system could be used for either manual or Békésy swept-frequency audiometry, compared these types of audiometry on two successive occasions. The test–retest correlations were between 0·8 and 0·9 between 1000 and 6000 Hz. Slightly lower coefficients were obtained at 500 Hz, but the probable reason was identified. The results gave standard deviations for the difference between the first and second test by manual audiometry of between 4 and 5 dB, except at 6000 Hz, when it was 7·6 dB. Delany, Whittle, Cook and Scott (1967) give values for threshold repeatability for experienced listeners, and the standard deviation is as low as 2 dB between 1000 and 2000 Hz, rising to about 4 and 5 dB respectively at the low and high audiometric frequencies. Rudmose (1963) describes his own work in which test–retest reliability using his ARJ 4 audiometer is investigated. His data describe the results of audiometric examinations numbering from 7–10, in seven subjects over a period of some six months, the intervals between audiometry being about two weeks or more. The average of each individual's audiograms, the standard deviations of each subject about his mean, and standard deviation of the whole group are given. The individual and group standard deviations are comparable at about 4 dB, being thus very similar to those of Knight (1962). Thus, on these values, for the range 500–4000 Hz 95% of the observations of an individual's threshold should fall within plus or minus 8 dB (and possibly a little more, perhaps 12 dB, at 6000 Hz) of his mean threshold. Delany (1971) examined the audiometric data of 33 young men in industry who were examined at an interval of 1 year, during which period their occupational noise was less than 82 dB(A) and thus they could be regarded as controls. They were subjects in Burns' and Robinson's (1970) study and so were otologically normal and the audiometry was of the pulsed Rudmose type. Their replication variance was about 20 dB^2 at 3 kHz and 76 dB^2 at 6 kHz. These are higher values than those of the laboratory investigations noted above, but similar to the industrial data of Burns, Hinchcliffe and Littler (1964). In the case of noise-exposed persons, the replication variances are, for some reason, appreciably larger than in persons not exposed; and they increase with increased noise exposure (Robinson, 1970). Thus with careful, well-conducted audiometry, in quiet surroundings, with audiometers maintained in a satisfactory state of calibration, changes of 10 dB between audiograms, in either manual or Rudmose audiometry, should be regarded as possibly significant. In any circumstances, greater accuracy can be obtained by repeating the audiogram and taking the average of the values of hearing level. Thus the accuracy could be increased two-fold by repeating the audiogram four times, so that in the case above, the mean so obtained would be expected to depart from the individual's true mean by not more than plus or minus 4 to 5 dB, instead of 8 to 9 dB, in 95% of occasions, for the range 0·5 to 4 kHz. It should be noted, however, that in repeated audiometry occasional large deviations may be found, the origin of which is not clear.

APPENDIX K[1]

Loudness: the sone and the phon; perceived noise

The great intrinsic and practical interest of the factors governing the sensation of loudness has prompted many studies of the loudness of tones or frequency bands. Kingsbury (1927) investigated the loudness of pure tones, and since then Fletcher and Munson (1933), Churcher and King (1937) and Robinson and Dadson (1956) have made investigations in the same field. The data on the relative loudness of pure tones of different frequency are expressed normally as equal loudness contours.

These curves state the relations between the frequency of a sound and its sound pressure for constant loudness. Such curves are shown in Fig. K.1, from ISO (1961), which is derived from the data of Robinson and Dadson (1956). In the curves, a pure tone of 1000 Hz at various specified levels is the reference and the SPL values necessary to achieve equal loudness at other frequencies are shown. The dashed line at the bottom of the family of curves represents a special case of equal loudness, specifically the threshold, in this case the free-field or MAF threshold. More recently the audibility of very low frequency sounds has been studied by Yeowart, Bryan and Tempest (1967) and by Whittle (1971). The latter data give threshold values down to 3·15 Hz where the threshold is in the region of 122 dB SPL. At 12·5 Hz the threshold is about 95 dB SPL. Equal loudness data were obtained which should enable the curves of Fig. K.1 to be extended downwards in frequency.

A scale of *loudness level* or *equivalent loudness* for general purposes is due to Fletcher and Munson (1933). The unit is the *phon*. In this system, the loudness level of a particular sound, in phons, is defined as the SPL of a pure tone of 1000 Hz that is judged to be of the same loudness as the sound.

---

[1] Refers to Chapters 8, 9 and 13.

Thus in Fig. K.1 each contour has a loudness in phons, of the value indicated at the 1000 Hz frequency. Recommended means of making a subjective comparison between a given sound and a pure tone of 1000 Hz are given by the British Standards Institution (1958). The judgement involves listening alternately to the 1000 Hz tone presented at different intensities, and to the sound, until the observer is satisfied that the loudness of the tone

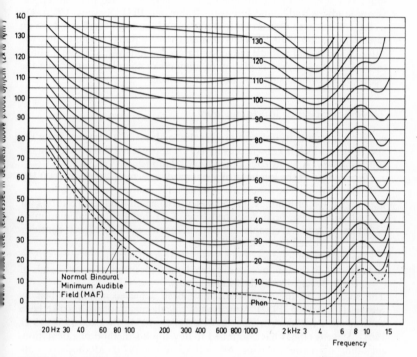

Fig. K.1.   Equal-loudness contours for pure tones and normal threshold of hearing for persons aged 18–25 years, using free-field listening. (ISO Recommendation R226.)

has been matched to the loudness of the sound. When this match has been achieved the phon value is the SPL of the 1000 Hz tone. The judgement of equal loudness between the 1000 Hz tone and a tone of a different frequency, or even a complex noise, is not particularly easy, but using the average of the judgements of a group of about 10 observers, this is quite a practical method and the accuracy is known. The method, however, is not a very convenient one. The computation of the loudness in phons of complex steady noises requires the construction of a numerical scale of loudness. This scale rests on the fact that when observers are asked to judge when one sound, for example, is 'twice as loud' or 'half as loud' as another, quite consistent differences in actual intensity are obtained in given conditions. For the purpose of such a scale, a unit of loudness, the *sone* was proposed by Stevens (1936). On this system, 1 sone is now defined as the loudness experienced by a typical listener when listening to a tone of frequency 1000 Hz and SPL 40 dB (Stevens, 1956). This tone would thus have a loudness level of 40 phons. A sound of loudness 2 sones would be twice as loud, and 0·5 sones half as loud, as 1 sone. Numbers of investigators have examined the relation between loudness and sound intensity (Robinson, 1953; Stevens, 1955). On average, for a pure tone a two-fold increase of loudness corresponds to an intensity increase of 10 dB. Thus an increase of 10 phons would double the value of sones, and a 10 dB reduction halves the sone value. These findings opened the way to the calculation of loudness of complex sounds, and the necessary additional steps can now be described.

The calculation of loudness level

To recapitulate, in order to predict the loudness of a complex noise the spectrum of which is known, Stevens (1957) approached this problem in the following way. A knowledge of the three relationships is required.
1. The relation of the loudness of a standard tone, for example at 1000 Hz, to its intensity. The sone scale describes this relation, as noted above.
2. The relation of loudness to frequency. The equal loudness relations for pure tones have been described (Fig. K.1), but for the calculation of loudness level of complex noises analogous equal loudness contours must be evolved for bands of noise.
3. The relation of loudness to bandwidth. This information is necessary in order to add the loudness of different bands, for instance octave bands of noise, in order to yield the total loudness.

The solutions to these problems have been achieved and described by Stevens (1955, 1956, 1957). A method of calculating loudness has also been evolved by Zwicker (Zwicker & Feldtkeller, 1955; Zwicker 1958, 1960, 1961). Stevens (1961) has continued to develop his method of calculation, including his newest version Mk VII (Stevens, 1970). Methods of both Stevens and Zwicker are now embodied in ISO Recommendation 532 (ISO, 1966), in BS 4198 (BSI, 1967), and in ANSI S3.4 (ANSI, 1968). The method of calculation will be illustrated by Stevens' procedure. The reader who wishes to calculate loudness by either method is strongly advised to consult the full text of the official ISO document, so that possible misapprehensions arising from the summarisation may be avoided.

Procedure for calculation

The calculation of loudness of a particular noise, in terms of a single figure in phons, can be performed from data of the SPL values of the noise in octave bands, one-half octave bands or one-third octave bands. The Stevens method to be described is applicable to any of these bandwidths, but will be described primarily as applying to octave bands, with a note on the use of one-third or one-half octave bands added. Having obtained the octave SPL values by analysis, these are converted into a loudness index and the total loudness in sones is then derived by an empirical summation

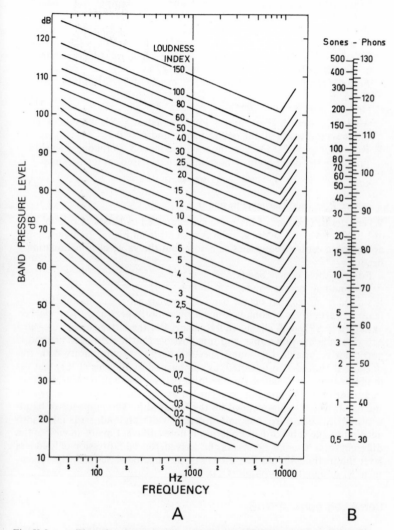

Fig. K.2. A: Chart for determination of loudness index from octave SPL values. B: Nomogram for the conversion of sones to phons. For details of use see text (p. 406). (ISO, 1966.) Band pressure level implies the sound pressure level in the part of the spectrum specified; in this case it may be $\frac{1}{3}$ octave, $\frac{1}{2}$ octave or one octave.

procedure. This value of total loudness is then converted into loudness level in phons by means of an equation.

In detail, the procedure is as follows.

1. Determine the SPL values in octave bands (for example, in the bands centred at 63 Hz to 8000 Hz).
2. Refer to Fig. K.2A. Enter the chart on the abscissa (frequency scale) at the centre frequency of each octave. From the SPL value of the octave (ordinate) read off the loudness index. (Example: octave centred at 500 Hz, SPL = 90; loudness index = 27.)
3. Find the total loudness in sones (S_t) by means of the formula

$$S_t = S_{max} + F(\sum S - S_{max}) \tag{1}$$

where S_{max} is the highest value of loudness index, $\sum S$ is the sum of the loudness indices of all the octave bands, and $F = 0.3$.

4. Refer to Fig. K.2B. Convert the total loudness S_t to loudness level in phons by the nomogram. This nomogram is based on the formula

$$S_t = 2^{(P-40)/10} \tag{2}$$

where P = loudness level in phons.

Where the SPL values of the noise are expressed in bands of one-third or one-half octave, the loudness index is read off from Fig. K.2A, by using the appropriate centre frequencies as before, but the value of F in Equation 1 becomes

For one-third octave bands: $F = 0.15$
For one-half octave bands: $F = 0.2$

It should be noted that octave band SPL values are most suitable for estimating the loudness of sounds which have smooth spectra, without marked variation in SPL with change in frequency. For the latter noises, one-third octave bands are preferable, and it is possible that the Zwicker method is particularly suitable for sound spectra sharply discontinuous with frequency. It should also be noted, as emphasised by Stevens (1961) that the loudness index is the same for one-third octave, one-half octave or octave bandwidth, and should not be regarded as equivalent to the loudness of the band, which would depend on the bandwidth. The influence of bandwidth for the one-third, one-half or octave width on loudness is taken into account by the value of F in Equation 1, noted above. Finally, in view of the different ways in which phons may be derived, for the expression of loudness level the method should be stated. For the method described, the recommended designation is phons (OD).

Perceived noise: PNdB

Turbojet engined commercial aircraft appeared in numbers in approximately their present form about 1959. The noise of these engines was different from that of piston-engined aircraft to which the public had become accustomed, and it was recognised that the potential annoyance from the new transports was much greater than anything previously experienced.

Kryter (1959) made a study of the relative 'noisiness' of types of jet trans-ports compared to piston-engined aircraft which were being superseded, as judged from the ground. The overall SPL values, recorded by sound level meter did not correlate well with judgements of noisiness. In order to achieve a more precise numerical description of noisiness, a method of computation from octave band levels was evolved, based on the work of Stevens on the calculation of loudness just described. This scale, which gives more weight to the high frequencies, has been called perceived noise level by Kryter, and the unit is the perceived noise decibel (PNdB). The computation of perceived noise level is entirely analogous to that of loud-ness level by Stevens' method (Kryter & Pearsons, 1963, 1964). The varia-tion in the Stevens method introduced by Kryter is in the stage of converting band sound pressures into a numerical index of loudness. The object of the perceived loudness scale being to quantify the noisiness or annoyance of a sound, Kryter proceeded to construct a scale of noisiness in a manner analogous to the sone scale of loudness. His unit, designated the *noy*, was defined originally as the noisiness of a band of noise from 910 to 1090 Hz of SPL 40 dB. Subsequently the bandwidth has been enlarged to one octave. The translation of one-third octave or octave sound pressure levels into noy values is based on data derived from equal noisiness contours, again analo-gous to equal loudness contours. The derivation of perceived noisiness in noys of the whole noise, followed by conversion into perceived noise level in PNdB, is performed as in Stevens' calculation of total loudness in sones and its conversion to loudness level in phons. The perceived noise level as a measure of the noisiness or annoyance value of noises has proved of great value in the specification of the noise experienced by a community from jet aircraft in flight or on the ground, and Kryter's procedure, using $\frac{1}{3}$ octave or octave bands of noise, has now been internationally agreed (ISO, 1970). For the purpose of calculation of perceived noise level, the reader is again advised to refer to the original publication. However, as an illustration of the principle, the procedure is briefly described. The steps are as follows.

Calculation procedure for perceived noise level in PNdB (for broad band noise without pronounced irregularities in spectrum, for example pure tones)
1. Determine the SPL values at any particular time in $\frac{1}{3}$ octave bands (centred at 50 to 10000 Hz).
2. Refer to Fig. K.3. Enter the chart on the abscissa (frequency scale) at the centre frequency of each $\frac{1}{3}$ octave. From the SPL value of each $\frac{1}{3}$ octave (ordinate) read off the noy value. (Example: $\frac{1}{3}$ octave centred at 500 Hz, SPL 90; noy value = 32.)
3. Find the total perceived noisiness N by combining the noy values of all the $\frac{1}{3}$ octaves by the formula

$$N = n_{max} + 0.15(\sum n - n_{max}) \qquad (3)$$

where n_{max} is the highest of the noy values and $\sum n$ is the sum of the noy values in all the $\frac{1}{3}$ octave bands. (Equation 3 corresponds to Equation 1.)

4. Refer to Table K. Convert the total perceived noisiness N to perceived noise level in PNdB. This table is based on the formula

$$N = 2^{(X-40)/10} \qquad (4)$$

where X = perceived noise level in PNdB. (Equation 4 corresponds to Equation 2.)

TABLE K

Perceived noise level as a function of total perceived noisiness from ISO (1970)

N (noy)			L_{PN}	N (noy)			L_{PN}
Lower	Mid	Upper	(PNdB)	Lower	Mid	Upper	(PNdB)
1·0	1·0	1·0	40	43·8	45·2	46·8	95
1·1	1·1	1·1	41	46·9	48·5	50·2	96
1·1	1·1	1·2	42	50·3	52·0	53·8	97
1·2	1·2	1·3	43	53·9	55·7	57·7	98
1·3	1·3	1·4	44	57·8	59·7	61·8	99
1·4	1·4	1·5	45	61·9	64·0	66·3	100
1·5	1·5	1·6	46	66·4	68·6	71·0	101
1·6	1·6	1·7	47	71·1	73·5	76·1	102
1·7	1·7	1·8	48	76·2	78·8	81·6	103
1·9	1·9	1·9	49	81·7	84·4	87·4	104
2·0	2·0	2·1	50	87·5	90·5	93·7	105
2·1	2·1	2·2	51	93·8	97·0	100·4	106
2·3	2·3	2·4	52	100·5	104·0	107·6	107
2·5	2·5	2·5	53	107·7	111·4	115·3	108
2·6	2·6	2·7	54	115·4	119·4	123·6	109
2·8	2·8	2·9	55	123·7	128·0	132·5	110
3·0	3·0	3·1	56	132·6	137·2	142·0	111
3·2	3·2	3·4	57	142·1	147·0	152·2	112
3·5	3·5	3·6	58	152·3	157·6	163·1	113
3·7	3·7	3·9	59	163·2	168·9	174·8	114
4·0	4·0	4·1	60	174·9	181·0	187·4	115
4·2	4·3	4·4	61	187·5	194·0	200·8	116
4·5	4·6	4·7	62	200·9	207·9	215·3	117
4·8	4·9	5·1	63	215·4	222·8	230·7	118
5·2	5·3	5·5	64	230·8	238·8	247·3	119
5·6	5·6	5·8	65	247·4	256·0	265·0	120
5·9	6·1	6·3	66	265·4	274·4	284·0	121
6·4	6·5	6·7	67	284·1	294·0	304·4	122
6·8	7·0	7·2	68	304·5	315·2	326·3	123
7·3	7·5	7·7	69	326·4	337·8	349·7	124

N (noy)			L_{PN}	N (noy)			L_{PN}
Lower	Mid	Upper	(PNdB)	Lower	Mid	Upper	(PNdB)
7·8	8·0	8·3	70	349·8	362·0	374·3	125
8·4	8·6	8·9	71	374·9	388·0	401·7	126
9·0	9·2	9·5	72	401·8	415·8	430·5	127
9·6	9·8	10·2	73	430·6	445·7	461·4	128
10·3	10·6	10·9	74	461·5	477·7	494·5	129
11·0	11·3	11·7	75	494·6	512·0	530·0	130
11·8	12·1	12·5	76	530·1	548·7	568·1	131
12·6	13·0	13·5	77	568·2	588·1	608·9	132
13·6	13·9	14·4	78	609·0	630·3	652·6	133
14·5	14·9	15·4	79	652·7	675·5	699·4	134
15·5	16·0	16·6	80	699·5	724·1	749·6	135
16·7	17·1	17·7	81	749·7	776·0	803·3	136
17·8	18·4	19·0	82	803·4	831·7	861·1	137
19·1	19·7	20·4	83	861·2	891·4	922·9	138
20·5	21·1	21·8	84	923·0	955·4	989·1	139
21·9	22·6	23·4	85	989·2	1024·0	1060·1	140
23·5	24·2	25·1	86	1060·2	1097·5	1136·1	141
25·2	26·0	26·9	87	1136·2	1176·2	1217·7	142
27·0	27·8	28·8	88	1217·8	1260·6	1305·1	143
28·9	29·8	30·9	89	1305·2	1351·1	1398·8	144
31·0	32·0	33·1	90	1393·9	1448·2	1490·1	145
33·2	34·3	35·5	91	1499·2	1552·1	1606·7	146
35·6	36·8	38·1	92	1606·8	1663·4	1722·1	147
38·2	39·4	40·8	93	1722·2	1782·8	1845·7	148
40·9	42·2	43·7	94	1845·8	1910·7	1978·2	149

In using Table K the values of N are given in ranges, with the appropriate PNdB values.

This procedure illustrates the basic principle of calculation. However, in practice, perceived noise level will probably be derived by methods either more complex or simpler than the above, and ISO (1970) should be consulted for details. As an example of practical elaborations of the basic procedure, perceived noise levels may be derived which (1) include an allowance for the presence of spectral irregularities such as pure tones if present; and (2) include an allowance for the duration of the noise. In the case of (1) the values are in *pure tone corrected perceived noise level* (L_{TPN}), but this form is not normally used in isolation, but combined with (2) the duration correction. In correcting for (2), the rationale is that the subjective effect depends on the time history of the noise, in addition to its level. Where such allowance is made (ISO, 1970) a correction is introduced whereby the duration of the flyover noise, as defined, modifies the perceived noise level, yielding *effective perceived noise level* (L_{EPN}) and expressed in EPNdB with the implication that the correction for pure tones has been made, if they occur.

Effective perceived noise level may in practice be derived from the *maximum perceived noise level* (L_{PNmax}) in the absence of pure tones, or maximum pure tone perceived noise level (L_{TPNmax}) if spectral irregularities are present. These maximum levels are read off a smooth curve against time of the values of L_{PN} or L_{TPN} calculated at intervals of 0·5 s or less and modified according to the time history of the noise. The practical impossibility of performing this operation (noted in Chapter 13) by any other means than automatic analysis and computation is obvious.

In view of these rigorous requirements for complex and expensive analytical facilities, it is not surprising that there also exists an alternative

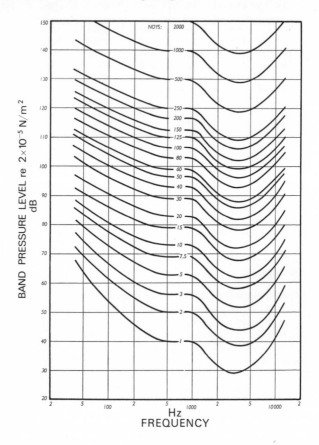

Fig. K.3.   Contours of perceived noisiness. (ISO, 1970.)

selection of simplified methods of indicating approximately, values of perceived noise level. These again are detailed in ISO (1970) but examples
include:

(1) Calculation of L_{PN} from octave band pressure levels instead of in
    $\frac{1}{3}$ octaves. In this case the factor 0·15 in Equation (3) becomes 0·3 (as
    it is in Equation (1)).
(2) Direct readings from a sound level meter. Two possibilities are
    available:

    (a) by using the A-weighting (Appendix G) of the sound level meter.
        Depending on the frequency characteristics of the noise, dB(A)
        levels should be increased by 7 to 14 dB to yield an approximation
        to PNdB values.
    (b) by using the D-weighting (Appendix G) of the sound level meter.
        This weighting is equal to the inverse of the 40 noy curve (Fig.
        K.3). Addition of 7 dB to the value will give an approximation to
        PNdB.

Pure tone corrections and duration allowances are neglected in these simple
methods.

TABLE L.1

Octave band pressure level values associated with preferred noise criterion (PNC) curves, from Beranek, Blazier and Figwer (1971)

Preferred noise criterion curves	31·5 Hz	63 Hz	125 Hz	250 Hz	500 Hz	1000 Hz	2000 Hz	4000 Hz	8000 Hz
PNC-15	58	43	35	28	21	15	10	8	8
PNC-20	59	46	39	32	26	20	15	13	13
PNC-25	60	49	43	37	31	25	20	18	18
PNC-30	61	52	46	41	35	30	25	23	23
PNC-35	62	55	50	45	40	35	30	28	28
PNC-40	64	59	54	50	45	40	36	33	33
PNC-45	67	63	58	54	50	45	41	38	38
PNC-50	70	66	62	58	54	50	46	43	43
PNC-55	73	70	66	62	59	55	51	48	48
PNC-60	76	73	69	66	63	59	56	53	53
PNC-65	79	76	73	70	67	64	61	58	58

[1] Refers to Chapter 9.

TABLE L.2

*Octave band pressure levels corresponding to noise rating number NR,
from ISO (1971)*

NR	Octave band sound pressure levels (dB)								
	Centre frequencies (Hz)								
	31·5	63	125	250	500	1000	2000	4000	8000
0	55·4	35·5	22·0	12·0	4·8	0	−3·5	−6·1	−8·0
5	58·8	39·4	26·3	16·6	9·7	5	+1·6	−1·0	−2·8
10	62·2	43·4	30·7	21·3	14·5	10	6·6	+4·2	+2·3
15	65·6	47·3	35·0	25·9	19·4	15	11·7	9·3	7·4
20	69·0	51·3	39·4	30·6	24·3	20	16·8	14·4	12·6
25	72·4	55·2	43·7	35·2	29·2	25	21·9	19·5	17·7
30	75·8	59·2	48·1	39·9	34·0	30	26·9	24·7	22·9
35	79·2	63·1	52·4	44·5	38·9	35	32·0	29·8	28·0
40	82·6	67·1	56·8	49·2	43·8	40	37·1	34·9	33·2
45	86·0	71·0	61·1	53·6	48·6	45	42·2	40·0	38·3
50	89·4	75·0	65·5	58·5	53·5	50	47·2	45·2	43·5
55	92·9	78·9	69·8	63·1	58·4	55	52·3	50·3	48·6
60	96·3	82·9	74·2	67·8	63·2	60	57·4	55·4	53·8
65	99·7	86·8	78·5	72·4	68·1	65	62·5	60·5	58·9
70	103·1	90·8	82·9	77·1	73·0	70	67·5	65·7	64·1
75	106·5	94·7	87·2	81·7	77·9	75	72·6	70·8	69·2
80	109·9	98·7	91·6	86·4	82·7	80	77·7	75·9	74·4
85	113·3	102·6	95·9	91·0	87·6	85	82·8	81·0	79·5
90	116·7	106·6	100·3	95·7	92·5	90	87·8	86·2	84·7
95	120·1	110·5	104·6	100·3	97·3	95	92·9	91·3	89·8
100	123·5	114·5	109·0	105·0	102·2	100	98·0	96·4	95·0
105	126·9	118·4	113·3	109·6	107·1	105	103·1	101·5	100·1
110	130·3	122·4	117·7	114·3	111·9	110	108·1	106·7	105·3
115	133·7	126·3	122·0	118·9	116·8	115	113·2	111·8	110·4
120	137·1	130·3	126·4	123·6	121·7	120	118·3	116·9	115·6
125	140·5	134·2	130·7	128·2	126·6	125	123·4	122·0	120·7
130	143·9	138·2	135·1	132·9	131·4	130	128·4	127·2	125·9

Calculation of equivalent continuous sound level for intermittent or fluctuating sounds

Tables M.1 and M.2 provide the information from which equivalent continuous sound level may be calculated from the duration and sound level A of noise exposure sustained during a period of 1 week. In Fig. M.1 a nomogram is provided to perform the same calculation on the basis of noise exposure over a period of 1 day.

A. *On a weekly basis* (*ISO, 1971*)

    The steps are:

    1. The total duration of each sound level is entered in the first column of Table M.1 and at the intersection with the appropriate noise level the *partial noise exposure index* is read off. If the total weekly duration is less than 10 min, the minimum value of 10 min should be used.

    2. Add the partial noise exposure indices; their arithmetical sum is the composite noise exposure index.

    3. Enter the value of composite noise exposure index in Table M.2 and read off the equivalent continuous sound level.

    (These tables can of course be used for finding the equivalent continuous sound level of one exposure, in which case the partial noise exposure index is entered directly into Table 2.)

B. *On a daily basis* (*Department of Employment, 1972*)

    The nomogram of Fig. M.1 is used in the same manner as Tables M.1 and M.2, the only difference being that it is calculated for exposure sustained during 1 day.

---

[1] Refers to Chapters 11 and 12.

*Partial noise exposure indices for sound levels 80 to 120 dB(A) and duration
10 minutes to 40 hours per week, from ISO (1971)*

Duration per week		Partial noise exposure indices								
		Sound level in dB(A) (Class mid-point)								
hours	min	80	85	90	95	100	105	110	115	120
	10					5	15	40	130	415
	12					5	15	50	160	500
	14					5	20	60	185	585
	16					5	20	65	210	665
	18					10	25	75	235	750
	20					10	25	85	265	835
	25				5	10	35	105	330	1040
	30				5	15	40	125	395	1250
	40				5	15	55	165	525	1670
	50				5	20	70	210	660	2080
	60			5	10	25	80	250	790	2500
	70			5	10	30	90	290	920	2920
	80			5	10	35	105	330	1050	3330
	90			5	10	40	120	375	1190	3750
	100			5	15	40	130	415	1320	4170
2				5	15	50	160	500	1580	5000
2·5				5	20	65	200	625	1980	6250
3				10	25	75	235	750	2370	7500
3·5			5	10	30	90	275	875	2770	8750
4			5	10	30	100	315	1000	3160	10000
5			5	15	40	125	395	1250	3950	12500
6			5	15	45	150	475	1500	4740	15000
7			5	20	55	175	555	1750	5530	17500
8			5	20	65	200	630	2000	6320	20000
9			5	25	70	225	710	2250	7110	22500
10		5	10	25	80	250	790	2500	7910	25000
12		5	10	30	95	300	950	3000	9490	30000
14		5	10	35	110	350	1110	3500	11100	
16		5	15	40	125	400	1260	4000	12600	
18		5	15	45	140	450	1420	4500	14200	
20		5	15	50	160	500	1580	5000	15800	
25		5	20	65	200	625	1980	6250	19800	
30		10	25	75	235	750	2370	7500	23700	
35		10	30	90	275	875	2770	8750	27700	
40		10	30	100	315	1000	3160	10000	31600	

The values are calculated from the formula:

$$E_i = \frac{\Delta t_i}{40} 10^{0.1(L_i - 70)}$$

where E_i = the partial noise exposure index

L_i = the sound level A in dB corresponding to the mid-point of the class i

Δt_i = the total duration in hours per week of sound levels within the class i.

TABLE M.2

Relation between composite noise exposure index and equivalent continuous sound level, from ISO (1971)

Composite noise exposure index	Equivalent continuous sound level, dB(A)
10	80
15	82
20	83
25	84
30	85
40	86
50	87
60	88
80	89
100	90
125	91
160	92
200	93
250	94
315	95
400	96
500	97
630	98
800	99
1000	100
1250	101
1600	102
2000	103
2500	104
3150	105
4000	106
5000	107
6300	108
8000	109
10000	110
12500	111
16000	112
20000	113
25000	114
31500	115

The values are calculated from the formula

$$L_{eq} = 70 + 10 \log_{10} \sum E_i$$

where L_{eq} = the equivalent continuous sound level in dB(A)
E_i = the partial noise exposure index (from Table M.1).

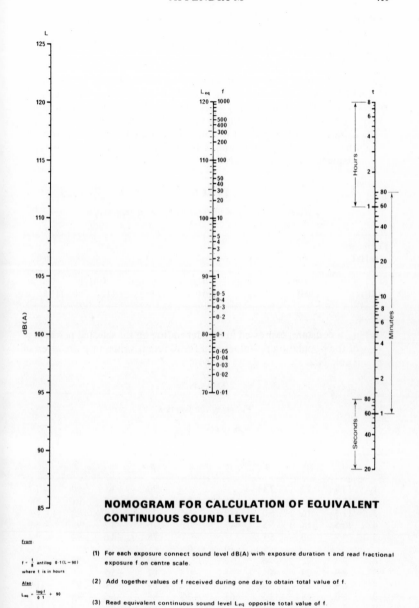

NOMOGRAM FOR CALCULATION OF EQUIVALENT CONTINUOUS SOUND LEVEL

From

$f = \frac{t}{8}$ antilog $0.1(L-90)$
where t is in hours

Also

$L_{eq} = \frac{\log f}{0.1} + 90$

(1) For each exposure connect sound level dB(A) with exposure duration t and read fractional exposure f on centre scale.

(2) Add together values of f received during one day to obtain total value of f.

(3) Read equivalent continuous sound level L_{eq} opposite total value of f.

Fig. M.1. (UK Department of Employment.)

Calculation of noise-induced threshold shift (Robinson, 1971)

The formulations of D. W. Robinson (1970) given in Chapter 11, Equations 11.2 and 11.3 may be conveniently handled numerically by the use of the necessary constants and partially-worked expressions. These constants are:
  (a) C_f, a coefficient depending on the audiometric frequency f.
  (b) λ_f, a constant, expressed in dB, depending on the audiometric frequency f. Values for C and λ for different values of audiometric frequency are given in Table N.1.

TABLE N.1

Values of C and λ for different audiometric frequencies

Frequency (kHz)	0·5	1	2	3	4	6
C	0·0040	0·0043	0·0060	0·0080	0·0120	0·0140
λ (dB)	130·0	126·5	120·0	114·5	112·5	115·5

  (c) u_p, a constant, expressed in dB, depending on the selected percentage of the population p. Values of u for different values of p are given in Table N.2.

TABLE N.2

Values of the function
$$u = 6\sqrt{2}.\mathrm{erf}^{-1}\left(\frac{p}{50} - 1\right)$$

p	u	p	u	p	u	p	u	p	u
1	13·9	12	7·1			70	−3·1	90	−7·7
2	12·3	14	6·5	35	2·3	72	−3·5	91	−8·0
3	11·3	16	6·0	40	1·5	74	−3·9	92	−8·4
4	10·5	18	5·5	45	0·8	76	−4·2	93	−8·9
5	9·9	20	5·1			78	−4·6	94	−9·3
				50	0				
6	9·3	22	4·6			80	−5·1	95	−9·9
7	8·9	24	4·2	55	−0·8	82	−5·5	96	−10·5
8	8·4	26	3·9	60	−1·5	84	−6·0	97	−11·3
9	8·0	28	3·5	65	−2·3	86	−6·5	98	−12·3
10	7·7	30	3·1			88	−7·1	99	−13·9

[1] Refers to Chapters 11 and 12.

(d) Tabulations for values for x and y in the expression

$$y = 27 \cdot 5 \left\{ 1 + \tanh \frac{E_A - \lambda_f + u_p}{15} \right\}$$

where $x = (E_A - \lambda_f + u_p)$

TABLE N.3

Values of the function

$$y = 27 \cdot 5 \left(1 + \tanh \frac{x}{15} \right)$$

x	y	x	y	x	y	x	y
−45	0·1	−25	1·9	−5	18·7	15	48·5
−44	0·2	−24	2·1	−4	20·3	16	49·2
−43	0·2	−23	2·4	−3	22·1	17	49·8
−42	0·2	−22	2·8	−2	23·9	18	50·5
−41	0·2	−21	3·2	−1	25·7	19	51·0
−40	0·3	−20	3·6	0	27·5	20	51·4
−39	0·3	−19	4·0	1	29·3	21	51·8
−38	0·4	−18	4·6	2	31·1	22	52·2
−37	0·4	−17	5·2	3	32·9	23	52·6
−36	0·4	−16	5·8	4	34·7	24	52·9
−35	0·5	−15	6·5	5	36·3	25	53·1
−34	0·6	−14	7·4	6	38·0	26	53·3
−33	0·7	−13	8·3	7	39·5	27	53·5
−32	0·8	−12	9·2	8	40·9	28	53·7
−31	0·9	−11	10·3	9	42·3	29	53·9
−30	1·0	−10	11·5	10	43·5	30	54·0
−29	1·1	−9	12·7	11	44·7	31	54·1
−28	1·3	−8	14·1	12	45·8	32	54·2
−27	1·5	−7	15·5	13	46·8	33	54·3
−26	1·7	−6	17·1	14	47·6	34	54·4

Example. What hearing level at 2 kHz would be attained or exceeded by 5% of a working population, without ear pathology, after 30 years of exposure to 90 dB(A), assuming age 50 years?

$E_A = 90 + 10 \log 30 = 105$ (Equation 11.1)
$\lambda = 120$ (Table N.1)
$p = 5$
therefore $u = 9 \cdot 9$ (Table N.2)
Thus $x = 105 - 120 + 9 \cdot 9$
$= -5 \cdot 1$
$y = 18 \cdot 5$ (Table N.3)

Noise-induced threshold shift $(y + u)$

$$= 18\cdot5 + 9\cdot9$$
$$= 28\cdot4 \text{ dB}$$

To add age allowance appropriate to age 50 and 2000 Hz audiometric frequency

$$C = 0\cdot006; N = 50 \qquad \text{Table N.1}$$

Thus $C(N - 20)^2$ \qquad (Equation 11.3)

$$= 0\cdot006 \times 30^2$$
$$= 5\cdot4 \text{ dB}$$

Thus expected hearing level is $28\cdot4 + 5\cdot4 = 33\cdot8$ dB.

References

(for abbreviations used see p. xii)

Chapter 1 INTRODUCTION pp. 4–10

Parry, C. H. (1825), *Collections from the unpublished medical writings of the late C. H. Parry,* **1**, 554. London, Underwood.

Chapter 2 PHYSICAL PROPERTIES OF SOUND pp. 11–27

Beranek, L. L. (1954), *Acoustics.* New York, McGraw-Hill.
—— (1971), *Noise and vibration control.* New York, McGraw-Hill.
BSI (1969), *BS 661. Glossary of acoustical terms.* London, BSI.
Hall, W., & Matthews, O. M. (1965), *Sound* (2nd Ed.). London, Edward Arnold.
Kinsler, L. E., & Frey, A. R. (1962), *Fundamentals of acoustics.* New York, John Wiley
Koffman, J. L., & Jarvis, R. G. (1964), 'Air springs as applied to multiple unit vehicles for heavy suburban services', *J. Inst. Locomotive Engineers,* **53**, 461.
Stephens, R. W. B., & Bate, A. E. (1966), *Acoustics and vibrational physics.* London, Edward Arnold.

Chapter 4 THE MEASUREMENT OF SOUND pp. 41–53

Beranek, L. L. (1960), *Noise reduction.* New York, McGraw-Hill.
—— (1971), *Noise and vibration control.* New York, McGraw-Hill.
Broch, J. T. (1969), *The application of the Brüel & Kjaer measuring systems to Acoustic Noise Measurements.* Naerum, Denmark, Brüel & Kjaer.
BSI (1963), *BS 3593. Recommendation on preferred frequencies for acoustical measurements.* London, BSI.
—— (1967), *BS 4196. Guide to the selection of methods of measuring noise emitted by machinery.* London, BSI.
DNA (1969), *DIN 45633 Blatt 2. Additional requirements for the extension of precision sound level meters to impulse sound level meters.* Berlin, Deutscher Normenausschuss.
IEC (1965), Publication 179, 1st Ed. *Precision sound level meters.* Geneva, IEC.
Peterson, A. P. G., & Gross, E. E. (1967), *Handbook of Noise Measurement.* West Concord, Massachusetts, USA, General Radio Company.

ISO (1970a), *R507. Procedure for describing aircraft noise around an airport* (2nd Ed.). ISO.

—— (1970b), *R1761. Monitoring aircraft noise around an airport.* ISO.

Chapter 5 MECHANISM OF HEARING pp. 54–80

Ades, H. W., Bredberg, G., & Engström, H. (1970), 'Scanning electron microscopy in the study of the inner ear'. In preparation.

Beagley, H. A. (1965a). 'Acoustic trauma in the guineapig I—Electrophysiology and histology', *Acta oto-laryng. (Stockh.)*, **60**, 437.

—— (1965b), 'Acoustic trauma in the guineapig II—Electron microscopy including the morphology of the cell junctions in the organ of Corti', *Acta oto-laryng. (Stockh.)*, **60**, 479.

Békésy, G. von (1960), *Experiments in hearing.* New York, McGraw-Hill.

Békésy, G. von, & Rosenblith, W. A. (1951), 'The mechanical properties of the ear', Chap. 27, *Handbook of experimental psychology*, ed. S. S. Stevens. New York, John Wiley.

Davis, H. (1957), 'Biophysics and physiology of the inner ear', *Physiological Reviews,* **37**, 1.

—— (1962), 'Advances in the neurophysiology and neuroanatomy of the cochlea', *J. Acoust. Soc. Am.,* **34**, 1377.

—— (1965), 'A model for the transducer action of the cochlea', *Cold Spring Harbor. Symp. on Quant. Bio.,* **30**, 181.

—— (1970), Chap. 3, 'Anatomy and physiology of the auditory system', in *Hearing and deafness*, ed. H. Davis and S. R. Silverman. New York, Holt, Rinehart & Winston.

Davis, H., & associates (1953), 'Acoustic trauma in the guineapig', *J. Acoust. Soc. Am.,* **25**, 1180.

Davis, H., Deatherage, B. H., Eldredge, D. H., & Smith, C. A. (1958), 'Summating potentials of the cochlea', *Am. J. Physiol.,* **195** (2), 251.

Davis, H., & Silverman, S. R. (1966), *Hearing and deafness*, 2nd Ed. New York, Holt, Rinehart & Winston.

—— (1970), *Hearing and deafness*, 3rd Ed. New York, Holt, Rinehart & Winston.

Eldredge, D. H., & Miller, James D. (1971), 'Physiology of hearing', *Ann. rev. Physiol.,* **33**, 281.

Engström, H., & Ades, H. (1960), 'Effect of high intensity noise on inner ear epithelia', *Acta oto-laryng. (Stockh.)*, Supp. 178, 217.

Engström, H., Ades, H. W., & Andersson, A. (1966), *Structural pattern of the organ of Corti.* Stockholm, Almqvist & Wiksell.

Engström, H., Ades, H. W., & Bredberg, G. (1970), 'Normal structure of the organ of Corti and the effect of noise-induced cochlear damage', in *Sensorineural Hearing Loss*, ed. G. E. W. Wolstenholme and Julie Knight, London, J. & A. Churchill.

Jepsen, O. (1963), Chap. 6, 'Middle ear muscle reflexes in man', *Modern developments in audiology*, ed. J. Jerger. New York & London, Academic Press.

Johnstone, B. M., Taylor, K. J., & Boyle, A. J. (1970), 'Mechanics of the guineapig cochlea', *J. Acoust. Soc. Am.*, **47**, 504.

Kiang, N. Y. S., Moxon, E. C., & Levine, R. A. (1970), 'Auditory-nerve activity in cats with normal and abnormal cochleas', in *Sensorineural hearing loss*, ed. G. E. W. Wolstenholme and Julie Knight, London, J. & A. Churchill.

Kiang, N. Y. S., Watanabe, T., Thomas, E. C., & Clark, L. F. (1965), *Discharge patterns of single fibers in the cat's auditory nerve.* Cambridge, Mass.: M.I.T. Press.

Littler, T. S. (1965), *The physics of the ear.* Oxford, Pergamon.

Møller, A. R. (1961), 'Bilateral contraction of the tympanic muscles in man examined by measuring acoustic impedance-change', *Ann. Otol. (St Louis)*, **70**, 735.

Reger, S. N. (1960), 'The effect of middle ear muscle action on certain psycho-physical measurements'. *Ann. Otol. (St. Louis)*, **69**, 1179.

Sachs, M. B., & Kiang, N. Y. S. (1968), 'Two-tone inhibition in auditory-nerve fibers', *J. Acoust. Soc. Am.*, **43**, 1120.

Smith, C. A. (1968), 'Ultrastructure of the Organ of Corti', *Advancement of Science*, Vol. 24, No. 122, p. 419.

Smith, C. A., & Sjöstrand, F. S. (1961), 'A synaptic structure in the hair cells of the guineapig cochlea', *J. Ultrastruct. Res.*, **5**, 184.

Spoendlin, H. (1966). *The organisation of the cochlear receptor.* Basel, Karger.

—— (1969), 'Innervation patterns in the organ of Corti of the cat', *Acta oto-laryng. (Stockh.)*, **67**, 239.

Stevens, S. S., & Davis, H. (1938), *Hearing, its psychology and physiology.* New York, John Wiley.

Tasaki, I. (1954), 'Nerve impulses in individual nerve fibres of guineapig', *J. Neurophysiol.*, **17**, 97.

Teas, D. C., Eldredge, D. H., & Davis, H. (1962), 'Cochlear responses to acoustic transients: an interpretation of whole-nerve action potentials', *J. Acoust. Soc. Am.*, **34**, 1438.

Whitfield, I. C. (1967), *The auditory pathway.* London, Edward Arnold.

Whitfield, I. C., & Ross, H. F. (1965), 'Cochlear-microphonic and summating potentials and the outputs of individual hair-cell generators', *J. Acoust. Soc. Am.*, **38**, 126.

Chapter 6 NORMAL HEARING AND ITS MEASUREMENT

pp. 81–108

ANSI (1951), *Z24.5. Specification for audiometers for general diagnostic purposes.* New York, ANSI.

—— (1969), *S3.6-1969. Specifications for audiometers.* New York, ANSI.

Beagley, H. A., & Knight, J. J. (1967), 'Application of computer averaging to evoked response audiology', Proceedings of the Second European Symposium on Medical Electronics. London, Hanover Press.

Békésy, G. von (1947), 'A new audiometer', *Acta oto-laryng. (Stockh.)*, **35**, 411.

—— (1960), *Experiments in hearing*, p. 145. New York, McGraw-Hill.

Bradford Hill, A. (1966, 1971), *Principles of medical statistics*. London, The Lancet.

BSI (1952), *BS 2042. An artificial ear for the calibration of earphones of the external type*. London, BSI.

—— (1954), *BS 2497. The normal threshold of hearing for pure tones by earphone listening*. London, BSI.

—— (1958), *BS 2980. Pure tone audiometers*. London, BSI.

—— (1966), *BS 4009. Specification for an artificial mastoid for the calibration of bone vibrators used in hearing aids and audiometers*. London, BSI.

—— (1968), *BS 2497*: Part 1, *Specification for a reference zero for the calibration of pure-tone audiometers*. Part 1, *Data for earphone coupler combinations maintained at certain standardising laboratories*. London, BSI.

—— (1969), *BS 2497*: Part 2, *Specification for a reference zero for the calibration of pure-tone audiometers*. Part 2, *Data for certain earphones used in commercial practice*. London, BSI.

Burns, W., & Hinchcliffe, R. (1957), 'Comparison of the auditory threshold as measured by individual pure tone and by Békésy audiometry', *J. Acoust. Soc. Am.*, **29**, 1274.

Chocholle, R. (1954), 'Measurement of the threshold of hearing', *Acustica*, **4**, 75.

Coles, R. R. A. (1972), 'Can present day audiology really help in diagnosis? An otologist's question', *J. Laryng.*, **86**, 191.

Dadson, R. S., & King, J. H. (1952), 'A determination of the normal threshold of hearing and its relation to the standardisation of audiometers', *J. Laryng.*, **66**, 366.

Davis, H., Hoople, G. D., & Parrack, H. O. (1958), 'Hearing level, hearing loss, and threshold shift', *J. Acoust. Soc. Am.*, **30**, 478.

Davis, H., & Silverman, S. R. (1970). *Hearing and Deafness*, 3rd Ed. New York, Holt, Rinehart & Winston.

Davis, Pauline A. (1939), 'The effects of acoustic stimuli on the waking human brain', *J. Neurophysiol.*, **2**, 494.

Delany, M. E. (1964), 'The acoustic impedance of human ears', *J. Sound Vib.*, **1**, 455.

—— (1971), 'Some sources of variance in the determination of hearing levels', in *Occupational hearing loss*, ed. D. W. Robinson. London & New York, Academic Press.

Delany, M. E., & Whittle, L. S. (1966), 'A new artificial ear and mastoid', *J. Sci. Instrum.*, **43**, 519.

—— (1967), 'Reference equivalent threshold sound pressure levels for audiometry', *Acustica*, **18**, 227.

Delany, M. E., Whittle, L. S., Cook, J. P., & Scott, V. (1967), 'Performance studies on a new artificial ear', *Acustica*, **18**, 231.

Delany, M. E., Whittle, L. S., & Knox, E. C. (1966), 'A note on the use of self-recording audiometry for children', *J. Laryng.*, **80**, 1135.

Glorig, A., & Davis, H. (1961), 'Age, noise and hearing loss', *Ann. Otol. (St Louis)*, **70**, 556.

Glorig, A., Wheeler, D., Quiggle, R., Grings, W., & Summerfield, A. (1957), *Wisconsin State Fair hearing survey 1954*. Los Angeles, Am. Acad. of Ophth. & Otolaryng.

Harrison, M. S. (1969), 'Audiological aspects of neuro-otology', *Sound*, 3, 61.

Hinchcliffe, R. (1958), 'The pattern of the threshold of perception for hearing and other special senses as a function of age', *Gerontologica*, 2, 311.

—— (1959a), 'The threshold of hearing as a function of age', *Acustica*, 9, 304.

—— (1959b), 'The threshold of hearing of a random sample rural population', *Acta oto-laryng. (Stockh.)*, 50, 411.

—— (1965), 'Deafness in Jamaica' (A pilot survey of a sample rural population), *West Indian Medical Journal*, 14, 4.

Hinchcliffe, R., & Littler, T. S. (1958), 'Methodology of air conduction audiometry for hearing surveys', *Ann. Occup. Hyg.*, 1, 114.

—— (1960), 'Auditory acuity of ex-coalminers', *Proc. 13th International congress on occup. health*, p. 712.

Hirsh, I. J. (1952), *The measurement of hearing*. New York, McGraw-Hill.

Hood, J. D. (1962), 'Bone conduction: a review of the present position with especial reference to the contributions of Dr Georg von Békésy, *J. Acoust. Soc. Am.*, 34, 1325,

IEC (1970a), Publication 303, *Provisional reference coupler for the calibration of earphones used in audiometry*. Geneva, IEC.

—— (1970b), Publication 318, *An IEC artificial ear of the wide-band type for the calibration of earphones used in audiometry*. Geneva, IEC.

ISO (1964), *R389. Standard reference zero for the calibration of pure tone audiometers*. ISO.

Knight, J. J., & Coles, R. R. A. (1960), 'Determination of hearing thresholds of naval recruits in terms of British and American standards', *J. Acoust. Soc. Am.*, 32, 800.

Knight, J. J., & Littler, T. S. (1953), 'The technique of speech audiometry and a simple speech audiometer with masking generator for clinical use', *J. Laryng.*, 67, 248.

Knox, Elizabeth C., & Lenihan, J. M. A. (1958), 'The Scottish Audiometer Calibration Service', *Ann. Occup. Hyg.*, 1, 104.

Littler, T. S. (1962), 'Techniques of industrial audiometry', National Physical Laboratory Symposium No. 12, London, HMSO.

McMurray, R. F., & Rudmose, W. (1956), 'An automatic audiometer for industrial medicine', *Noise Control*, 2, 33.

Moroney, M. J. (1962), *Facts from figures*. London, Penguin.

Naunton, R. F. (1963), Chap. 1, 'The measurement of hearing by bone conduction', in *Modern developments in audiology*, ed. J. Jerger. New York & London, Academic Press.

Robinson, D. W. (1960), 'Variability in the realisation of the audiometric zero', *Ann. Occup. Hyg.*, 2, 107.

—— (1971), 'A review of audiometry', *Phys. Med. and Biol.*, 16, 1.

Rosen, S. (1962), 'Presbycusis study of a relatively noise-free population in the Sudan', *Ann. Otol. (St Louis)*, 71, 727.

Rudmose, W. (1963), Chap. 2, 'Automatic audiometry', in *Modern developments in audiology*, ed. J. Jerger. New York & London, Academic Press.

Sivian, L. J., & White, S. D. (1933), 'On minimum audible sound fields', *J. Acoust. Soc. Am.*, **4**, 288.

Ward, W. D., Glorig, A., & Sklar, Diane L. (1959), 'Susceptibility and sex', *J. Acoust. Soc. Am.*, **31**, 1138.

Watson, L. A., & Tolan, T. (1949), *Hearing tests and hearing instruments*. Baltimore, Williams & Wilkins.

Weissler, P. G. (1968), 'International standard reference zero for audiometers', *J. Acoust. Soc. Am.*, **44**, 264.

Wheeler, L. J., & Dickson, E. D. D. (1952), 'The determination of the threshold of hearing', *J. Laryng.*, **66**, 379.

Chapter 7 DEAFNESS pp. 109–114

Ballantyne, J. C. (1970), *Deafness*. London, J. & A. Churchill.

Beagley, H. A. (1965), 'Acoustic trauma in the guinea pig. I. Electrophysiology and histology. II. Electron microscopy, including morphology of cell junctions in the organ of Corti', *Acta oto-laryng. (Stockh.)*, **60**, 437, 479.

Cawthorne, T. E., & Hewlett, A. B. (1954), 'Ménière's disease', *Proc. Roy. Soc. Med.*, **47**, 663.

Davis, H., & Silverman, S. R. (1966), *Hearing and deafness*. (2nd Ed.). New York, Holt, Rinehart & Winston.

—— (1970), *Hearing and deafness* (3rd Ed.). New York, Holt, Rinehart & Winston.

Williams, H. L. (1952), *Ménière's disease*. Springfield, Ill., C. C. Thomas.

Chapter 8 DISTURBANCE pp. 115–147

Abey-Wickrama, I., a'Brook, M. F., Gattoni, F. E. G., & Herridge, C. F. (1969), 'Mental hospital admissions and aircraft noise', *Lancet*, **2**, 1275.

Bartlett, F. C. (1934), *The problem of noise*. Cambridge University Press.

Beranek, L. L. (1971), *Noise and vibration control*, Chap. 18. New York, McGraw-Hill.

Broadbent, D. E. (1954), 'Some effects of noise on visual performance', *Quart. J. Exp. Psychol.*, **6**, 1.

—— (1955), 'Some clinical implications of recent experiments on the psychology of hearing', *Proc. Roy. Soc. Med.*, **48**, 961.

—— (1957a), 'Noise and behaviour', *Proc. Roy. Soc. Med.*, **50**, 525.

—— (1957b), 'Effects of noises of high and low frequency on behaviour', *Ergonomics*, **1**, 21.

—— (1958), 'The effects of noise on an intellectual task', *J. Acoust. Soc. Am.*, **30**, 824.

—— (1971), *Decision and stress*. New York & London, Academic Press.

Broadbent, D. E., & Gregory, M. (1963), 'Vigilance as a statistical decision'. *Brit. J. Psychol.,* **54**, 309.

Broadbent, D. E., & Little, E. A. J. (1960), 'Effects of noise reduction in a work situation', *Occup. Psychol.,* **34**, 133.

BSI (1958), *BS 3045. The relation between the sone scale of loudness and the phon scale of loudness level.* London, BSI.

—— (1967), *BS 4198. Method for calculating loudness.* London, BSI.

—— (1969), *BS 661. Glossary of acoustical terms.* London, BSI.

Burns, W., Hinchcliffe, R., & Littler, T. S. (1964), 'An exploratory study of hearing and noise exposure in textile workers', *Ann. Occup. Hyg.,* **7**, 323.

Burns, W., & Robinson, D. W. (1970), *Hearing and noise in industry.* London, HMSO.

Carpenter, A. (1959), 'Effects of noise on work', *Ann. Occup. Hyg.,* **1**, 42.

—— (1962), 'Effects of noise on performance and productivity', National Physical Laboratory Symposium No. 12. London, HMSO.

Committee on the Problem of Noise (1963). *Noise, final report.* London, HMSO.

Corcoran, D. W. J. (1962), 'Noise and loss of sleep', *Quart. J. Exp. Psychol.,* **14**, 178.

Davis, H. (1958), ed., 'Auditory and non-auditory effects of high-intensity noise'. Project ANEHIN: Final report, Joint Project 13 01, Subtask 1, Report No. 7. Central Institute for the Deaf, St Louis, Mo., and US Naval School of Aviation Medicine, Pensacola, Fla.

Davis, R. C., Buchwald, A. M., & Frankman, R. W. (1955), 'Autonomic and muscular responses and their relation to simple stimuli', *Psychol. Monographs* 69, No. 405.

Fletcher, H. (1940), 'Auditory patterns', *Rev. Mod. Physics,* **12**, 47.

—— (1953), *Speech and hearing in communication.* Princeton, N.J., D. Van Nostrand.

French, N. R., & Steinberg, J. C. (1947), 'Factors governing the intelligibility of speech sounds', *J. Acoust. Soc. Am.,* **19**, 90.

Galloway, W. J., & Clark, W. E. (1962). 'Prediction of noise for motor vehicles in freely flowing traffic', Fourth International Congress on Acoustics, Paper L28. Copenhagen.

Greater London Council (1970), Urban design bulletin No. 1, *Traffic noise.* London, Greater London Council, County Hall, London, S.E.1.

Harrison, M. S. (1969), 'Audiological aspects of neuro-otology', *Sound,* **3**, 61.

Hawkins, J. E., & Stevens, S. S. (1950), 'The masking of pure tones and speech by vehicle noise', *J. Acoust. Soc. Am.,* **22**, 6.

Hood, J. D. (1968), 'Observations upon the relationship of loudness discomfort level and auditory fatigue to sound-pressure level and sensation level', *J. Acoust. Soc. Am.,* **44**, 959.

Hood, J. D., & Poole, J. P. (1966), 'Tolerable limit of loudness: its clinical and physiological significance', *J. Acoust. Soc. Am.,* **40**, 47.

ISO (1959), *R131. Expression of the physical and subjective magnitudes of sound or noise.* ISO.

ISO (1961), *R226. Normal equal loudness contours for pure tones and normal threshold of hearing under free-field listening conditions.* ISO.
—— (1966), *R532. Method for calculating loudness level.* ISO.
—— (1970), *R507. Procedure for describing aircraft noise around an airport.* 2nd Ed. ISO.

Jansen, G. (1969). 'Effects of noise on physiological state', National Conference on Noise as a Public Health Hazard. American Speech and Hearing Association, Washington, D.C.

Johnson, D. R., & Saunders, E. G. (1968), 'The evaluation of noise from freely flowing road traffic', *J. Sound Vib., 7*, 287.

Kosten, C. W., & van Os, G. J. (1962), 'Community reaction criteria for external noises', National Physical Laboratory Symposium No. 12. London, HMSO.

Kryter, K. D. (1970), *The effects of noise on man.* New York & London, Academic Press.

Lehmann, G. (1965), 'Autonomic reactions to hearing impressions', *Stud. Gen. (Berlin), 18*, 700.

Licklider, J. C. R. (1951), Chap. 25 and 26 (with Miller, G. A.) in *Handbook of experimental psychology*, ed. S. S. Stevens. New York, John Wiley.

Mackworth, N. H. (1950), *Medical Research Council Report No. 268.*

McKennell, A. C., & Hunt, E. A. (1966), *Noise annoyance in central London.* The Government Social Survey.

Meister, F. J. (1964), 'Protection against traffic noise (Die Lärmdampfung im Strassenverkehr)', *VDI-Zeitschrift, 106*, 1165.

Morgan, C. T., Cook, J. S., Chapanis, A., & Lund, M. W. (1963), *Human engineering guide to equipment design.* New York, McGraw-Hill.

Oppliger, G., & Grandjean, E. (1957), 'Vasomotor reactions of the hand to noise stimuli', *Helv. Physiol. Pharmacol. Acta, 17*, 275.

Parkin, P. H. (1962), 'Propagation of sound in air', NPL Symposium No. 12. London, HMSO.

Parkin, P. H., Purkis, H. J., Stephenson, R. J., & Schlaffenberg, B. (1968), *London Noise Survey.* London, HMSO.

Road Research Laboratory (1970), 'A review of road traffic noise', RRL Report LR 357, Road Research Laboratory, Crowthorne, Berks.

Robinson, D. W. (1957), 'The subjective loudness scale', *Acustica, 7*, 217.
—— (1970a), 'An outline guide to criteria for the limitation of urban noise', Aeronautical Research Council, C.P. 1112. London, HMSO.
—— (1970b), 'Relations between hearing loss and noise exposure', Appendix 10, Burns & Robinson (1970, above).

Robinson, D. W., & Dadson, R. S. (1956), 'A re-determination of the equal-loudness relation for pure tones', *Brit. J. appl. Phys., 7*, 166.

Robinson, D. W., & Whittle, L. S. (1964), 'The loudness of octave-bands of noise', *Acustica, 14*, 24.

Rucker, A., & Glück, K. (1964), 'Die Ausbreitung und Dampfung des Strassenverkehrslärms in Bebauungsgebieten', *StrBau StrVerk., Heft 32.*

Sanders, A. F. (1961), 'The influence of noise on two discrimination tasks', *Ergonomics, 4*, 235.

Stephenson, R. J., & Vulkan, G. H. (1968), 'Traffic noise', *J. Sound Vib.,* **7**, (2), 247.

Stevens, S. S. (1955), 'The measurement of loudness', *J. acoust. Soc. Am.,* **27**, 815.

—— (1956), 'Calculation of the loudness of complex noise', *J. Acoust. Soc. Am.,* **28**, 807.

—— (1957), 'Calculating loudness', *Noise Control,* **3**, 11.

—— (1961), 'Procedure for calculating loudness, Mk VI', *J. Acoust. Soc. Am.,* **33**, 1577.

Taylor, W., Pearson, J., Mair, A., & Burns, W. (1965), 'Study of noise and hearing in jute weaving', *J. Acoust. Soc. Am.,* **38**, 113.

Teichner, W. H., Arees, E., & Reilly, R. (1963), 'Noise and human performance: a psychophysiological approach', *Ergonomics,* **6**, 83.

Wegel, R. L., & Lane, C. E. (1924), 'The auditory masking of one sound by another and its probable relations to the dynamics of the cochlea', *Phys. Rev.,* **23**, 266.

Weston, H. C., & Adams, S. (1932), *Industrial Health Research Board Report No. 65.* London, HMSO.

—— (1935), *Industrial Health Research Board Report No. 70.* London, HMSO.

Wilkinson, R. T. (1963), 'Interaction of noise with knowledge of results and sleep deprivation', *Exp. Psychol.,* **66**, 332.

Zwicker, E. (1960), 'Ein Verfahren zur Berechnung der Lautstärke', *Acustica,* **10**, 304.

—— (1961), 'Subdivision of the audible frequency range into critical bands (Frequenzgruppen)', *J. Acoust. Soc. Am.,* **33**, 248.

Zwicker, E., Flottorp, G., & Stevens, S. S. (1957), 'Critical band width in loudness summation', *J. Acoust. Soc. Am.,* **29**, 548.

Chapter 9 MEASURES TO REDUCE INTERFERENCE EFFECTS
pp. 148–188

ANSI (1967), *S1.6. Preferred frequencies and band numbers for acoustical measurements.* New York, ANSI.

ASHRAE (1967), Chap. 31, Sound and vibration control, Fig. 2, p. 377, in *Guide and data book: systems and equipment.* ASHRAE.

Bazley, E. N. (1966), *The airborne sound insulation of partitions.* London, HMSO.

Beranek, L. L. (1956), 'Criteria for office quietening based on questionnaire rating studies', *J. Acoust. Soc. Am.,* **28**, 833.

—— (1957), 'Revised criteria for noise in buildings', *Noise Control,* **3**, No. 1, 19.

—— (1960), *Noise reduction,* Chap. 20. New York, McGraw-Hill.

—— (1971), Chap. 18, *Noise and vibration control.* New York, McGraw-Hill.

Beranek, L. L., Blazier, W. E., & Figwer, J. J. (1971), Preferred noise criteria curves and their application. (Personal communication.)

Bor, W. (1971), 'Urban transport and environment', *J. Sound Vib.*, **15**, 23.

Bottom, C. G., & Waters, D. M. (1971), 'A social survey into annoyance caused by the interaction of aircraft noise and traffic noise'. Report TT7102. Department of Transport Technology, Loughborough University of Technology.

BSI (1960), C.P. 3: Chap. III, 'Sound insulation and noise reduction'. London, BSI.

—— (1961), *BS 3425* (with amendments issued April 1963 and June 1966): *Measurement of noise emitted by motor vehicles*. London, BSI.

—— (1962), *BS 3539. Specification for sound level meters for the measurement of noise emitted by motor vehicles*. London, BSI.

—— (1963), *BS 3593. Preferred frequencies for acoustical measurements*. London, BSI.

—— (1966), *BS 3425. Method for the measurement of noise emitted by motor vehicles*. London, BSI.

—— (1967), *BS 4142. Method of rating industrial noise affecting mixed residential and industrial areas*. London, BSI.

Broadbent, D. E. (1966), Personal communication.

Broadbent, D. E., & Little, E. A. J. (1960), 'Effects of noise reduction in a work situation', *Occup. Psychol.*, **34**, 133.

Building Research Station (1956), *Digests* 88 and 89, London, HMSO.

—— (1968), *London Noise Survey*. London, HMSO.

Burns, W., & Robinson, D. W. (1970), *Hearing and noise in industry*. London, HMSO.

Burt, M. E. (1971), 'Aspects of highway design and traffic management', *J. Sound Vib.*, **15**, 23.

Coblentz, A., Xydias, N., & Alexandre, A. (1967), 'Enquête sur le bruit autour des aéroports', *Anthropologie Appliquée,* document AA 16/67, 5.

Committee on the Problem of Noise (1963), *Noise, final report*. London, HMSO.

Denby, W. (1967), 'Effective control of road vehicle noise', IEE Conference publication No. 26, *Acoustic noise and its control*. London, Instn. Elec. Engrs.

Dennington, D. (1971), 'Aspects of highway design and traffic management', *J. Sound Vib.*, **15**, 35.

Evans, Margaret J., & Tempest, W. (1972), 'Some effects of infrasonic noise in transportation', *J. Sound Vib.*, **22**, 19.

Fog, H., Jonsson, E., Kajland, A., Nilsson, A., & Sörensen, S. (1968), Report 36E. *Traffic noise in residential areas*. National Swedish Institute for Building Research.

Galloway, W. J., & Bishop, D. E. (1970), *Noise exposure forecasts: evolution, extensions, and land use interpretations*. Bolt, Beranek & Newman, Final Report No. FAA-NO-70-9, Contract Fa68WA-1900, FAA.

Greater London Council (1970), Urban design bulletin No. 1, *Traffic noise*. London, Greater London Council, County Hall, London, S.E.1.

Griffiths, I. D., & Langdon, F. J. (1968), 'Subjective responses to road traffic noise', *J. Sound Vib.*, **8**, 16.

Hardy, A. C., & Lewis, P. T. (1971), 'Sound insulation standards for buildings adjacent to urban motorways', *J. Sound Vib.,* **15**, 53.

ISO (1962), *R266. Preferred frequencies for acoustical measurements.* ISO.

―― (1964), *R362. Measurement of noise emitted by vehicles.* ISO.

―― (1971), *R1996. Assessment of noise with respect to community response.* ISO.

Keighley, E. C. (1966), 'The determination of acceptability criteria for office noise', *J. Sound Vib.,* **4**, 73.

Klumpp, R. G., & Webster, J. C. (1963), 'Physical measurements of equally interfering navy noises', *J. Acoust. Soc. Am.,* **35**, 1328.

Koffman, J. L. (1967), 'Development of British Railways coach insulation', *Railway Gazette,* June 2, 415.

Koffman, J. L., & Jeffs, D. C. (1967), 'Reducing noise from locomotive cooling fans', *Railway Gazette,* March 17, 229.

Kosten, C. W., & van Os, G. J. (1962), 'Community reaction criteria for external noises', NPL Symposium No. 12. London, HMSO.

Kryter, K. D. (1946), 'Effects of ear protective devices on the intelligibility of speech in noise', *J. Acoust. Soc. Am.,* **18**, 413.

―― (1962), 'Methods for the calculation and use of the articulation index', *J. Acoust. Soc. Am.,* **39**, 1689.

Kryter, K. D., Licklider, J. C. R., Webster, J. C., & Hawley, M. (1963), Chap. 4, in *Human engineering guide to equipment design,* ed. C. T. Morgan, J. S. Cook, A. Chapanis, & M. W. Lund. New York, McGraw-Hill.

Kryter, K. D., & Pearsons, K. S. (1963), 'Some effects of spectral content and duration on perceived noise level', *J. Acoust. Soc. Am.,* **35**, 866.

Langdon, F. J., & Scholes, W. E. (1968), 'The traffic noise index: a method of controlling noise nuisance'. BRS Current Papers SfB Ac, UDC 711.73.

Meister, F. J. (1968), VDI Zeitschrift, Research report series 11, No. 5.

McKennell, A. C. (1963), *Aircraft noise annoyance around London (Heathrow) Airport.* COI Report S.S. 337. London, Central Office of Information.

Mills, C. H. G., & Robinson, D. W. (1963), 'The subjective rating of motor vehicle noise', Committee on the Problem of Noise, Appendix IX. London, HMSO.

Moser, H. M., & Bell, G. E. (1955), Joint United States–United Kingdom Report AFCRC-TN-55-56, Air Force Cambridge Research Center, Cambridge, Mass., USA.

Parkin, P. H., & Humphreys, H. R. (1969), *Acoustics, noise and buildings,* 3rd Ed. London, Faber.

Parrack, H. O. (1957), *Handbook of noise control,* ed. C. M. Harris, New York, McGraw-Hill.

Pearsons, K. S. (1966), 'The effects of duration and background noise level on perceived noisiness', Report FAA-ADS-78. Bolt, Beranek and Newman Inc., Van Nuys, Calif.

Priede, T. (1967), 'Noise and vibration problems in commerce vehicles', *J. Sound Vib.,* **5**, 129.

Priede, T. (1971), 'Origins of automotive vehicle noise', *J. Sound Vib.*, **15**, 61.

Road Research Laboratory (1970), *A review of Road Traffic Noise*, Report LR 357. Road Research Laboratory, Crowthorne, Berkshire.

Robinson, D. W. (1969), *The concept of noise pollution level.* NPL Aero Report Ac 38.

—— (1970), *An outline guide to criteria for the limitation of urban noise.* C.P. No. 1112. London, HMSO.

—— (1971), 'Towards a unified system of noise assessment', *J. Sound Vib.*, **14**, 279.

Robinson, D. W., & Whittle, L. S. (1964), 'The loudness of octave-bands of noise', *Acustica*, **14**, 24.

Rosenblith, W. A., & Stevens, K. N. (1953), *WADC Technical Report 52-204*, Chap. 18. Wright Air Development Centre.

Scholes, W. E., & Sargent, J. W. (1971), 'Designing against noise from road traffic', *J. appl. Acoust.*, **4**, 203.

Schultz, T. J. (1968), 'Noise-criterion curves for use with the USASI preferred frequencies', *J. Acoust. Soc. Am.*, **43**, 637.

Stevens, S. S. (1956), 'Mark I. Calculation of the loudness of complex noise', *J. Acoust. Soc. Am.*, **28**, 807.

Waters, P. E., Lalor, N., & Priede, T. (1970), 'The diesel engine as a source of commercial vehicle noise'. Proc. Inst. mech. Engrs., **184**, 63.

Webster, J. C. (1965), 'Speech communications as limited by ambient noise', *J. Acoust. Soc. Am.*, **37**, 692.

—— (1969), Personal communication.

Chapter 10   TEMPORARY EFFECTS OF NOISE ON HEARING
pp. 189–213

Atherley, G. R. C. (1964), 'Monday morning auditory threshold in weavers', *Brit. J. Ind. Med.*, **21**, 150.

Békésy, G. von (1949a), 'On the resonance curve and the decay period at various points on the cochlear partition', *J. Acoust. Soc. Am.*, **21**, 245.

—— (1949b), 'The structure of the middle ear and the hearing of one's own voice by bone conduction', *J. Acoust. Soc. Am.*, **21**, 217.

Bronstein, A. (1936), 'Sensibilization of the auditory organ by acoustic stimuli', *Byull eksp. Biol. Med.*, **1**, 274.

Burns, W. (1971), 'The relation of temporary to permanent threshold shift in individuals', in *Occupational hearing loss*, ed. D. W. Robinson. London & New York, Academic Press.

Burns, W., & Robinson, D. W. (1970). *Hearing and noise in industry.* London, HMSO.

Burns, W., Stead, J. C., & Penney, H. W. (1970), Appendix 13 'The relations between temporary threshold shift and occupational hearing loss', in

Hearing and noise in industry, W. Burns & D. W. Robinson. London, HMSO.

Carter, N., & Kryter, K. D. (1962), *Equinoxious contours for pure tones and some data on the 'critical band' for TTS*. Bolt, Beranek & Newman, Inc. Rept. No. 948, Contract No. DA49-007-MD-985, OSG, US Army, Washington, D.C.

Caussé, R., & Chavasse, P. (1947), 'Etudes sur la fatigue auditive', *Ann. psychol.*, **43/44**, 265.

Chisman, J. A., & Simon, J. R. (1961), 'Protection against impulse-type industrial noise by utilising the acoustic reflex', *J. appl. Psychol.*, **45**, 402.

Davis, H., Hoople, G., & Parrack, H. O. (1958a), 'Hearing level, hearing loss and threshold shift', *J. Acoust. Soc. Am.*, **30**, 478.

—— (1958b), 'Medical principles of monitoring audiometry', A.M.A. *Archives of Indust. Health*, **17**, 1.

Davis, H., Morgan, C. T., Hawkins, J. E., Galambos, R., & Smith, F. W. (1950), 'Temporary deafness following exposure to loud tones and noise', *Acta oto-laryng. (Stockh.)*, Supp. 88.

Ewing, A., & Littler, T. S. (1935), 'Auditory fatigue and adaptation', *Brit. J. Psychol.*, **25**, 284.

Fletcher, J. L., & Riopelle, A. J. (1960), 'Protective effect of the acoustic reflex for impulsive noises', *J. Acoust. Soc. Am.*, **32**, 401.

Glorig, A., Ward, W. D., & Nixon, J. (1961), 'Damage risk criteria and noise-induced hearing loss', *Arch. otolaryng.*, **74**, 413.

Hinchcliffe, R. (1957), 'Threshold changes at 4 Kc/s produced by bands of noise', *Acta oto-laryng. (Stockh.)*, **47**, 496.

Hirsh, I. J., & Bilger, R. C. (1955), 'Auditory threshold recovery after exposure to pure tones', *J. Acoust. Soc. Am.*, **27**, 1186.

Hirsh, I. J., & Ward, W. D. (1952), 'Recovery of the auditory threshold after strong acoustic stimulation', *J. Acoust. Soc. Am.*, **24**, 131.

Hood, J. D. (1950), 'Studies in auditory fatigue and adaptation', *Acta oto-laryng. (Stockh.)*, Supp. 92.

Jerger, J. F., & Carhart, R. (1956), 'Temporary threshold shift as an index of noise susceptibility', *J. Acoust. Soc. Am.*, **28**, 611.

Jepsen, O. (1963), Chap. 6, 'Middle ear muscle reflexes in man', in *Modern developments in audiology*, ed. J. Jerger. New York & London, Academic Press.

Kiang, N. Y. S., Moxon, E. C., & Levine, R. A. (1970), 'Auditory-nerve activity in cats with normal and abnormal cochleas', in *Sensorineural hearing loss*, ed. G. E. W. Wolstenholme and Julie Knight. London, J. & A. Churchill.

Kryter, K. D. (1950), 'The effects of noise on man'', *J. Speech Dis.*, monograph, Supp. 1.

Kryter, K. D., Ward, W. D., Miller, J. D., & Eldredge, D. H. (1966), 'Hazardous exposure to intermittent and steady-state noise', *J. Acoust. Soc. Am.*, **39**, 451.

Kylin, B. (1960), 'Temporary threshold shift and auditory trauma following exposure to steady state noise', *Acta oto-laryng. (Stockh.)*, Supp. 152.

Loeb, M., Fletcher, J. L., & Benson, R. W. (1965), 'Some preliminary studies of temporary threshold shift with an arc discharge impulse noise generator', *J. Acoust. Soc. Am.*, **37**, 313.

Miller, J. D., Eldredge, D. H., & Bredberg, G. (1970), Personal communication.

Mills, J. H., Gengel, R. W., Watson, C. S., & Miller, J. D. (1970), 'Temporary changes of the auditory system due to exposure to noise for one or two days', *J. Acoust. Soc. Am.*, **48**, 524.

Murray, N. E., & Reid, G. (1946), 'Temporary deafness due to gunfire', *J. Laryng.*, **61**, 92.

Perlman, H. B., & Case, T. J. (1939), 'Latent period of the crossed stapedius reflex in man, *Ann. Otol. (St Louis)*, **48**, 663.

Reger, S. N. (1960), 'Effect of middle ear muscle action on certain psychophysical measurements', *Ann. Otol. (St Louis)*, **69**, 1179.

Smith, R. P., & Loeb, M. (1969), 'Recovery from temporary threshold shifts as a function of test and exposure frequencies', *J. Acoust. Soc. Am.*, **45**, 238.

Spieth, W., & Trittipoe, W. J. (1958), 'Intensity and duration of noise exposure and temporary threshold shifts', *J. Acoust. Soc. Am.*, **30**, 710.

Terkildsen, K. (1960), 'The intra-aural muscle reflexes in normal persons and in workers exposed to intense industrial noise', *Acta oto-laryng. (Stockh.)*, **52**, 384.

Ward, W. D. (1960), 'Recovery from high values of temporary threshold shift', *J. Acoust. Soc. Am.*, **32**, 497.

—— (1961), 'Noninteraction of temporary threshold shifts', *J. Acoust. Soc. Am.*, **33**, 512.

—— (1962a), 'Damage-risk criteria for line spectra', *J. Acoust. Soc. Am.*, **34**, 1610.

—— (1926b), 'Studies on the aural reflex. II. Reduction of temporary threshold shift from intermittent noise by reflex activity; implications for damage-risk criteria', *J. Acoust. Soc. Am.*, **34**, 234.

—— (1963), Chap. 7, 'Auditory fatigue and masking', in *Modern developments in audiology*, ed. J. Jerger. New York & London, Academic Press.

—— (1965), 'The concept of susceptibility to hearing loss', *J. occup. Med.*, **7**, 595.

—— (1966), 'The use of TTS in the derivation of damage risk criteria for noise exposure', *Internat. Audiol.*, **5**, 309.

—— (1968), 'Susceptibility to auditory fatigue', in *Contributions to Sensory Physiology*, ed. W. D. Neff. New York & London, Academic Press.

—— (1970), 'Temporary threshold shift and damage-risk criteria for intermittent noise exposures', *J. Acoust. Soc. Am.*, **48**, 561.

Ward, W. D., & Glorig, A. (1961), 'A case of fire cracker-induced hearing loss', *Laryngoscope*, **71**, 1590.

Ward, W. D., Glorig, A., & Sklar, D. L. (1958), 'Dependence of temporary threshold shift at 4 Kc on intensity and time', *J. Acoust. Soc. Am.*, **30**, 944.

—— (1959a), 'Temporary threshold shift from octave-band noise: applications to damage-risk criteria', *J. Acoust. Soc. Am.*, **31**, 522.

Ward, W. D., Glorig, A., & Sklar, D. L. (1959b), 'Relation between recovery from temporary threshold shift and duration of exposure', *J. Acoust. Soc. Am.,* **31**, 600.

—— (1959c), 'Threshold shift produced by intermittent exposure to noise', *J. Acoust. Soc. Am.,* **31**, 791.

Ward, W. D., Selters, W., & Glorig, A. (1961), 'Exploratory studies on temporary threshold shift from impulses', *J. Acoust. Soc. Am.,* **33**, 781.

Chapter 11   PERMANENT EFFECTS OF NOISE ON HEARING

pp. 214–251

ANSI (1954), *The relations of hearing loss to noise exposure,* Report Z24-X-2. New York, ANSI.

Atherley, G. R. C., & Martin, A. M. (1971), 'Equivalent continuous noise level as a measure of injury from impact and impulsive noise', *Ann. Occup. Hyg.,* **14**, 11.

Atherley, G. R. C., & Noble, W. G. (1968), 'A review of studies of weavers' deafness', *Applied Acoustics,* **1**, 3.

Baughn, W. L. (1966), 'Noise control: percent of population protected', *Int. Audiol.,* **5**, 331.

Burns, W. (1958), 'Noise and hearing: physiological aspects', *Proceedings of the symposium on engine noise and noise suppression.* London, Instn. Mech. Engrs.

Burns, W., Hinchcliffe, R., & Littler, T. S. (1964), 'An exploratory study of hearing and noise exposure in textile workers', *Ann. Occup. Hyg.,* **7**, 323.

Burns, W., & Littler, T. S. (1960), Chap. 17, *Modern trends in occupational health,* ed. R. S. F. Schilling. London, Butterworth.

Burns, W., & Robinson, D. W. (1970a), *Hearing and noise in industry.* London, HMSO.

—— (1970b), 'An investigation of the effects of occupational noise on hearing', in *Sensorineural hearing loss,* ed. G. E. W. Wolstenholme & Julie Knight. London, J. & A. Churchill.

Burns, W., Stead, J. C., & Penney, H. W. (1970), 'The relations between temporary threshold shift and occupational hearing loss', in W. Burns & D. W. Robinson (1970a, above), Appendix 13.

Burns, W., Wood, B. E., Stead, J. C., & Penney, H. W. (1970), 'Noise-induced hearing loss in pathological cases', in Burns & Robinson (1970a, above), Appendix 14.

Copeland, W. C. T., Whittle, L. S., & Saunders, E. G. (1970), 'The mobile audiometric laboratory', in Burns & Robinson (1970a, above), Appendix 4.

Crowden, G. P. (1933), *13th Annual report.* London, Industrial Health Research Board.

Dickson, E. D. D., Ewing, A., & Littler, T. S. (1939), 'The effects of aeroplane noise on the auditory acuity of aviators: some preliminary remarks', *J. Laryng.,* **54**, 531.

Gallo, R., & Glorig, A. (1964), 'Permanent threshold shift changes produced by noise exposure and aging', *Amer. ind. Hyg. Ass. J.,* **25**, 237.

Glorig, A., & Davis, H. (1961), 'Age, noise and hearing loss', *Ann. Otol. (St Louis)*, **70**, 556.

Glorig, A., Ward, W. D., & Nixon, J. (1961), 'Damage-risk criteria and noise-induced hearing loss', National Physical Laboratory Symposium No. 12. London, HMSO.

Hickish, D. E., & Challen, P. J. R. (1966), 'A serial study of noise exposure and hearing loss in a group of small and medium size factories', *Ann. Occup. Hyg., 9*, 113.

ISO (1971), *R1999. Assessment of occupational noise exposure for hearing conservation purposes.* ISO.

Jerger, J. F., & Carhart, R. (1956), 'Temporary threshold shift as an index of noise susceptibility', *J. Acoust. Soc. Am., 28*, 611.

Kryter, K. D. (1963), 'Exposure to steady-state noise and impairment of hearing', *J. Acoust. Soc. Am., 35*, 1515.

Kylin, B. (1960), 'Temporary threshold shift and auditory trauma following exposure to steady-state noise', *Acta oto-laryng. (Stockh.), 51*, Supp. 152.

Littler, T. S. (1958), 'Noise measurement, analysis and evaluation of harmful effects', *Ann. Occup. Hyg., 1*, 11.

Miller, J. D., Watson, C. S., & Covell, W. P. (1963), 'Deafening effects of noise on the cat', *Acta oto-laryng. (Stockh.)*, Supp. 176.

Nixon, J. C., & Glorig, A. (1961), 'Noise-induced permanent threshold shift at 2000 cps and 4000 cps', *J. Acoust. Soc. Am., 33*, 904.

Robinson, D. W. (1970), 'Relations between hearing loss and noise exposure', in Burns & Robinson (1970a, above), Appendix 10.

—— (1971), ed. *Occupational hearing loss.* London & New York, Academic Press.

Robinson, D. W., & Burdon, Lynda A. (1970), 'Serial audiometry and the prospective study', in Burns & Robinson (1970a, above), Appendix 12.

Robinson, D. W., & Cook, Judith P. (1970), 'Experimental basis for the concept of noise immission level', in Burns & Robinson (1970a, above), Appendix 11.

Rosenwinkel, N. E., & Stewart, K. C. (1957), 'The relationship of hearing loss to steady-state noise exposure', *Ann. Ind. Hyg. Assoc. Quart., 18*, 227.

Rudmose, W. (1957), Chap. 7, *Handbook of noise control*, ed. C. M. Harris. New York, McGraw-Hill.

Taylor, W., Pearson, J., Mair, A., & Burns, W. (1965), 'Study of noise and hearing in jute weaving', *J. Acoust. Soc. Am., 38*, 113.

Ward, W. D. (1963), Chap. 7, 'Auditory fatigue and masking', in *Modern developments in audiology*, ed. J. Jerger. New York & London. Academic Press.

—— (1965), 'The concept of susceptibility to hearing loss', *J. occup. Med., 7*, 595.

Ward, W. D., Glorig, A., & Sklar, D. L. (1959), 'Temporary threshold shift from octave band noise: applications to damage-risk criteria', *J. Acoust. Soc. Am., 31*, 522.

Chapter 12   PRESERVATION OF HEARING   pp. 252–292

Adam, J. A., & Thomas, J. M. (1967), 'Noise in the iron and steel industry', *J. Iron Steel Inst.*, **205**, 701.

American Academy of Ophthalmology and Otolaryngology (1959), *Guide for conservation of hearing in noise*. Los Angeles, AAOO.

—— (1969), *Guide for conservation of hearing*. Los Angeles, AAOO.

ANSI (1954), *The relations of hearing loss to noise exposure*. Report Z24-X-2. New York, ANSI.

Atherley, G. R. C., & Martin, A. M. (1971), 'Equivalent continuous noise level as a measure of injury from impact and impulsive noise', *Ann. Occup. Hyg.*, **14**, 11.

Baughn, W. L. (1966), 'Noise control: percent of population protected', *Int. Audiol.*, **5**, 331.

Bazley, E. N. (1966), *The airborne sound insulation of partitions*. London, HMSO.

Beranek, L. L. (1971), *Noise and vibration control*. New York, McGraw-Hill.

British Occupational Hygiene Society (1971), 'Hygiene standards for wideband noise', *Ann. Occup. Hyg.*, **14**, 57.

Broch, J. T. (1969), *The application of the Brüel & Kjaer measuring systems to acoustic noise measurements*. Naerum, Denmark, Brüel & Kjaer.

BSI (1960), C.P. 3: Chap. III. 'Sound insulation and noise reduction'. London, BSI.

Burns, W. (1965), 'Noise as an environmental factor in industry', *Trans. Ass. industr. med. Offrs.*, **15**, 2.

Burns, W., & Littler, T. S. (1960), Chap. 17, *Modern trends in occupational health*, ed. R. S. F. Schilling. London, Butterworth.

Burns, W., & Robinson, D. W. (1970), *Hearing and noise in industry*. London, HMSO.

Burns, W., Wood, B. E., Stead, J. C., & Penney, H. W. (1970), 'Noise-induced hearing loss in pathological cases', in Burns & Robinson (1970, above), Appendix 14.

Committee on the Problem of noise (1963), *Noise, final report*. London, HMSO.

Copeland, W. C. T., & Saunders, E. G. (1970), 'The mobile acoustical laboratory', in Burns & Robinson (1970, above), Appendix 5.

Copeland, W. C., Whittle, L. S., & Saunders, E. G. (1964), 'A mobile audiometric laboratory', *J. Sound Vib.*, **1**, 388.

Davis, H. (1962), Opening address, *Report of the Royal National Institute for the Deaf 1962 Conference*, p. 4. London RNID.

—— (1970), 'A historical introduction', in *Occupational hearing loss*, ed. D. W. Robinson. London & New York, Academic Press.

Department of Employment (1972), *Code of practice for reducing the exposure of employed persons to noise*. London, HMSO.

Fleming, N., & Copeland, W. C. (1958), 'Principles of noise suppression', *Ann. Occup. Hyg.*, **1**, 55.

Forrest, M. R., & Coles, R. R. A. (1970), 'Problems of communication and ear protection in the Royal Marines', *J. Royal Naval Medical Service*, **56**, No. 1-2, 162.

Glorig, A., Ward, W. D., & Nixon, J. (1962), 'Damage-risk criteria and noise-induced hearing loss', National Physical Laboratory Symposium No. 12. London, HMSO.

Harris, C. M. (1957), *Handbook of noise control*. New York, McGraw-Hill.

Hinchcliffe, R. (1959), 'Correction of pure-tone audiograms for advancing ages', *J. Laryng.*, **73**, 830.

Hood, J. D., & Poole, J. P. (1971), 'Speech audiometry in conductive and sensorineural hearing loss', *Sound*, **5**, 30.

Institution of Mechanical Engineers (1958), *Symposium on engine noise and noise suppression*. London, Instn. Mech. Engrs.

ISO (1964), *R389. Standard reference zero for the calibration of pure-tone audiometers*. ISO.

—— (1971), *R1999. Assessment of occupational noise exposure for hearing conservation purposes*. ISO.

King, A. J. (1958), 'Reduction of internal-combustion engine noise by enclosure', *Proceedings of the Symposium on engine noise and noise suppression*. London, Instn. Mech. Engrs.

Kosten, C. W., & van Os, G. J. (1962), 'Community reaction criteria for external noises', NPL Symposium No. 12. London, HMSO.

Kryter, K. D. (1946), 'The effects of ear protection devices on the intelligibility of speech in noise', *J. Acoust. Soc. Am.*, **18**, 413.

—— (1950), 'The effects of noise on man', *J. Speech Dis.*, monograph, Supp. 1.

Kryter, K. D., Ward, W. D., Miller, J. D., & Eldredge, D. H. (1966), 'Hazardous exposure to intermittent and steady-state noise', *J. Acoust. Soc. Am.*, **39**, 451.

Littler, T. S. (1958), 'Noise measurement analysis and evaluation of harmful effects', *Ann. Occup. Hyg.*, **1**, 11.

Mosko, J. D., & Fletcher, J. L. (1971), 'Evaluation of the Gundefender earplug: temporary threshold shift and speech intelligibility', *J. Acoust. Soc. Am.*, **49**, 1732.

NPL (1962), Symposium No. 12. London, HMSO.

Parfitt, G. G. (1958), 'The analysis and control of vibration', *Ann. Occup. Hyg.*, **1**, 68.

Parkin, P. H., & Humphreys, H. R. (1969), *Acoustics, noise and buildings*, 3rd Ed. London, Faber and Faber.

Peterson, A. P. G., & Gross, E. E. (1967), *Handbook of noise measurement*. West Concord, Massachusetts, USA, General Radio Company.

Piesse, R. A. (1957), *Report No. 21*. Commonwealth Acoustic Laboratories, Sydney, Australia.

Richards, E. J. (1965), 52nd Thomas Hawksley Lecture, *Proc. Instn. mech. Engrs.*, **180**, Part 1, 1099.

Robinson, D. W. (1970), 'Relations between hearing loss and noise exposure', in Burns & Robinson (1970, above), Appendix 10.

Robinson, D. W. (1971), ed. *Occupational hearing loss*. London & New York, Academic Press.

Robinson, D. W., & Burdon, Lynda A. (1970), 'Serial audiometry and the prospective study', in Burns & Robinson (1970, above), Appendix 12.

Rosenblith, W. A., & Stevens, K. N. (1953), *Handbook of acoustic noise control*. Office of technical services, Washington, D.C.

Shaw, E. A. G., & Thiessen, G. J. (1958), 'Improved cushion for ear defenders', *J. Acoust. Soc. Am., 30*, 24.

Svenska Elektriska Kommissionen (1969), 'Estimation of risk of hearing damage from noise: measuring methods and acceptable values', Proposal SEC 590111, Stockholm. (In Swedish.)

Taylor, W., Burns, W., & Mair, A. (1964), 'A mobile unit for the assessment of hearing', *Ann. Occup. Hyg., 7*, 343.

Taylor, W., Pearson, J., Mair, A., & Burns, W. (1965), 'Study of noise and hearing in jute weaving', *J. Acoust. Soc. Am., 38*, 113.

Thiessen, G. J. (1962), 'Reduction of noise at the listener's ear', NPL Symposium No. 12, p. 153. London, HMSO.

United States Air Force (1956), USAF Medical Service, *Hazardous noise exposure*, AF Reg. No. 160-3. Washington, D.C.

Van Atta, F. A. (1971), 'Hearing conservation through regulation', in *Occupational hearing loss,* ed. D. W. Robinson. New York & London, Academic Press.

von Gierke, H. E. (1956), 'Personal protection', *Noise control, 2*, No. 1, 37.

Waters, P. E., Lalor, N., & Priede, T. (1970), 'The diesel engine as a source of commercial vehicle noise', *Proc. Inst. mech. Engrs., 184*, Part 3P.

Chapter 13 AIRCRAFT NOISE pp. 293–335

Bell, G. E. (1963), 'The noise problem at airports', Chap. 9 in *Noise measurement and control*, ed. P. Lord and F. L. Thomas. London, Heywood & Co.

Bitter, C. (1968), 'Noise nuisance due to aircraft', paper presented at Colloque sur la définition des exigences humaines a l'égard du bruit, Paris.

British Airports Authority Annual Report and Accounts for year ended March 1, 1971. London, HMSO.

Bürck, W., Grützmacher, M., Meister, F. J., & Müller, E. A. (1965), 'Fluglärm'. Göttingen.

Central Office of Information (1969), *Action against aircraft noise*. London, HMSO.

Cmnd. Paper No. 3259 (1967), *The Third London Airport*. London, HMSO.

Commission on The Third London Airport: Report (1971), (Chairman, The Hon. Mr Justice Roskill). London, HMSO.

Committee on Public Engineering Policy, National Academy of Engineering, USA (1969), *A Study of Technology Assessment*, Committee on Science and Astronautics, US House of Representatives.

Committee on the Problem of Noise (1963). *Noise, final report*. London, HMSO.

Cook, R. H. (1960), *Aviation Week,* **73**, No. 40.

Davies, D. P. (1971), *Handling the big jets.* Cheltenham, England, Air Registration Board.

FAA (1970), Regulations, Volume III, Part 36: Noise standards: aircraft type certification. FAA.

Flight International (1966a), No. 3012, p. 942.

—— (1966b), No. 3013, p. 975.

—— (1970), No. 3211, p. 485.

—— (1971), No. 3265, p. 559.

Galloway, W. J., & Bishop, D. E. (1970), *Noise exposure forecasts: evolution, evaluation, extensions, and land use interpretations.* Bolt, Beranek & Newman, Final Report No. FAA-NO-70-9, Contract Fa68WA-1900, FAA.

Greatrex, F. B., & Bridge, R. (1967), 'Evolution of the engine noise problem', *Flight International*, No. 3021, p. 165.

Greatrex, F. B., & Brown, D. M. (1958), 'Progress in jet engine noise reduction'. First Congress, International Council of Aeronautical Sciences. Madrid, Sept. 1958.

Hargest, T. J. (1962), 'Jet engine noise reduction'. NPL Symposium No. 12. London, HMSO.

ICAO (1969a), 'Correlation of surveys with the determination of noise areas around aerodromes', ICAO Special Meeting on Aircraft Noise in the vicinity of aerodromes, Noise 1969-WP/15, Item 2, Paper No. 2. Montreal, ICAO.

—— (1969b), Report of the Special Meeting on Aircraft Noise in the vicinity of aerodromes. ICAO Doc. 8857 Noise. Montreal, ICAO.

—— (1971), Annex 16 to the convention on international civil aviation. Montreal, ICAO.

ISO (1970), *R507. Procedure for describing aircraft noise around an airport,* 2nd Ed. ISO.

Knowler, A. E. (1971), 'The second noise and social survey around Heathrow, London Airport', Seventh International Congress on Acoustics, Budapest.

Lighthill, M. J. (1952), 'On sound generated aerodynamically', *Proc. Roy. Soc.,* **221A**, 564.

Lloyd, P. (1959), 'Some aspects of engine noise', *J. Roy. Aero Soc.,* **63**, 541.

McKennell, A. C. (1963), *Aircraft noise annoyance round London Heathrow Airport.* London, UK Central Office of Information.

Parnell, J. E., Nagel, D. G., & Cohen, A. (1972), *Evaluation of hearing levels of residents living near a major airport.* Report FAA-RD-72-72. Washington, DC, Department of Transportation.

Robinson, D. W. (1970), *An outline guide to criteria for the limitation of urban noise.* Aeronautical Research Council, CP 1112. London, HMSO.

—— (1971), 'Towards a unified system of noise assessment', *J. Sound Vib.,* **14**, 279.

Smith, M. J. T. (1970), *Aero gas turbine noise.* Derby, England, Rolls-Royce (1971) Limited.

Smith, M. J. T. (1971), *Progress towards quieter aircraft—British developments in aero engines.* Derby, England, Rolls-Royce (1971) Limited.

Statutory Instrument (1970), No. 823. Civil aviation: the air navigation (noise certification) order 1970. London, HMSO.

UK Department of Trade and Industry (1970), Noise measurement for aircraft design purposes including noise certification purposes. CAP 335.

—— (1971), *Second survey of aircraft noise annoyance around London (Heathrow) Airport.* London, HMSO.

van Niekerk, C. G., & Muller, J. L. (1969), 'Assessment of aircraft noise disturbance', *J. R. Ae. Soc.,* **73**, 383.

Chapter 14 IMPULSE NOISE; INTENSE NOISE; ULTRA,

INFRASOUND; SONIC BOOM; VIBRATION

pp. 336–360

Acoustical Society of America (1966), 'Proceedings of the Sonic Boom Symposium', *J. Acoust. Soc. Am.,* **39**, No. 5, Part 2.

—— (1972), 'Sonic Boom Symposium', *J. Acoust. Soc. Am.,* **51**, 671.

Acton, W. I., & Carson, M. B. (1967), 'Auditory and subjective effects of airborne noise from industrial ultrasonic sources', *Brit. J. industr. Med.,* **24**, 297.

Baals, D. D., & Foss, W. E., Jr. (1966), 'Assessment of sonic boom problem for future air-transport-vehicles', *J. Acoust. Soc. Am.,* **39**, 573.

Beranek, L. L. (1960), *Noise reduction.* New York, McGraw-Hill.

Carlson, H. W., Mack, R. J., & Morris, O. A. (1966), 'Sonic Boom Pressure Field Estimation Techniques', *J. Acoust. Soc. Am.,* **39**, S10.

CHABA (1968), Proposed damage-risk criterion for impulse noise (gunfire). Report of Working Group 57, ed. W. D. Ward. National Academy of Sciences National Research Council, Committee on Hearing, Bioacoustics and Biomechanics, Washington, D.C.

Coles, R. R. A. (1962), 'Some considerations concerning the effects on hearing of noise of small arms', *J. Roy. Naval Med. Service,* **49**, 18.

Coles, R. R. A., Garinther, G. R., Hodge, D. C., & Rice, C. G. (1968), 'Hazardous exposure to impulse noise', *J. Acoust. Soc. Am.,* **43**, 336.

Coles, R. R. A., & Knight, J. J. (1965), 'The problem of noise in the Royal Navy and Royal Marines', *J. Laryng.,* **79**, 131.

Coles, R. R. A., & Rice, C. G. (1970), 'Towards a criterion for impulse noise in industry', *Ann. Occup. Hyg.,* **13**, 43.

—— (1971), 'Assessment of risk of hearing loss due to impulse noise', in *Occupational Hearing Loss,* p. 71, ed. D. W. Robinson. London & New York, Academic Press.

Gavreau, V., Condat, R., & Saul, H. (1966), 'Infra-sons: générateurs, détecteurs, propriétés physiques, effets biologiques', *Acustica,* **17**, 1.

Goldman, D. E. (1957), 'Effects of vibration on man', in *Handbook of noise control,* ed. C. M. Harris. New York, McGraw-Hill.

Goldman, D. E., & von Gierke, H. E. (1961), 'Effects of shock and vibration on man', in *Shock and vibration handbook*, vol. 3, Chap. 44, ed. C. M. Harris and C. E. Crede. New York, McGraw-Hill.

Guignard, J. C., & Guignard, Elsa (1970), 'Human response to vibration—a critical survey of published work'. University of Southampton, ISVR Memorandum 373 September 1970.

Hilton, D. A., & Newman, J. W. (1966), 'Instrumentation techniques for measurement of sonic boom signatures', *J. Acoust. Soc. Am., 39*, S31.

Hubbard, H. H. (1966), 'Nature of the sonic boom problem', *J. Acoust. Soc. Am., 39*, S1.

Hubbard, H. H., & Maglieri, D. J. (1967), 'The nature, measurement and control of sonic booms', IEE Conference publication No. 26, *Acoustic noise and its control.* London, Instn. Elec. Engrs.

Johnson, D. R., & Robinson, D. W. (1969), 'Procedure for calculating the loudness of sonic bangs', *Acustica, 21*, 307.

Kryter, K. D. (1969), 'Sonic booms from supersonic transport', *Science, 163*, 359.

Kryter, K. D., Ward, W. D., Miller, J. D., & Eldredge, D. H. (1966), 'Hazardous exposure to intermittent and steady-state noise', *J. Acoust. Soc. Am., 39*, 451.

McLean, F. E. (1965), 'Some nonasymptotic effects on the sonic boom of large airplanes', *NASA Tech. Note D 2877.*

McLean, F. E., & Shrout, B. L. (1966), 'Design methods for minimization of sonic boom pressure field disturbances', *J. Acoust. Soc. Am., 39*, S19.

Maglieri, D. J. (1966), 'Some effects on airplane operations and the atmosphere of sonic boom signatures', *J. Acoust. Soc. Am., 39*, S36.

Mohr, G. C., Cole, J. N., Guild, Elizabeth, & von Gierke, H. E. (1965), 'Effects of low frequency and infrasonic noise on man', *Aerospace Medicine, 36*, No. 9, 817.

Murray, N. E., & Reid, G. (1946), 'Temporary deafness due to gunfire', *J. Laryng., 61*, 92.

Nixon, C. W. (1965), 'Sonic boom, people and SST operation: a status report', *Aerospace Med., 36*, 170(A).

Nixon, C. W., & Borsky, P. N. (1966), 'Effects of sonic boom on people: St Louis, Missouri, 1961–1962', *J. Acoust. Soc. Am., 39*, S51.

Parrack, H. O. (1966), 'Effect of airborne ultrasound on humans', *International Audiology, 5*, 294.

Rylander, R. (1972), ed. 'Sonic boom exposure effects: report from a workshop on methods and criteria', Stockholm 1971, *J. Sound Vib., 20*, 477.

Shaw, E. A. G., & Thiessen, G. J. (1958), 'Improved cushion for ear defenders', *J. Acoust. Soc. Am., 20*, 24.

Stephens, R. W. B. (1969), 'Infrasonics', *Ultrasonics, 7*, 30.

Stephens, R. W. B., & Bate, A. E. (1966), *Acoustics and vibrational physics.* London, Edward Arnold.

von Gierke, H. E. (1964), 'Biodynamic response of the human body', *Applied Mechanics Reviews, 17*, No. 12, 951.

von Gierke, H. E. (1966), 'Effects of sonic boom on people. Review and outlook', *J. Acoust. Soc. Am., 39, S43.

von Gierke, H. E., & Nixon, C. W. (1972), 'Human response to sonic boom in the laboratory and the community', *J. Acoust. Soc. Am., 51, 766.

Warren, C. H. E. (1966), 'Experience in the United Kingdom on the effects of sonic bangs', *J. Acoust. Soc. Am., 39, S59.

—— (1972), 'Recent sonic-bang studies in the United Kingdom', *J. Acoust. Soc. Am., 51, 783.

Whittle, L. S. (1971), 'The audibility of low-frequency sounds', British Acoustical Society Spring meeting 1971, communication No. 71SBB6.

Appendix B   Units and equivalents   pp. 379–381

Aitken, A. C., & Connor, R. D. (1965), *Physical and mathematical tables* prepared by the late John B. Clark. Edinburgh & London, Oliver & Boyd.

BSI (1959), *BS 350:* Part I, *Conversion factors and tables.* London, BSI.

—— (1962), *BS 350:* Part 2, *Conversion factors and tables.* London, BSI.

—— (1964), *BS 3763: The International System ('SI) units.* London, BSI.

—— (1967a), *The use of SI units* PD 5686, April 1967. London, BSI.

—— (1967b), *BS 1991:* Part I, *Letter symbols, signs and abbreviations.* London, BSI.

Appendix I   Effects of background noise on audiometry
pp. 398–400

BSI (1954), *BS 2497. The normal threshold of hearing for pure tones by earphone listening.* London, BSI. Note: This is the original version of BS 2497 and as such is no longer available in precisely this form.

Copeland, W. C., Whittle, L. S., & Saunders, E. G. (1964), 'A mobile audiometric laboratory', *J. Sound Vib., 1, 388.

Fletcher, H. (1940), 'Auditory patterns', *Rev. Mod. Phys., 12, 47.

French, N. R., & Steinberg, J. C. (1947), 'Factors governing the intelligibility of speech sounds', *J. Acoust. Soc. Am., 19, 90.

Hawkins, J. E., & Stevens, S. S. (1950), 'The masking of pure tones and of speech by white noise', *J. Acoust. Soc. Am., 22, 6.

Royal Air Force Central Medical Establishment Acoustics Laboratory (1963), *Report J3/6/RI.*

Scharf, B. (1959), 'Critical bands and the loudness of complex sounds near threshold', *J. Acoust. Soc. Am., 31, 365.

Taylor, W., Burns, W., & Mair, A. (1964), 'A mobile unit for the assessment of hearing', *Ann. Occup. Hyg., 7, 343.

Zwicker, E., Flottorp, G., & Stevens, S. S. (1957), 'Critical band width in loudness summation', *J. Acoust. Soc. Am., 29, 549.

Appendix J Sources of error in audiometry pp. 400–404

Brown, R. E. C. (1948), 'Experimental studies on the reliability of audiometry', *J. Laryng.*, **62**, 487.

Burns, W., & Hinchcliffe, R. (1957), 'Comparison of the auditory threshold as measured by individual pure tone and by Békésy audiometry', *J. Acoust. Soc. Am.*, **29**, 1274.

Burns, W., Hinchcliffe, R., & Littler, T. S. (1964), 'An exploratory study of hearing and noise exposure in textile workers', *Ann. Occup. Hyg.*, **7**, 323.

Burns, W., & Morris, G. A. (1955), 'A recording audiometer', *J. Physiol.*, **131**, 4P.

Burns, W., & Robinson, D. W. (1970), *Hearing and noise in industry*. London, HMSO.

Dadson, R. S., & King, J. H. (1952), 'A determination of the normal threshold of hearing and its relation to the standardisation of audiometers', *J. Laryng.*, **66**, 366.

Delany, M. E. (1971), 'Some sources of variance in the determination of hearing level', in *Occupational hearing loss*, ed. D. W. Robinson. London & New York, Academic Press.

Delany, M. E., Whittle, L. S., Cook, J. P., & Scott, V. (1967), 'Performance studies on a new artificial ear', *Acustica,* **18**, 231.

High, W. S., & Glorig, A. (1962), 'The reliability of industrial audiometry', *J. Auditory Res.,* **2**, 56.

Knight, J. J. (1962), Personal communication.

Robinson, D. W. (1960), 'Variability on the realisation of the audiometric zero', *Ann. Occup. Hyg.,* **2**, 107.

—— (1970), Appendix 10, 'Relations between hearing loss and noise exposure', in *Hearing and noise in industry*, W. Burns & D. W. Robinson. London, HMSO.

Robinson, D. W., & Dadson, R. S. (1956), 'A re-determination of the equal-loudness relation for pure tones', *Brit. J. appl. Phys.,* **7**, 166.

Rudmose, W. (1963), Chap. 2, 'Automatic audiometry', *Modern developments in audiology*, ed. J. Jerger. New York, Academic Press.

Appendix K Loudness: the sone and the phon; perceived noise
pp. 404–413

ANSI (1968), S3.4, *Procedure for the computation of loudness of noise.* New York, ANSI.

BSI (1958), *BS 3045. The relation between the sone scale of loudness and the phon scale of loudness level.* London, BSI.

—— (1967), *BS 4198. Method for calculating loudness.* London, BSI.

Churcher, B. G., & King, A. J. (1937), 'The performance of noise meters in terms of the primary standard', *J. Instn. Elec. Engrs.,* **81**, 927.

Fletcher, H., & Munson, W. A. (1933), 'Loudness, its definition, measurement and calculation', *J. Acoust. Soc. Am.,* **24**, 80.

ISO (1961), *R226. Normal equal loudness contours for pure tones and normal threshold of hearing under free-field listening conditions*. ISO.

—— (1966), *R532. Method for calculating loudness level*. ISO.

—— (1970), *R507. Procedure for describing aircraft noise around an airport*, 2nd Ed., June 1970. ISO.

Kingsbury, B. A. (1927), 'A direct comparison of the loudness of pure tones', *Phys. Rev., 21*, 84.

Kryter, K. D. (1959), 'Scaling human reactions to the sound from aircraft', *J. Acoust. Soc. Am., 31*, 1415.

Kryter, K. D., & Pearsons, K. S. (1963), 'Some effects of spectral content and duration on perceived noise level', *J. Acoust. Soc. Am., 35*, 866.

—— (1964), 'Modifications of noy table', *J. Acoust. Soc. Am., 36*, 394.

Robinson, D. W. (1953), 'The relation between the sone and phon scale of loudness', *Acustica, 7*, 344.

Robinson, D. W., & Dadson, R. S. (1956), 'A re-determination of the equal-loudness relation for pure tones', *Brit. J. appl. Phys., 7*, 166.

Stevens, S. S. (1936), 'A scale for the measurement of a psychological magnitude: loudness', *Psychol. Rev., 43*, 405.

—— (1955), 'The measurement of loudness', *J. Acoust. Soc. Am., 27*, 815.

—— (1956), 'Calculation of the loudness of complex noise', *J. Acoust. Soc. Am., 28*, 807.

—— (1957), 'Calculating loudness', *Noise Control, 3*, 11.

—— (1961), 'Procedure for calculating loudness, Mk VI', *J. Acoust. Soc. Am., 33*, 1577.

—— (1970), 'Assessment of noise: calculation procedure Mark VII', Personal communication.

Whittle, L. S. (1971), 'The audibility of low frequency sounds', British Acoustical Society Spring Meeting 71S BB6.

Yeowart, N. S., Bryan, M. E., & Tempest, W. (1967), 'The monaural M.A.P. threshold of hearing at frequencies from 1·5 to 100 c/s', *J. Sound Vib., 6*, 335.

Zwicker, E., & Feldtkeller, R. (1955), 'Über die Lautstärke von gleichförmigen Geräuschen', *Acustica, 5*, 303.

Zwicker, E. (1958), 'Über psychologische und methodische Grundlagen der Lautheit', *Acustica, 8*, 237.

—— (1960), 'Ein Verfahren zur Berechnung der Lautstärke', *Acustica, 10*, 304.

—— (1961), 'Subdivision of the audible frequency range into critical bands (Frequenzgruppen)', *J. Acoust. Soc. Am., 33*, 248.

Appendix L   Data on noise rating   pp. 414–415

Beranek, L. L., Blazier, W. E., & Figwer, J. J. (1971), 'Preferred noise criteria curves and their application'. Personal communication. This material is now available in L. L. Beranek (1971), Chap. 18, *Noise and vibration control*, ed. L. L. Beranek. New York, McGraw-Hill,

ISO (1971), *R1996. Assessment of noise with respect to community response*. ISO.

Appendix M   Calculation of equivalent continuous sound level for intermittent or fluctuating sounds   pp. 416–419

Department of Employment (1972). *Code of practice for reducing the exposure of employed persons to noise.* London, HMSO.

ISO (1971), *R1999. Assessment of occupational noise exposure for hearing conservation purposes.* ISO.

Appendix N   *Calculation of noise-induced threshold shift* (Robinson, 1971)   pp. 420–422

Robinson, D. W. (1970), Appendix 10, 'Relations between hearing loss and noise exposure', in *Hearing and noise in industry*, W. Burns & D. W. Robinson. London, HMSO.

—— (1971), *Occupational hearing loss,* ed. D. W. Robinson. London & New York, Academic Press.

Index of authors

Index of subjects